Analytical and Chromatographic Techniques in Radiopharmaceutical Chemistry

Analytical and Chromatographic Techniques in Radiopharmaceutical Chemistry

Edited by
Donald M. Wieland
Michael C. Tobes
Thomas J. Mangner

With 115 Figures

Springer-Verlag
New York Berlin Heidelberg Tokyo

Donald M. Wieland
Division of Nuclear Medicine
Department of Internal Medicine
The University of Michigan
Ann Arbor, MI 48109
USA

Michael C. Tobes
Division of Nuclear Medicine
Department of Internal Medicine
The University of Michigan
Ann Arbor, MI 48109
USA

Thomas J. Mangner
Division of Nuclear Medicine
Department of Internal Medicine
The University of Michigan
Ann Arbor, MI 48109
USA

Library of Congress Cataloging in Publication Data
Main entry under title:
Analytical and chromatographic techniques in radio-pharmaceutical chemistry.

Updated and expanded versions of presentations at a symposium held June 4, 1984, in Los Angeles, Calif., under the sponsorship of the Radiopharmaceutical Science Council of the Society of Nuclear Medicine.

Includes bibliographies and index.

1. Radiopharmaceuticals—Analysis—Congresses. 2. Thin layer chromatography—Congresses. 3. Liquid chromatography—Congresses. 4. Chemistry, Analytic—Congresses. I. Wieland, Donald M. II. Tobes, Michael C. III. Mangner, Thomas J. IV. Society of Nuclear Medicine (1953–). Radiopharmaceutical Science Council. [DNLM: 1. Chemistry, Analytical—methods—congresses. 2. Chromatography, High Pressure Liquid—congresses. 3. Chromatography, Thin Layer—congresses. 4. Radiochemistry—methods—congresses.
QD 605 A532] RS190.R34A53 1985 615.84 85-14707

Softcover reprint of the hardcover 1st edition 1986

Typeset by BiComp, Inc., York, Pennsylvania.

9 8 7 6 5 4 3 2 1

ISBN-13: 978-1-4612-9331-6 e-ISBN-13: 978-1-4612-4854-5
DOI:10.1007/978-1-4612-4854-5

Foreword

In 1906, Michael T. Sweet first developed the chromatographic method by using an adsorbant to separate pigments. Since that time, the technological advances in TLC and HPLC have brought about new definitions of purity in parallel with the advances. Radiopharmaceutical chemistry is especially dependent on the chromatographic technique because of the relatively small amount of material in most radiopharmaceuticals—often so small that the usual physical methods of analytical chemistry cannot be used. As a result, this collection of papers represents the key to successful radiopharmaceutical development by setting the standard for the present-day definition of radiochemical purity.

William C. Eckelman, Ph.D.
Diagnostics Associate Director
The Squibb Institute for Medical Research
New Brunswick, New Jersey

Preface

The chapters herein are updated and expanded versions of presentations that the authors made at a symposium held on June 4, 1984 in Los Angeles, California under the sponsorship of the Radiopharmaceutical Science Council of the Society of Nuclear Medicine. All manuscripts were refereed.

The intent of the symposium organizers was to enlist participants who work on a day-to-day basis with the analytical and chromatographic techniques to be discussed at the symposium. We feel confident that this distillation of hands-on experience will be of value to graduate students as well as experienced researchers in radiopharmaceutical chemistry and related fields which use radiotracer methodology.

The short history of radiopharmaceutical chemistry has been marked by vivid examples of the value of conscientious use of analytical and chromatographic techniques. Nearly a decade ago, radio-TLC revealed the presence of a radioactive "impurity" in preparations of the adrenal cortex imaging agent I-131-19-iodocholesterol. Identification of this "impurity" showed in fact that it was the active agent I-131-6β-iodomethyl-19-norcholesterol. The recent renaissance in Tc-99m radiopharmaceuticals, especially the cationic heart agents, has been accompanied by successful application of reverse phase radio-HPLC to purity analyses of these agents. Application of similar radio-HPLC techniques to clinical mainstays such as Tc-99m bone agents, many of which were first synthesized in the "HPLC-less" days of the early 1970s, has revealed that certain of these agents are complex mixtures. Just as radio-HPLC was being embraced as the definitive technique for purity confirmation, a 1984 report using fluorine nuclear magnetic resonance revealed that F-18-2-fluorodeoxyglucose, a radiotracer of major importance in determining regional glucose metabolism by positron emission tomography, was in certain cases contaminated with varying amounts of its isomer, F-18-2-fluorodeoxymannose. To date no radio-HPLC system has been reported that distinguishes these fluoro isomers.

The historical lesson is clear—no single analytical technique should be relied on. Working at nanogram levels, radiopharmaceutical chemists have always had cause to be light sleepers. Hopefully, reading this book will keep them up all night.

Donald M. Wieland, Ph.D.
Michael C. Tobes, Ph.D.
Thomas J. Mangner, Ph.D.

Acknowledgments

The editors and the Radiopharmaceutical Science Council would like to thank the speakers, panel participants, poster presentors, and commercial exhibitors who participated in the symposium. A special thanks to Joanna Fowler of Brookhaven National Laboratories for moderating the panel discussion and to Richard Chamberlain of the Society of Nuclear Medicine Central Office for organizational assistance. The editors are grateful to Linder Markham for secretarial assistance.

Research at the University of Michigan Medical Center in Radiopharmaceutical Chemistry and Nuclear Medicine is supported by the Department of Internal Medicine and the following grants from the National Institutes of Health: Contract No. HL-27555, NCI Training Grant, Contract No. 5-T32-CA09015-09 and the U.S. Department of Energy, Contract No. DE-AC02-76EV02031.

Contents

Contributors

Jorge R. Barrio, Division of Nuclear Medicine and Biophysics, UCLA School of Medicine, Los Angeles, CA 90024, USA

Gerald Bida, Division of Nuclear Medicine and Biophysics, UCLA School of Medicine, Los Angeles, CA 90024, USA

Hal T. Butler, Department of Chemistry, Wayne State University, Detroit, MI 48202, USA

Alan P. Carpenter, Jr., Division of Analytical Chemistry, Radiopharmaceutical Research, Dupont/New England Nuclear Corporation, North Billerica, MA 01845, USA

Joseph H. Chasko, Donner Laboratory, Lawrence Berkeley Laboratory, University of California, Berkeley, CA 94720, USA

Diane C. Chugani, Division of Nuclear Medicine and Biophysics, UCLA School of Medicine, Los Angeles, CA 90024, USA

Myra E. Coddens, Department of Chemistry, Wayne State University, Detroit, MI 48202, USA

Carmen S. Dence, Division of Radiation Sciences, The Edward Mallinckrodt Institute of Radiology, Washington University, St. Louis, MO 63110, USA

Heinz Filthuth, Laboratorium Prof. Dr. Berthold, D-7547 Wildbad 1, West Germany

Alan R. Fritzberg, NeoRx Corporation, Seattle, WA 98119, USA

Sheryl J. Hays, Warner-Lambert Company, Pharmaceutical Research Division, Ann Arbor, MI 48105, USA

Richard D. Hichwa, Division of Nuclear Medicine, University of Michigan Medical Center, Ann Arbor, MI 48109, USA

Donald J. Hnatowich, Department of Nuclear Medicine, University of Massachusetts Medical School, Worcester, MA 01605, USA

Reese M. Jones, Donner Laboratory, Lawrence Berkeley Laboratory, University of California, Berkeley, CA 94720, USA

Randy E. Keen, Division of Nuclear Medicine and Biophysics, UCLA School of Medicine, Los Angeles, CA 90024, USA

Michael J. Kessler, Radiomatic Instruments & Chemical Co., Tampa, FL 33611, USA

Michael R. Kilbourn, Division of Radiation Sciences, The Edward Mallinckrodt Institute of Radiology, Washington University, St. Louis, MO 63110, USA

Keith R. Lechner, Division of Radiation Sciences, The Edward Mallinckrodt Institute of Radiology, Washington University, St. Louis, MO 63110, USA

Chester A. Mathis, Donner Laboratory, Lawrence Berkeley Laboratory, University of California, Berkeley, CA 94720, USA

Thomas J. Mangner, Division of Nuclear Medicine, University of Michigan Medical Center, Ann Arbor, MI 48109, USA

Adrian D. Nunn, Division of Radiopharmaceutical Research and Development, The Squibb Institute for Medical Research, New Brunswick, NJ 08903, USA

Michael E. Phelps, Division of Nuclear Medicine and Biophysics, UCLA School of Medicine, Los Angeles, CA 90024, USA

Colin F. Poole, Department of Chemistry, Wayne State University, Detroit, MI 48202, USA

Nagichettiar Satyamurthy, Division of Nuclear Medicine and Biophysics, UCLA School of Medicine, Los Angeles, CA 90024, USA

Sheila A. Schuette, Department of Chemistry, Wayne State University, Detroit, MI 48202, USA

Michael J. Welch, Division of Radiation Sciences, The Edward Mallinckrodt Institute of Radiology, Washington University, St. Louis, MO 63110, USA

D. Scott Wilbur, NeoRx Corporation, Seattle, WA 98119, USA

Thin-Layer Chromatography

CHAPTER 1

Instrumental Evaluation of Thin-Layer Chromatograms

COLIN F. POOLE, HAL T. BUTLER, MYRA E. CODDENS, and SHEILA A. SCHUETTE

INTRODUCTION

Recent changes in the practice of thin-layer chromatography (TLC) have created a renaissance of interest in this technique and led to its wider acceptance as a powerful tool for qualitative and quantitative analysis of mixtures. The performance breakthrough in TLC, referred to as high performance thin-layer chromatography (HPTLC), was not a result of any specific advances in instrumentation or materials, but was rather a culmination of improvements in practically all of the operations comprising TLC, [1–6]. Improvements in the quality of the sorbent layer, methods of sample application, new development techniques, and the availability of scanning densitometers for *in situ* quantitative analysis were all important developments in the evolution of HPTLC. The new HPTLC plates were prepared from sorbents of smaller average particle diameter and of narrower particle size distribution than those used to prepare conventional TLC plates. The performance of these plates improved an order of magnitude over those used for conventional TLC. Because of the lower sample capacity of the HPTLC sorbent layer, the amount of sample applied to it had to be scaled down from those commonly used in conventional TLC. Typical sample volumes applied to the plates are in the range 100–200 nl, sample amounts less than ca. 0.1 μg, and starting spot sizes of 1.0–2.0 mm. Separations are performed using shorter migration distance 3–6 cm, providing faster analysis times (in spite

---

Abbreviations: Thin-Layer Chromatography, TLC; High Performance Thin-Layer Chromatography, HPTLC

TABLE 1.1. A Comparison of the Techniques of Conventional and High Performance Thin-Layer Chromatography

Parameter	*Conventional TLC*	*HPTLC*
Plate size	20 × 20 cm	10 × 10 cm 20 × 5 cm
Particle size		
Average	20 μm	5 μm
Distribution	10–60 μm	Tight
Adsorbent layer thickness	100–250 μm	200 μm
Plate height	30 μm	12 μm
Total number of usable theoretical plates (conventional single development)	<600	<5000
Separation number	7–10	10–20
Sample volume	1–5 μl	0.1–0.2 μl
Starting spot diameter	3–6 mm	1–1.5 mm
Diameter of separated spots	6–15 mm	2–5 mm
Solvent migration distance	10–15 cm	2–6 cm
Separation time	30–200 min	3–20 min
Detection limits		
Absorption	1–5 ng	0.1–0.5 ng
Fluorescence	0.05–0.1 ng	0.005–0.01 ng
Tracks per plate	10	18 or 36

of the lower mobile phase velocities of HPTLC plates), and producing more compact spots, which are easier to detect. Sample detectability is improved by approximately an order of magnitude over values for conventional TLC.

In summary, HPTLC is a more rapid, efficient, and sensitive technique than conventional TLC. As such, it should be useful in radiopharmaceutical chemistry where the ability to separate and detect substances in low concentrations is very important. Important experimental parameters for TLC and HPTLC are summarized in Table 1.1. Because of these advantages, HPTLC represents a considerable advance in the practice of TLC.

SAMPLE APPLICATION

Samples are applied to HPTLC plates as spots or bands such that maximum component resolution and detectability are obtained. Samples may be applied as bands in two general ways, by using HPTLC plates with concentrating zones, or by using a band applicator. HPTLC plates with concentrating zones, available from Merck and Whatman, are prepared from two layers of silica gel having different properties [5,7]. The two layers abut each other, parallel to one edge, forming a very narrow interface. The concentrating zone is the narrower of the two zones, typically about 2 cm wide. The separating zone is prepared from the same silica gel used to coat normal HPTLC plates, while a silica gel of large pore diameter and extremely low surface area (ca. 0.5 $m^2 \cdot g^{-1}$) is used for the concentrating zone. Sample application is simplified because microliter volumes as spots or bands can be applied to any area of the concentrating zone. HPTLC plates with concentrating zones are particularly useful when large sample sizes are spotted, or crude samples (e.g., biological fluids) are directly applied to the plate. In the latter case, salts, biological polymers and other interferents that would bind with or change the homogeneity of the separation layer are retained in the concentrating zone and do not reach the analytical layer. During development, the sample migrates out of the concentrating zone and is focused at the interface as a narrow band. The sample may be developed several times just to the interface, to assist in the refocusing mechanism. The sample band at the interface should be as narrow as possible, typically about 1 mm wide, and the sample should be distributed homogeneously throughout the band for quantitative analysis. A homogeneous sample band distribution may not be obtained unless the sample is carefully applied to the concentrating zone and, thus, HPTLC plates with concentrating zones are more frequently used in qualitative than quantitative analysis.

HPTLC plates with concentrating zones can be used in laboratories lacking the sophisticated equipment necessary for spotting nanoliter volumes onto normal HPTLC plates. This is the approach that many laboratories have used to judge the value of HPTLC for applications of interest to them prior to purchasing equipment for HPTLC. It should be noted, however, that purposely designed spotting and scanning equipment is necessary to obtain the full resolution available in HPTLC and for the greatest quantitative accuracy.

Samples can be applied to HPTLC plates as narrow bands using the Camag Linomat (Applied Analytical Industries, Wilmington, NC). With this applicator, bands of any desired length can be spotted by mechanically moving the plate on a stage beneath a fixed sample syringe. A controlled nitrogen atomizer sprays the sample from the syringe, forming narrow, homogeneous bands on the plate surface. Resolution is not affected by whether the sample is applied as a band or a spot, as bands and spots of the same initial starting diameter will undergo the same broadening in the direction of migration. Different scan settings will be required to avoid distortion of the recorded chromatogram (i.e., loss of resolution), depending on whether the sample is applied as a band or a spot. The advantage of applying samples as bands is that they are easier to scan for, maximizing quantitative accuracy. The disadvantages are that relatively large sample volumes are needed, the application process is slow, and if multiple samples of various concentrations are to be applied, then flushing the syringe between applications adds considerably to the sampling time. An automated version of the Linomat for applying spots or bands has recently become available. This device can apply samples to the plate from a number of sequentially addressed vials, and can solvent-flush the applicator between samples. This expensive equipment might be useful for busy analytical laboratories with a high routine workload.

For most quantitative work in HPTLC, the sample is applied directly to the sorbent layer as a spot from a fixed volume dosimeter. The dosimeter consists of a platinum-iridium capillary of 100 or 200 nl volume sealed into a glass support capillary of larger bore. The capillary tip is polished to provide a smooth planar surface of small area (ca. 0.05 mm^2), which, when used with a mechanical applicator, does not seriously damage the plate surface. Spotting by hand is to be discouraged, as this invariably damages the plate layer, causing the chromatogram to develop irregularly, and the dosimeter to become blocked by sorbent particles from the plate layer. Mechanical application of the sample is made possible by attaching a metal collar to the glass support capillary so that it can be held by a magnet and lowered to make contact with the plate surface. The simplest form is the *rocker* applicator (W + W Electronics): An arm houses a magnet at one end to hold the dosimeter and a counterweight at the other to control the force with which the dosimeter strikes the plate surface, as shown in Figure 1.1. The dosimeter is both lowered to and removed from the plate's surface by a tipping action of the applicator arm about its fulcrum. A somewhat more sophisticated sample applicator is the Nanomat (Camag), which holds and lowers the dosimeter electromagnetically, Figure 1.1. The force with which the dosimeter hits the plate, the time it spends in contact with the layer, and the number of repetitions needed for complete sample transfer can be controlled electronically. Both applicators use a click-stop grid mechanism to aid in the even spacing of the samples on the plate to provide a frame of reference for

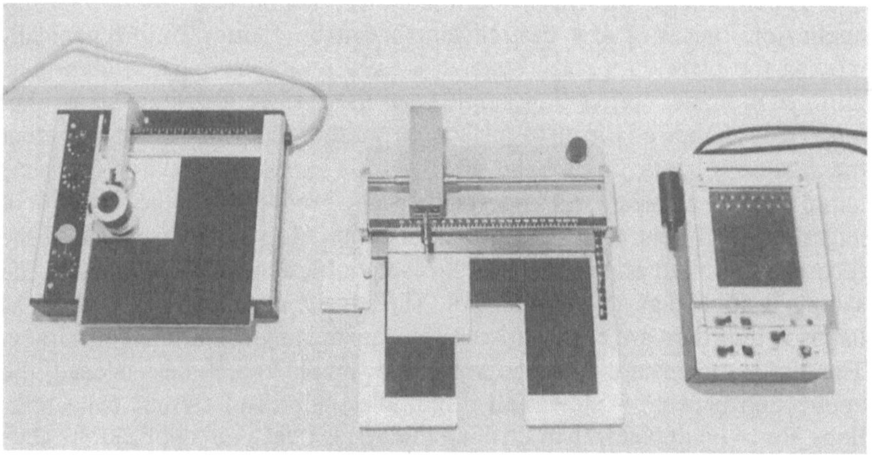

FIG. 1.1. Sample applicators for high performance thin-layer chromatography. From left to right: Nanomat, *rocker* applicator, and Transpot (from [5] with permission).

sample location during scanning densitometry. Additionally, the Nanomat is equipped with an accessory for spotting circular chromatograms. This enables the samples to be spotted in a tight circle in the center of the plate for circular development or as a larger circle toward the plate edge for anticircular development.

As an alternative to the dosimeter, samples can be spotted with a microsyringe, for example, the Camag Nano-Applicator. For accurate dispension of nanoliter volumes, the syringe is controlled by a micrometer screw gauge. This micrometer microsyringe can be operated to provide various precisely selected volumes (50–230 nl), or via a fixed level mechanism, a repetitive constant sample volume. The microsyringe delivers the sample volume by displacement rather than capillary action and, therefore, does not deform the plate surface. The microsyringe needle is brought only close enough to the plate surface for the convex sample drop of the ejected liquid to touch the plate surface.

Any of the methods discussed above are suitable for spotting solutions of low viscosity, or some viscous solutions, by repetitive application after dilution. However, samples from environmental and biological sources invariably yield viscous residues; dilution and repetitive sample application can be tedious and time consuming, or impossible, for these samples. The Transpot (Clarke Analytical Systems) provides a simple means of spotting viscous samples or large sample volumes of up to 100 microliters onto HPTLC plates, Figure 1.1. This apparatus is designed for the solventless sample application of evaporated residues of several samples simultaneously at precise locations on the thin-layer plate [8]. The transfer medium is a fluorinated ethylene-propylene film coated with perfluorokerosene, positioned over a series of depressions in a metal platform. With application of vacuum through orifices in the center of each depression, the film conforms to the contour of the platform's surface. The sample solutions are pipetted into these depressions, and the solvent is evaporated by gentle heat and a flow of nitrogen. The HPTLC plate is then positioned over the film, adsorbent side down, and with slight pressure replacing the vacuum, the spots are all transferred simultaneously to the plate. The various steps in the sample application process are summarized in Figure 1.2. Problems can arise from samples wandering out of the depressions before all the solvent has been evaporated and from the nonquantitative transfer of highly crystalline samples. Sample displacement occurs because of a buildup of static charges on the perfluorocarbon film; this is corrected for by wiping the film with antistatic paper moistened with ethanol or by using a variety of antistatic guns, radiation emitters, etc., available from record stores. Crystalline samples do not penetrate the sorbent layer, and they must be transferred in a small volume of a nonvolatile solvent such as octanol, dodecane, methyl myristate, acetophenone, etc. These carrier solvents may be added to the sample solution prior to evaporation to give a final volume of a few nanoliters, typically 2–10 nl, when the residue is transferred. A small amount of a colored dye that does

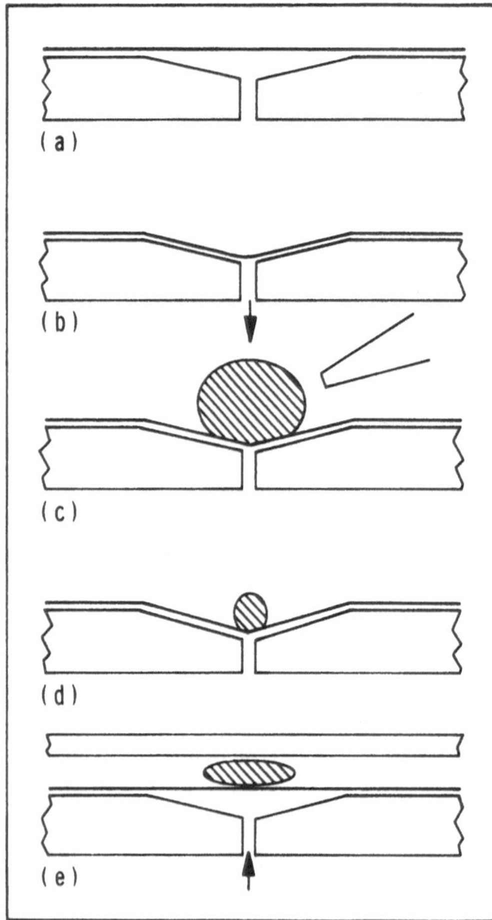

FIG. 1.2. Sample application by contact spotting using the Transpot. A specially treated fluoropolymer film is pulled into a series of depressions in a metal plate by application of vacuum (a) and (b). Sample solution is delivered by pipette (c), and after evaporation, a residue remains (d), which is transferred to the HPTLC plate by replacing the vacuum with slight pressure (e) (from [3] with permission).

not migrate in the chromatographic system can be added to each sample prior to evaporation to mark the exact position of the sample tracks for scanning densitometry [9].

The choice of sample solvent is also an important consideration in sample application. Ideally, it should be a good solvent for the sample, be nonviscous and sufficiently volatile to be easily evaporated from the plate, it must wet the sorbent layer, and it should be a weak chromatographic solvent. These considerations cannot always be met in practice. Nearly all solvents wet silica gel, so this is not normally a problem. Sorbents with chemically bonded layers, however, are poorly wet by many solvents, restricting solvent selection. If the sample solvent does not wet the sorbent layer, it will transfer to the sorbent surface as a bead without penetrating the layer. This frequently occurs with weak chromatographic solvents using reversed-phase layers. In this case, the sample may have to be applied repetitively in small volumes of a strong chromatographic solvent. For reversed-phase plates, methanol, acetonitrile, methylene chloride, and acetone are often used. Aqueous solutions of water-miscible organic solvents are frequently unsuitable, although they are more desirable to minimize sample predevelopment.

Predevelopment, itself a miniaturized chromatogram, produces enlarged starting spots, often with an uneven sample distribution (Figure 1.3). Predevelopment compromises sample resolution and detectability and can also result in the formation of elliptical spots. The extent of predevelopment can be minimized by spotting small sample volumes in a volatile and weak chromatographic solvent.

Intuitively, one might anticipate that a strong correlation should exist between the diameter of the initial starting spot and resolution. If so, decreasing the starting spot diameter to very small sizes should result in a concomitant increase in resolution. Experimental observations reveal that this is not the case (Figure 1.4). The spot size after development seems to correlate more strongly with the quality of the sorbent layer than the initial starting spot width within the normal spot size ranges used in HPTLC [10]. This does not support the use of large starting spots, for these will certainly

SPOT SIZE IS A FUNCTION OF

1. Sample volume
2. R_f value of most mobile component in spotting solvent

SPOTTING

Solvent

1. must wet the sorbent layer
2. must have sufficient solvent strength to ensure quantitative transfer of sample from the dosimeter
3. preferable volatile

FIG. 1.3. Guidelines for selecting a suitable solvent for sample application.

SPOT SIZE

Minimum starting spot required

1. For Optimum Resolution
2. Improved Detectability

size of developed spot depends more on the quality of the Chromatographic System than starting spot width

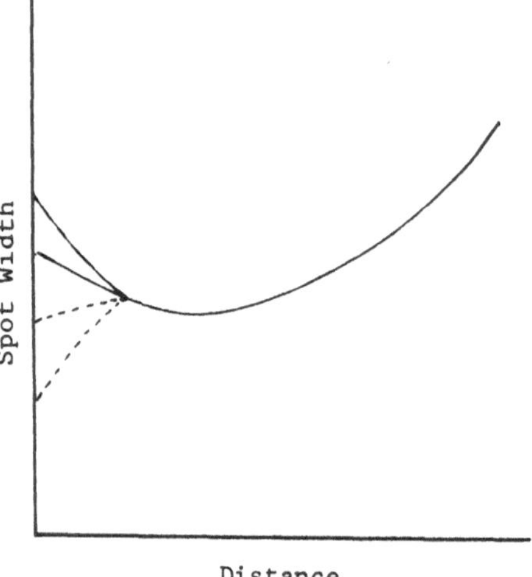

There is no particular advantage to be gained from applying spots <1.0 mm on HPTLC plates

Spots >>1.0 mm tend to develop with an eliptical shape

FIG. 1.4. Factors influencing the choice of the starting spot size in HPTLC.

cause a loss in resolution, but merely reflects the fact that without an improvement in sorbent quality, there is little to be gained by applying spots with a diameter less than 1.0 mm.

MODE SELECTION FOR SCANNING DENSITOMETRY

Commercial instruments for performing quantitative evaluation of thin-layer chromatograms first appeared in about 1967 [4,11–13]. Such instruments have played an important role in the evolution of modern TLC; without such equipment, the exquisite resolution obtained by HPTLC would be to no avail and TLC would have remained a semiquantitative technique. At best, inspection by eye of a TLC plate is capable of detecting about 1–10 μg of colored components with a reproducibility of not better than 10–30%. Excising the separated spots, eluting the substances from the sorbent material, and measuring by solution photometry are time consuming and fairly insensitive. Difficulties in accurately locating the edge of the spot by eye, incomplete elution of the sample from the sorbent, and nonspecific background absorption resulting from colloidal sorbent particles in the analytical solution add to the problem. *In situ* detection is essential for the accurate measurement of both spot size and location, for a true measure of resolution, and for rapid, accurate quantitation.

In situ measurements of substances on HPTLC plates can be made by a variety of methods: reflectance, transmission, simultaneous reflectance and transmission, fluorescence quenching, and fluorescence. Light striking the plate surface is both transmitted and diffusely scattered by the layer. Light striking a spot on the plate will undergo absorption so that the light transmitted or reflected is diminished in intensity at those wavelengths forming the absorption profile of the spot. The measurement of the signal diminution in the transmission or reflectance mode attributable to this absorption by the spot provides the mechanism for *in situ* quantitation.

In the fluorescence mode, the light striking the plate represents a source of excitation, and the sample spot can be considered as a point source of light defined by the spot boundary and the concentration profile of emitting molecules throughout the spot. Fluorescence emission occurs at a longer wavelength than excitation, so in this case the densitometer views a dark background (in the optical sense), superimposed upon which are the sample components as emission sources. This situation differs fundamentally from the conditions pertaining to absorption measurements.

Absorption measurements may also be made by fluorescence quenching. In this case, a special HPTLC plate incorporating a fluorescent indicator is used. When such a plate is exposed to ultraviolet (UV) light of short wavelength, the UV-absorbing spots appear dark against the brightly fluorescing background of lighter color. In this instance, the sample spots could be imagined as acting as UV filters: The spots absorb some of the excitation radiation, thus diminishing the intensity of the fluorescence emission eminating from that area of the plate. Hence, they appear darker against a lighter background. Only those substances whose absorption spectra overlap the excitation spectra of the fluorescent indicator will be visualized as previously discussed. Fluorescence quenching measurements are less specific and less sensitive than absorption measurements. Severe background fluctuations resulting from the inhomogeneous distribution of fluorescent indicator in the sorbent layer are the principal reason for low sensitivity. Fluorescence quenching should primarily be considered as a visualization technique. If a densitometer is available, then the measurement of absorption is preferred to fluorescence quenching.

An exact theoretical description of the optical behavior of a thin-layer plate would be very complex. Since TLC plates are opaque and scatter light strongly, absorption measurements cannot be expected to obey the Beer–Lambert law. The properties of light absorption by a sample in an optically dense medium are reasonably well described by the Kubelka–Munk equation (1) [11,13–15].

$$\frac{(1-R)^2}{2R} = \varepsilon \frac{C}{S} \tag{1}$$

R = reflectance

ε = absorption coefficient

S = scatter coefficient

C = spot concentration

The Kubelka–Munk equation predicts that calibration curves will be linear initially, and pass through the origin, but curve toward the concentration axis at higher concentrations.

When applicable, fluorescence detection is the measurement technique of choice. Compared with absorption measurements, it provides higher sensitivity, selectivity, wider linear calibration range, and a signal that is virtually independent of the spot shape. For compounds that fluoresce strongly, detection limits 100 to 1000 times lower than for absorption are possible. In common with solution fluorometry, in TLC there is an exponential correlation between the fluorescence intensity, I_{FL}, and the substance quantity, C:

$$I_{FL} \simeq I_{0\lambda}(1 - 10^{-\varepsilon_\lambda C}) \tag{2}$$

ε_λ = absorption coefficient at the excitation wavelength

$I_{0\lambda}$ = intensity of the excitation wavelength

For low concentrations, the product $\varepsilon_\lambda C$ is small, and the linear approximation may be applied

$$I_{FL} \simeq I_{0\lambda}\,\varepsilon_\lambda C \tag{3}$$

In fluorescence scanning densitometry, linear calibration curves passing through or close to the origin are obtained. The linearity of the calibration curve usually extends over two or three orders of magnitude in sample amount.

There is much confusion in the literature concerning whether absorption or fluorescence measurements should be made in the reflectance or transmission mode [6,16]. For measurements in the UV region, the transmission mode may be inappropriate owing to the strong absorption of wavelengths shorter than ca. 320 nm by the glass backing plate and by silica gel itself. Sample resolution is not influenced by the measurement method, and, therefore, sample detectability at wavelengths greater than 320 nm is the determinant factor. Normally, transmission measurements provide larger signals than those observed by reflectance, but this is usually accompanied by an increase in the background noise. In the reflectance mode, most of the scattered light arises from layers close to the surface and is less significantly influenced by changes in the thickness of the sorbent layer, which contributes much of the background noise in the transmission mode. On an instrument-to-instrument basis, sample detectability in the reflectance and transmission modes will depend largely on the provision made for controlling background noise. In many instances, the difference is quite small, and either mode will provide acceptable analytical data.

INSTRUMENTATION FOR SCANNING DENSITOMETRY

Commercial instruments for scanning densitometry share many features in common. Halogen or tungsten lamps are used as sources for the visible, and deuterium lamps for the UV region. High intensity mercury or xenon sources are required for fluorescence measurements. Some low cost densitometers use filters and a mercury source for absorption measurements in the UV range. Filter densitometers provide only a limited number of measuring wavelengths, determined by the characteristic line spectra of the source. More frequently, grating or prism monochromators and wide-spectrum sources are used. For fluorescence measurements, a monochromator or filter is used to select the excitation wavelength. A cutoff filter, which transmits the wavelength of emission but attenuates the wavelength of excitation, is placed between the detector plates. Interference filters can be substituted if better selectivity is required; however, sensitivity will decrease because emission spectra usually consist of broad bands. Photomultipliers or photodiodes are used to measure the light reflected or transmitted by the plate.

Three optical arrangements are commonly used in contemporary scanning densitometers, as shown in Figure 1.5. The single-beam arrangement is capable of producing excellent quantitative results, but spurious background noise resulting from fluctuations in the source output, inhomogeneity in the distribution of extraneous absorbed impurities, and irregularities in the plate surface can be troublesome. Background disturbances are compensated for in the double-beam operating mode by exposing the plate to two beams and recording the difference between the two signals. The two beams can either be separated in time at the same point on the plate, or separated in space and recorded simultaneously by two detectors.

The double beam in space optical arrangement divides a single beam of monochromatic light into two beams that scan different positions on the plate. One beam scans the sample lane while the other traverses the blank regions between sample lanes. The two beams are subsequently detected by matched photomultipliers, and a difference signal is fed to the recorder; fluctuations in the output of the source are corrected in this way. As the two beams impinge on different areas of the plate, however, small irregularities in the plate surface and undesired background contributions from impurities in the sorbent layer may still pose problems.

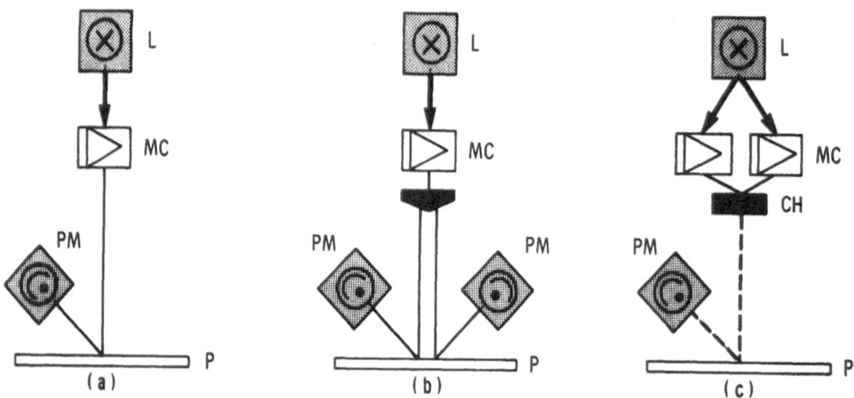

FIG. 1.5. Schematic diagrams showing the optical arrangements of different kinds of scanning densitometers: (a) single beam, (b) double beam in space, and (c) double beam in time (from [3] with permission).

In the single-beam dual-wavelength mode shown in Figure 1.6, fluctuations caused by scattering at a light-absorbing wavelength (λ_1) are compensated for by subtracting the fluctuations at a different wavelength (λ_2) at which the spot exhibits no absorption but experiences the same scatter. The two beams are altered by a chopper and recombined into a single beam to provide the difference signal at the detector. Because the scatter coefficient

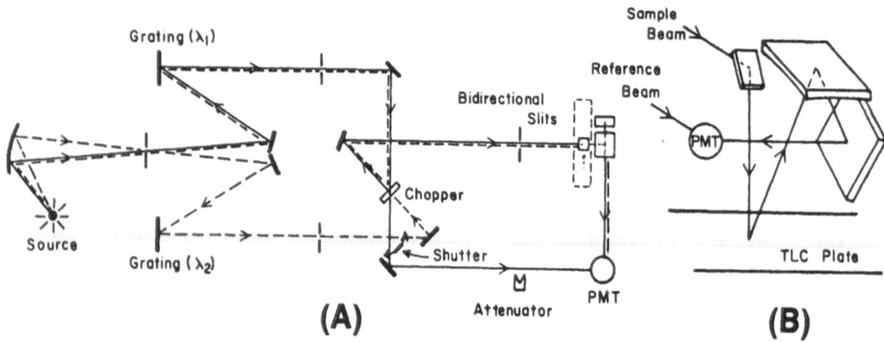

FIG. 1.6. Optical path of single-beam dual-wavelength scanning densitometer (Shimadzu CS-910). (A) top view; (B) perspective of collection optics. Single-wavelength (———) and dual-wavelength (----) modes. The source beam exiting the monochromator is chopped by a twin-lobed mirror, rotating at 60 Hz, yielding two beams modulated at 120 Hz. These beams are referred to as the sampling and reference beams. The reference beam passes through a V-shaped shutter, to a frosted glass scatter plate, and then to the photomultiplier. The sampling beam passes through the bidirectional slits and is focused onto the scanning stage. The reflected light from the TLC plate is collected through a 5-cm orifice on the stage lid by a 7 × 7 cm spherical mirror focused 18° off axis from the sampling beam. The collected light is then directed to the photomultiplier.

is wavelength dependent to some extent, the correction is better when λ_1 and λ_2 are as close as possible. Background correction for spurious absorbing impurities can be very good in this mode, as indicated in Figure 1.7 [17].

In practice, the quality and surface homogeneity of HPTLC plates are generally very good, and single-beam single-wavelength operation is predominantly used. The lower quality of conventional and home-made TLC plates generally requires the use of background correction to obtain a stable baseline. Background correction may also be required if sorbent impurities interfere in recording the chromatogram. Of the methods discussed for background correction, the single-beam dual-wavelength mode is the most reliable, but it is often difficult to meet the requirement that λ_1 and λ_2 should be similar, and yet sample absorption should occur at only one of the wavelengths.

In all contemporary densitometers, the position of the sample beam is fixed. The plate is scanned by mounting it on a movable stage controlled by

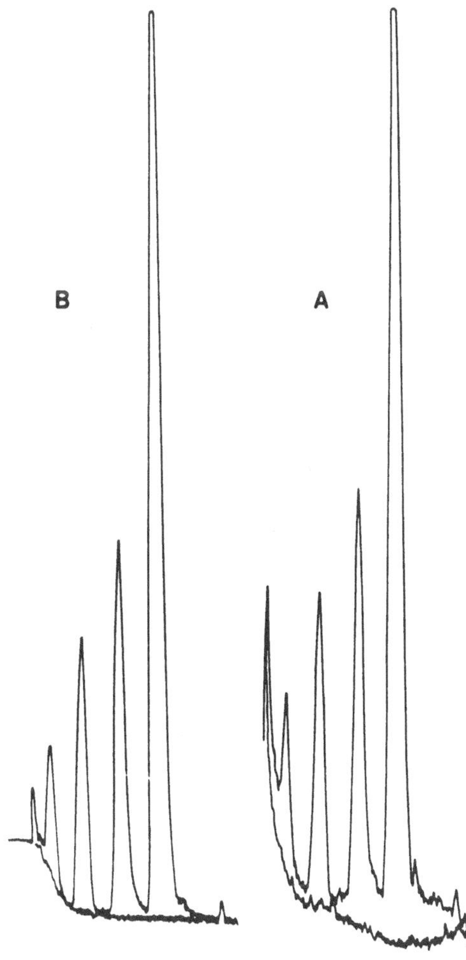

FIG. 1.7. Use of background correction to improve the baseline stability in the analysis of a mixture containing metoprolol and some anticipated contaminants. (A) was measured using the single-wavelength mode: λ = 280 nm, and (B) by the single-beam dual-wavelength mode: λ_1 = 280 nm and λ_2 = 300 nm (from [17] with permission).

stepping motors. The motor-driven scanning stage transports the plate through the beam in the direction perpendicular to the slit; the stage is manually operated or motor driven in the orthogonal direction. Each scan represents a lane whose length is defined by the sample migration distance and whose width is specified by the slit dimensions. Several modern densitometers completely automate the scan sequence, including lane changing, through commands entered into computer memory.

Linear scanning is the most frequently used scan mode. Some instruments can perform zig-zag, radial, and peripheral scanning as well. Radial and peripheral scanning are used for recording circular or anticircular chromatograms. Neither circular nor anticircular chromatography has become popular and, therefore, the need for radial or peripheral scanning has declined. Zig-zag scanning improves the quantitative accuracy with which spots of an irregular shape can be recorded. Zig-zag scanning is more time consuming than linear scanning and produces a signal output that is not readily handled by standard laboratory recorders and integrators without modification. It is more useful in conventional TLC where spots are large and frequently distorted compared with the compact symmetrical spots obtained by HPTLC.

INSTRUMENT PARAMETERS THAT AFFECT THE PERFORMANCE OF SLIT-SCANNING DENSITOMETERS

The faithfulness with which a scanning densitometer can convert a separation on a TLC plate into a strip chart chromatogram is dependent on several instrument parameters. The most important of these are the dimensions of the measuring beam, the scan rate, and the total electronic time constant of the instrument and recording device. Recorders and integrators used for open tubular column gas chromatography are generally suitable for HPTLC. At high scan rates, peak profiles may become distorted as a result of inadequate response times of the amplifier and recording device. At normal scan rates, 2–5 $cm \cdot min^{-1}$, signal distortion should be absent in a well-designed densitometer interfaced to a reliable recorder or integrator. The dimensions of the measuring beam can influence both resolution and sensitivity. For slit-scanning densitometers, the image of the scanning beam on the plate is a rectangle defined by the slit width and the slit height. The slit width is the parameter that fixes the dimensions of the beam in the direction of scanning; the slit height defines the beam dimensions in the orthogonal direction. Resolution is generally independent of the slit height, but shows some dependence on the slit width [4,18]. As shown in Figure 1.8, as the slit width is increased, the resolution of the two partially separated PTH-amino acids, leucine and isoleucine, declines. At small slit-width settings, the baseline noise increases and the optimum slit-width setting is generally the largest value that does not compromise the observed resolution. The ratio of the slit height to spot width affects mainly the signal-to-noise ratio, Figure 1.9. The

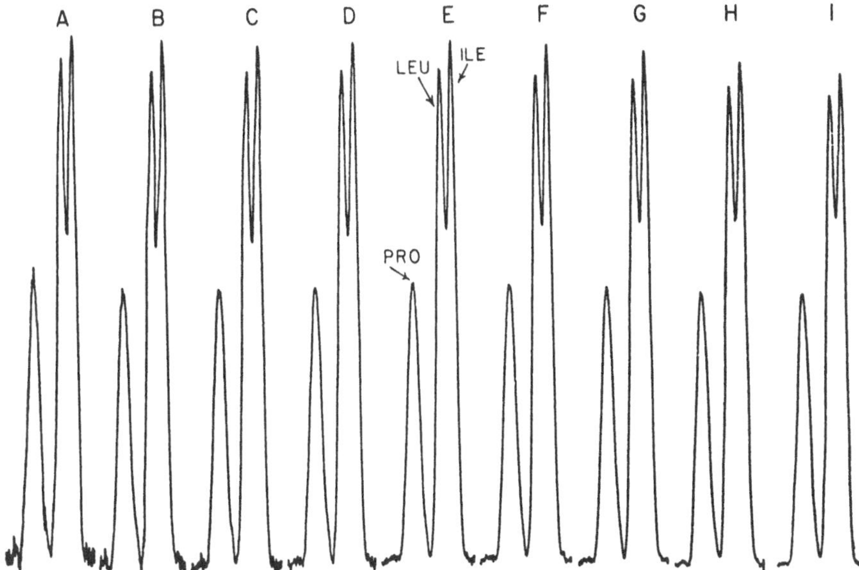

FIG. 1.8. Resolution of PTH-derivatives of proline, leucine, and isoleucine measured at a constant slit height 1.50 mm and scan rate 24 mm·min^{-1}. Slit width: (A) 0.05 mm; (B) 0.10 mm; (C) 0.20 mm; (D) 0.40 mm; (E) 0.60 mm; (F) 0.80 mm; (G) 1.0 mm; (H) 1.20 mm; (I) 1.40 mm (from [18] with permission).

detector signal represents the diminution of scattered or transmitted light at the detector as the measuring beam moves from a blank area of the plate over the area occupied by the sample. If the slit height is large compared with the spot dimensions, then the signal will be weak as the detector sees mainly light scattered or transmitted by the plate, with only a small contribution from absorption by the spot. With beam dimensions similar to the size of the spot, the signal increases as the amount of scattered or transmitted light is reduced, but the amount of absorbed light remains constant. As the sample concentration across the diameter of the spot is not constant, the signal continues to increase as the slit height is reduced to less than the spot diameter. The contribution of absorbing molecules at the fringe of the spot is small compared with those at the center. At very small slit heights, the signal levels off as the slit height approaches the diameter of the core of the spot where sample concentration is approximately constant. At small slit heights, though, instrumental noise increases such that an optimum value for the slit height signal-to-noise ratio exists. A value slightly larger than the diameter of the spot is generally used.

A contrary trend is observed in the measurement of fluorescence [18,19]. This is to be expected because the signal originates in a manner different from that for absorption. Here the signal is a true emission signal, and the spot can be considered as a point source. The slit-width value can affect both

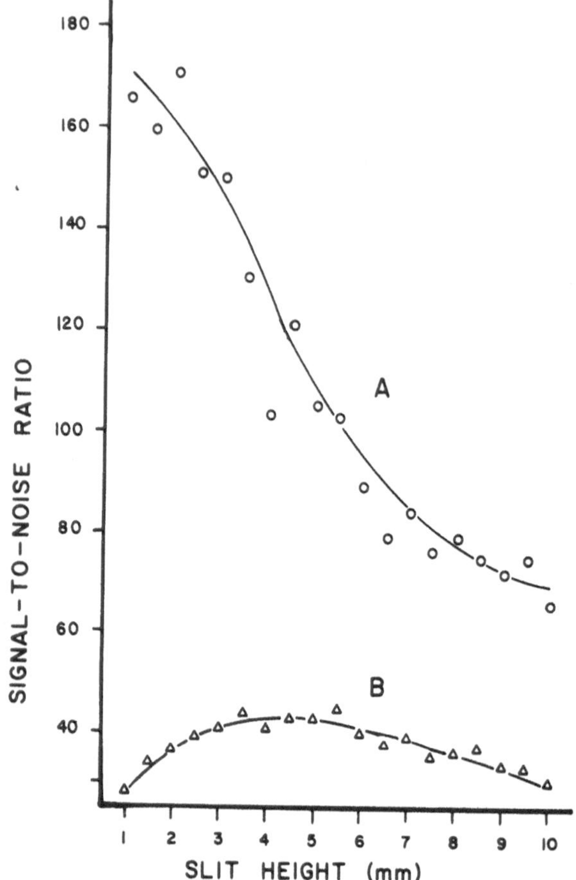

FIG. 1.9. Relationship between signal-to-noise and slit-height setting at a constant slit width (0.5 mm) for (A) a compact spot (3.0 mm) and (B) a diffuse spot (6.6 mm).

resolution and sensitivity. The signal declines dramatically as the slit width is reduced, as Figure 1.10 indicates. The optimum slit-width value is thus the largest that can be used without compromising the observed resolution. The slit height does not affect resolution but does influence sensitivity. The signal increases linearly with increasing slit-height values for slit heights smaller than the diameter of the spot, as shown in Figure 1.11. When the spot diameter is reached, the signal intensity levels off and only increases slightly with larger slit heights. Thus, the optimum slit height is a value slightly larger than the spot diameter.

PROTOCOL FOR MEASURING THE SENSITIVITY OF A SLIT-SCANNING DENSITOMETER

The performance of a slit-scanning densitometer can be judged from its ability to record a chromatogram without loss of resolution or dynamic signal range, and by its ability to determine spots of low concentration [20,21]. No

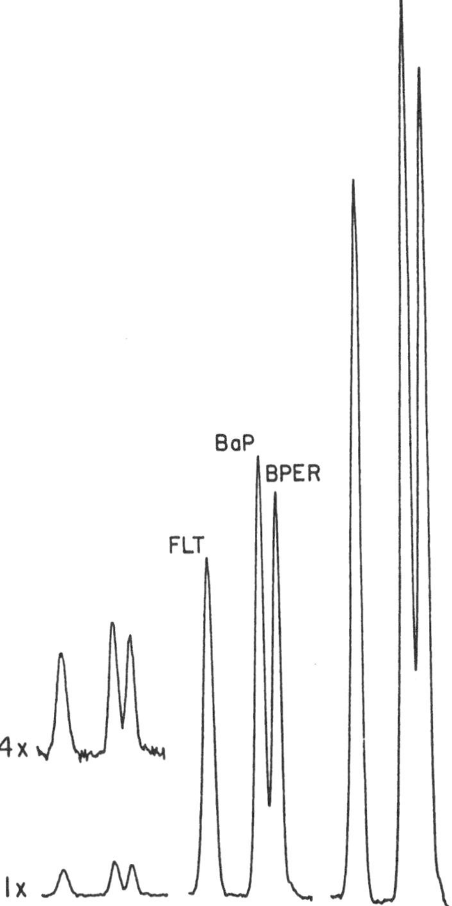

FIG. 1.10. The effect of measuring beam dimensions on fluorescence emission intensity for a mixture of benzo[a]pyrene (BaP), benzo[ghi]perylene (BPer), and fluoranthene (Flt). The separation was scanned with a slit height of 6.0 mm and slit width of 1.40 mm, 0.60 mm, and 0.05 mm, respectively from right to left.

absolute measure of resolution is available, and this makes a comparison of individual densitometers difficult except for side-by-side comparison of the same separation with both instruments set to optimum measuring conditions. Some manufacturers indicate the spatial resolution of their instruments as measured by scanning a photographic test pattern of adjacent equal-density lines. Typical values quoted for spatial resolution are 10–100 nm. Since real spots do not have discrete edges, are not regularly shaped, and are generally larger than the lines in a photographic test pattern, it is not clear how quantitatively useful spatial resolution is as an indicator of chromatographic resolving power.

Instrument sensitivity is a more accessible parameter and can be measured by a standardized protocol [20,21]. For single-wavelength densitometers, azobenzene is used as the reference standard for the UV and visible range. Diphenylacetylene is used as a secondary standard for dual-wavelength instruments. The protocol specifies the concentration of the standards, measuring wavelengths, spot size, and dimensions of the measuring beam. An outline of the protocol is given in Table 1.2, and some typical

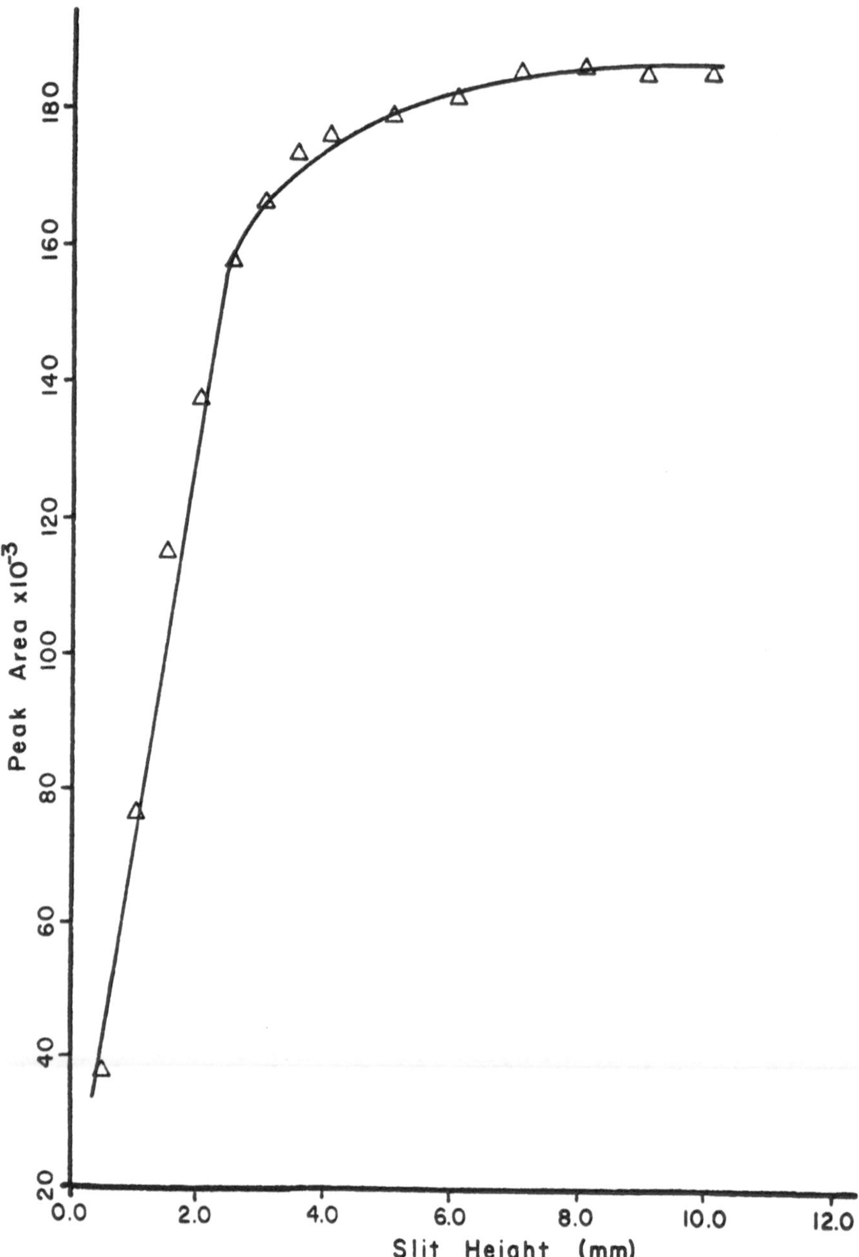

FIG. 1.11. The effect of slit height on observed signal for fluoranthene. Spot size was ca. 4.0 mm and slit width 0.50 mm (from [19] with permission).

TABLE 1.2. Sequence of Operations for Determining the Sensitivity and Detectability Values of a Scanning Densitometer

Step	*Procedure*	*Explanation*
1	Preclean all HPTLC plates	Development in acetonitrile reduces background levels
2	Spot several plates and develop for different times	Suggested times: 2, 3, 4, 5, and 10 min
3	Plot development time vs. spot width	Spot width is taken as the width of the peak at baseline; the slit height and width are not specified for this step, but it is suggested that a slit height be chosen that is large enough to include the entire spot; plot should have linear region
4	Determine appropriate development time	From the graph of development time vs. spot width or by calculation, determine the development time necessary to obtain a 4.0-mm spot
5	Spot plate and develop for appropriate time	Check to be sure that the resulting spot is indeed 4.0 mm in diameter; some fine tuning of the system may be necessary
6	Scan plate	Use slit height of 4.4 mm (peak width plus 10%), slit width of 0.8 mm, and wavelength 320 nm (UV) or 430 nm (vis) for azobenzene, and 290 nm and 350 nm (reference) for diphenylacetylene. The sample concentration is 1.0 mg/ml (UV) and 5.0 mg/ml (vis) azobenzene or 0.1 mg/ml diphenylacetylene. Sample size 200 nl
7	Measure peak height from chromatogram	Should be constant when all necessary parameters are controlled; recorder sensitivity at 10 mV
8	Calculate sensitivity (S)	$\frac{\text{peak height (mm)}}{\text{sample weight (ng)}} = S \text{ m/g} \times 10^6$
9	Measure noise level	Using all parameters as stated, measure the noise level at a blank area of the plate in the nonscan mode. Average recorder pen deflections over a 5-min period, measured at 5 mV and scaled to match peak height measurement (10 mV)
10	Calculate detectability (D)	$\frac{2 \times \text{noise (mm)}}{S \text{ (mm/ng)}} = D \text{ (g} \times 10^9)$

TABLE 1.3. Summary of Sensitivity and Detectability Data for a Shimadzu CS-910 Scanning Densitometer

Mode	*Standard*	*Concn. (mg/ml)*	*Measuring Wavelength (nm)*	*Noise (mm)*	*Ave. Peak Height (mm)*	*Sensitivity (mm/ng)*	*Detectability (ng)*
Transmission Single wavelength	Azobenzene	5	430	0.43	152 (±3.3%)	0.152 (±3.2%)	5.65 (±3.4%)
Reflectance Single wavelength		1	430	0.45	141 (±3.3%)	0.140 (±3.3%)	6.37 (±3.5%)
			320	5.2	321 (±2.8%)	1.61 (±2.8%)	6.42 (±2.8%)
	Diphenylacetylene	0.1	290	3.0	66 (±4.99%)	3.27 (±4.9%)	1.84 (±5.5%)
Reflectance Dual wavelength			290	2.5	56 (±7.2%)	2.79 (±7.2%)	1.8 (±8.1%)

values obtained with a Shimadzu CS-910 scanning densitometer appear in Table 1.3.

QUALITATIVE SAMPLE IDENTIFICATION BY HPTLC AND SCANNING DENSITOMETRY

Separated sample components are identified in HPTLC by coincidence of their migration properties with standards spotted on the same plate and by measuring the *in situ* absorption or fluorescence spectra of components for comparison with standard spectra. Most densitometers make some provision for either manually or automatically recording the *in situ* spectra of any desired spot. For automatic spectrum recording, a motor-driven monochromator controlled by the central microprocessor is used. For manual recording of spectra, the spot is scanned repetitively while the monochromator position (or fluorescence emission filter for recording the fluorescence spectra) is changed by fixed-wavelength increments between scans. An example of a manually scanned spectrum is shown in Figure 1.12. A line drawn through the peak maxima gives the characteristic absorption or fluorescence emission envelope of the compound as a function of wavelength. For absorption studies, the solution spectra and *in situ* plate spectra of a compound are frequently similar. In the case of fluorescence studies, the solution and plate spectra may differ substantially from one another. The solution spectra are thus a reasonable guide for optimizing measuring conditions in the absorption mode, but may be misleading for fluorescence studies.

It is less time consuming to scan a separation sequentially at different wavelengths than to record the full spectra of each spot [22–25]. The ratios of the response values at different wavelengths are characteristic of the sample and can be used to confirm identity or indicate contamination of the spot with other components. The wavelengths selected for identification should emphasize the spectral characteristics of the substance of interest (λ values for peaks and troughs, etc.). If standards are run on the same plate and used to measure the absorbance response ratios, then the reproducibility of the ratios is very good (RSD = 1–3%). Otherwise, the reproducibility of the absorbance ratios will depend on how accurately the monochromator can be reset to a particular wavelength between measurements. Most densitometers do not allow wavelengths to be reset with sufficient accuracy to make the use of absolute response ratios meaningful.

Sequential wavelength scanning may also be used to enhance the resolution of unresolved components on the HPTLC plate. For this purpose, wavelengths must be chosen such that neither unresolved component has significant absorption or fluorescence where the other component is to be determined. An example is shown in Figures 1.13 and 1.14 for the determination of penicillic acid, patulin, and luteoskyrin. Patulin and luteoskyrin slightly overlap each other, but as patulin does not show any absorption at the absorption maximum for luteoskyrin (440 nm), the determination of lu-

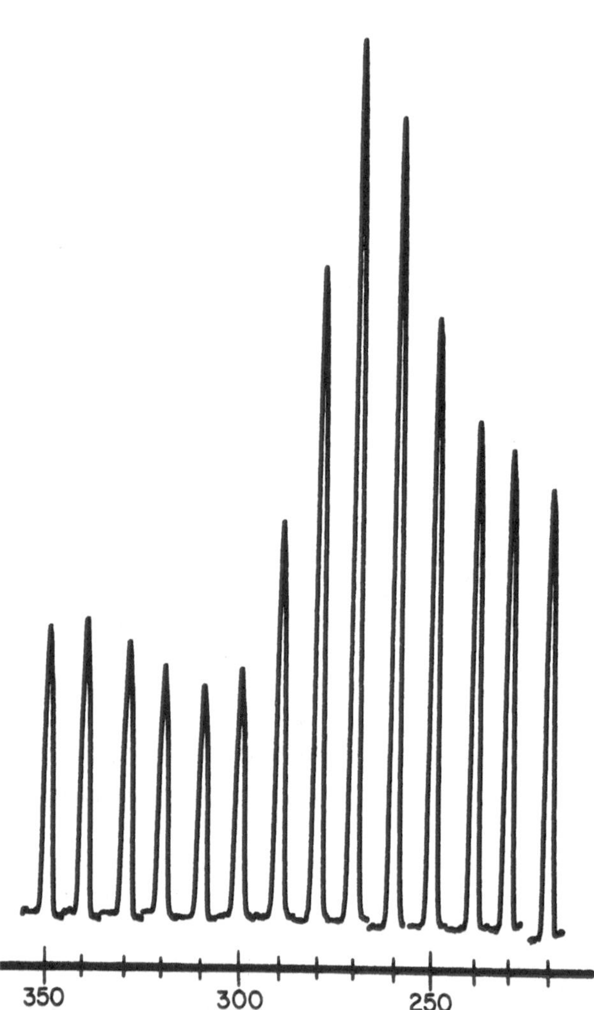

FIG. 1.12. UV absorption spectra of PTH-L-asparagine recorded *in situ*. The spot was scanned repetitively at 10 nm intervals and the absorption maxima measured. A line drawn through the peak apices provides a diagram of the conventional absorption envelope (from [5] with permission).

teoskyrin in the presence of patulin does not present a problem. Patulin can be determined at 240 nm where the contribution from luteoskyrin is very small.

Spectral selectivity is much higher in the fluorescence mode than in the absorption mode. This arises for two reasons. First, of all known organic compounds, only a small percentage are naturally fluorescent; and second, two wavelengths, an excitation wavelength and an emission wavelength, are

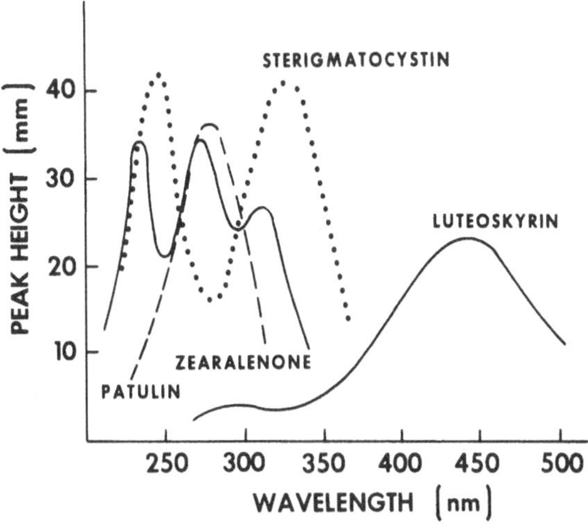

FIG. 1.13. *In situ* UV-visible absorption spectra for some mycotoxins.

used for measurement instead of the single wavelength used for absorption. Figure 1.15 shows an example of the use of sequential wavelength scanning at two different emission wavelengths to characterize a mixture of polycylic aromatic hydrocarbons [26]. The fluorescence emission properties of the polycyclic aromatic hydrocarbons (PAHs) can be expressed on a standard scale using excitation wavelengths of 254, 266, 313, and 365 nm [24,25]. These values are known as the emission response ratios (ERR) and are formally defined by equation (4).

$$\mathrm{ERR(ex,em)} = \frac{[\mathrm{PAH(ex,em)}]/[\mathrm{ng}]}{[\mathrm{Per(ex,em)}/[\mathrm{ng}^1]} \times 100 \qquad (4)$$

ERR(ex,em) = Emission Response Ratio at a given excitation and emission wavelength

PAH(ex,em) = Peak area of test compound at the given excitation and emission wavelength

Per(ex,em) = Peak area for perylene at the given excitation and reference emission wavelength

ng = Nanograms of PAH

ng^1 = Nanograms of perylene

Perylene is used as a reference standard to account for minor changes in instrumental parameters on a day-to-day basis. Perylene was selected for this purpose as it is readily available in high purity, is chemically and photolytically stable on the plate, and provides a suitable emission signal at all excitation wavelengths. The emission intensity of perylene per nanogram at

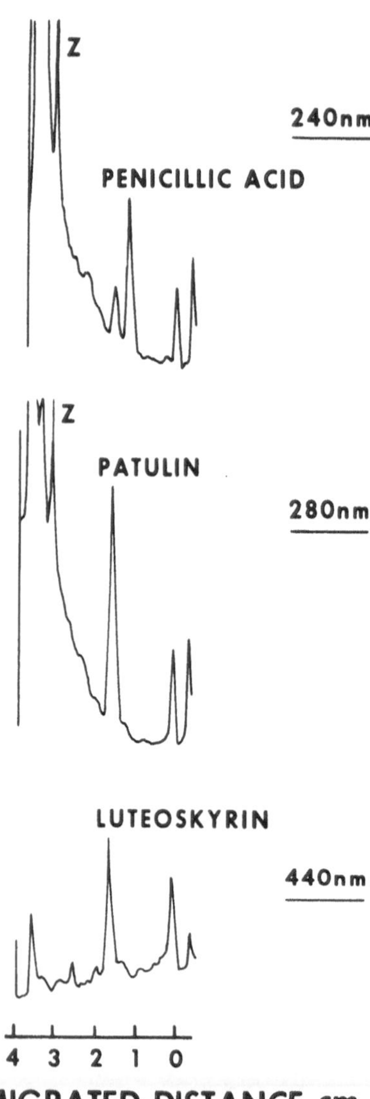

FIG. 1.14. The use of sequential wavelength scanning to determine the poorly resolved mixture of penicillic acid, luteoskyrin, and patulin (from [22] with permission).

450 nm was assigned a value of 100 at each excitation wavelength to provide a scale for comparison. For sample amounts that remain within the linear response range of the detector, the ERR values can be reproduced with a RSD = 3% on a day-to-day basis.

The ERR values can be used in several ways. Plotted in diagrammatic form, for example, as shown in Figure 1.16, they can be used to select the optimum emission wavelength for detection at any particular excitation wavelength. They also provide a convenient means of selecting combinations of excitation and emission wavelengths to maximize discrimination in the detection of specific PAHs in unresolved spots. The product obtained by

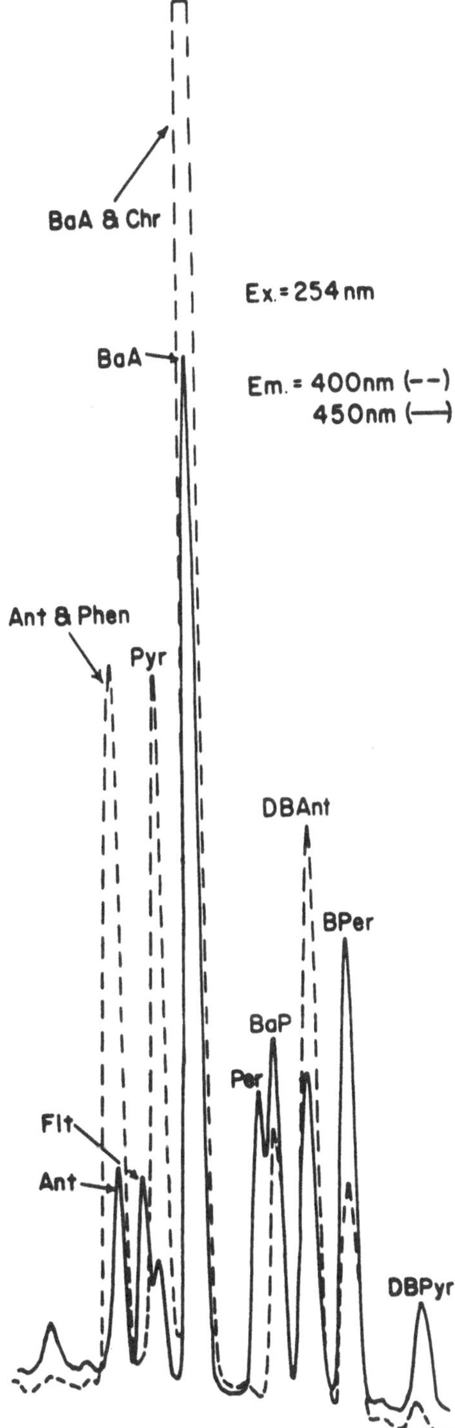

FIG. 1.15. Separation of a mixture of PAHs by HPTLC. Sequential wavelength scanning at two emission wavelengths provides complementary data on the complexity of the mixture. Ex = 254 nm, and Em = 400 nm (----) and 450 nm (———). DBPyr = dibenzo[a,i]pyrene, BPer = benzo[ghi]perylene, DBAnt = dibenzo[a,h]anthracene, BaP = benzo[a]pyrene, Per = perylene, BaA = benzo[a]anthracene, Chr = chrysene, Pyr = pyrene, Flt = fluoranthene, Ant = anthracene, and Phen = phenanthrene.

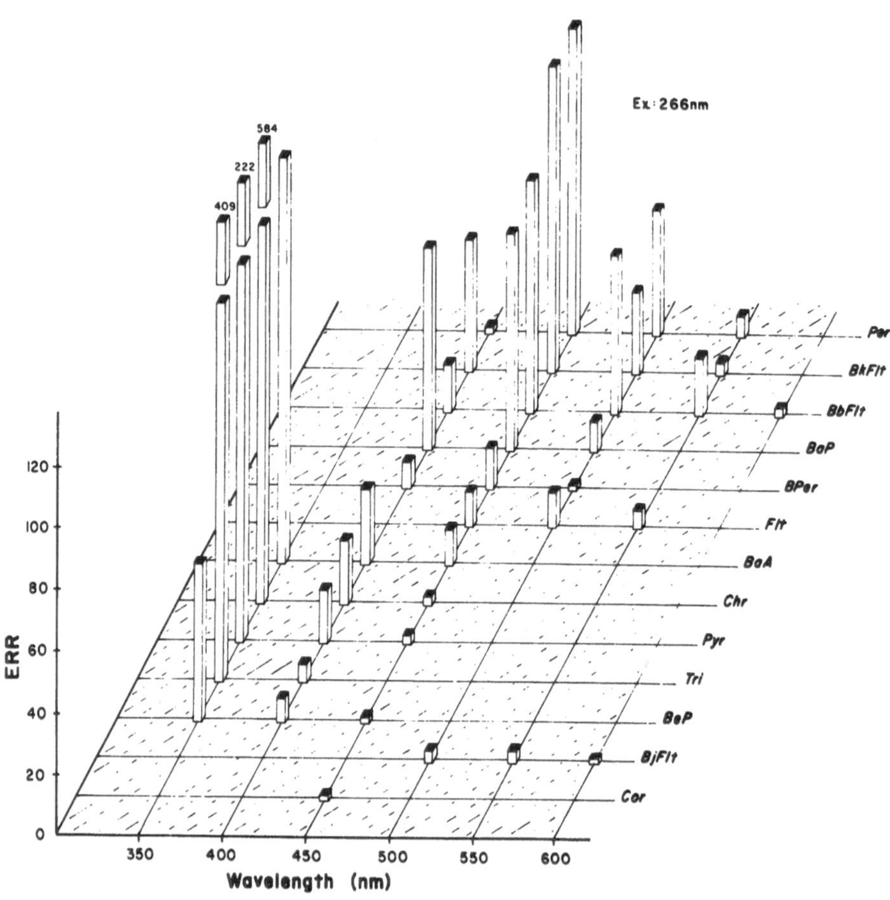

FIG. 1.16. Three-dimensional plot of emission response ratios at 266 nm excitation (from [24] with permission).

dividing any combination of ERR values by each other at a fixed excitation wavelengths provides a substance-specific ratio suitable for qualitative identification of PAHs in environmental samples [24,26]. The ERR values were designed for use with PAHs, but there is no fundamental reason why the method could not be extended to include any fluorescent compound.

QUANTITATION OF SEPARATED COMPONENTS BY SCANNING DENSITOMETRY

Calibration is the method used to convert detector response data into sample concentration. Samples and a sufficient number of standards to construct a calibration curve are spotted on the same plate and developed simultaneously. Each lane is then scanned in turn to produce the response data for the samples and to generate the calibration curve. In the absorption mode, calibration curves are often nonlinear, as shown, for example, in Figure 1.17. This is not an artifact, as absorption of light by samples on a thin-layer plate obeys the Kubelka–Munk equation discussed previously (equation 1). The curved upper branch of the calibration graph can often be linearized by a parabolic approximation

$$A^2 = f(C) \tag{5}$$

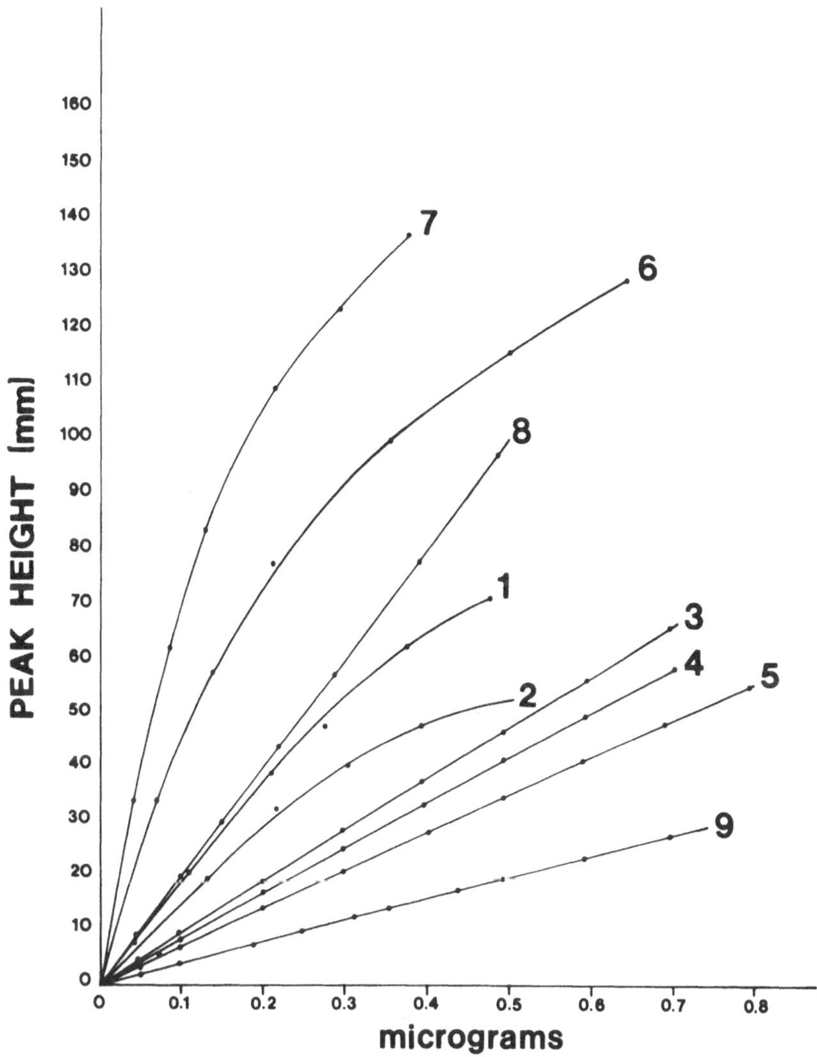

FIG. 1.17. Calibration curves for a series of β-blocking drugs measured at their absorption maximum by scanning densitometry. 1 = metoprolol (220 nm), 2 = propranolol (310 nm), 3 = oxprenolol (270 nm), 4 = atenolol (205 nm), 5 = alprenolol (270 nm), 6 = timolol (310 nm), 7 = practolol (250 nm), 8 = pamatolol (220 nm), and 9 = prenalterol (220 nm) (from [23] with permission).

or a twofold logarithmic approximation

$$\log A = f(\log C) \tag{6}$$

where A is the area enclosed by the reflectance curve. If an internal standard with similar absorption properties to the substance or substances of interest is added to the sample, then the ratio of the peak areas of these substances to the internal standard plotted as a function of concentration is often linear over one to two orders of magnitude [27]. Electronic linearization of the signal response is available on some densitometers. These devices usually employ some approximate solution of the Kubelka–Munk equation. However, electronic linearization often results in the introduction of large errors into the substance calibration and is of dubious value [15,28]. In many instances, linearization is used to fulfill a psychological rather than a scientific need. Why else increase the error associated with an analysis for the comfort of drawing a straight line through the calibration data? Restricting calibration to the pseudolinear region of the calibration curve or the use of nonlinear regression techniques would be most preferable [28].

For normal development conditions, the reproducibility of absorbance and fluorescence measurements by scanning densitometry is 1–3% RSD. The most important source of error for absorption measurements is the poor reproducibility of the spot shape or size as a result of inadequate experimental technique [29]. It is very difficult to maintain a constant spot size by applying variable sample volume to the plate unless the Transpot sample applicator is used. Thus, to prepare a calibration curve using dosimeters, etc., similar volumes of different concentrations of standards should be spotted and not different volumes of a fixed concentration. Spots will also be distorted and poorly quantified if the plate is developed under conditions that cause a sagging or ragged solvent front.

External calibration using samples and standards developed on the same plate is used for quantitation in fluorescence scanning densitometry. Calibration curves are usually linear over two or three orders of magnitude and largely independent of small changes in the spot size or shape. Calibration is thus quite straightforward.

A novel method of calibration in fluorescence scanning densitometry, which requires only a single standard, has recently been described [30]. In fluorescence scanning densitometry, a linear relationship exists between the detector response and the slit width used for scanning. Also, the slope of the slit width vs. detector response curve is proportional to the sample amount, as indicated in Figure 1.18. The linear relationship between the slit width and detector response can be represented by equation (7)

$$A = MW + B \tag{7}$$

where:

$$A = \text{area of the peak}$$

$$M = \text{slope of the line}$$

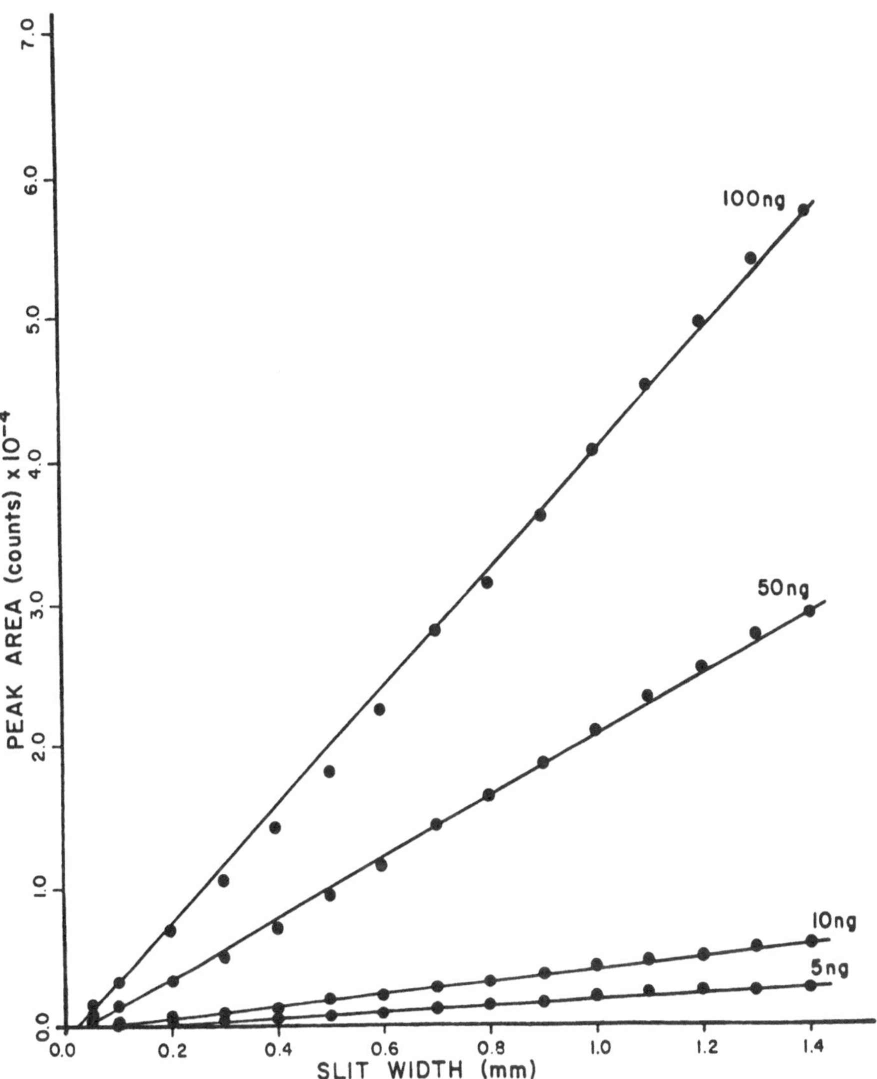

FIG. 1.18. Relationship between detector response and slit width for fluoranthene standards of different amounts (from [30] with permission).

$$W = \text{slit width}$$

$$B = \text{constant}$$

The slope M of equation (7) is proportional to the sample amount C. Making this substitution into equation (7) for a standard and an unknown, leads to equation (8)

$$\frac{M_s}{M_u} = \frac{C_s}{C_u} \tag{8}$$

M_s = slope of the detector response vs. slit width curve for a standard of known amount

M_u = slope of the detector response vs. slit width curve for the unknown

C_s = amount of standard

C_u = amount of unknown

The values for M_s and M_u can be determined by linear regression analysis. As C_s is known, C_u is calculated from equation (9)

$$C_u = \frac{C_s M_u}{M_s} \tag{9}$$

Values for M_s and M_u can be obtained by scanning the standard and unknown twice with slit-width values of 0.4 mm and 0.8 mm. Compared with conventional calibration, the accuracy of the method is not as good, with typical errors of 5–14% for values of C_u. This is sufficiently accurate, however, for screening purposes and for estimating the range of sample concentrations for conventional calibration. Most importantly, it preserves the high sample throughput of the HPTLC method by requiring only a single track for calibration standards.

REAGENTS USED TO ENHANCE FLUORESCENCE OF ORGANIC COMPOUNDS ON THIN-LAYER PLATES

Some fluorescent compounds show an unusually low response when adsorbed onto sorbents compared with the value anticipated from solution behavior. This is frequently observed for compounds adsorbed on silica gel and less frequently for chemically bonded sorbents. It has been observed that the fluorescence signal can be restored to its anticipated value by dipping or spraying the HPTLC plate with a solution of a viscous liquid in a volatile organic solvent. Suitable fluorescence-enhancing reagents are liquid paraffin [31], triethanolamine [31,32], dodecane [6,33], Triton X-100 [34], and Fomblin Y-Vac [6,33]. Fluorescence enhancement values of 10- to 200-fold have been observed in favorable cases. An example is shown in Figure 1.19 for some polycyclic aromatic hydrocarbons adsorbed on silica gel using dodecane as the fluorescence-enhancing reagent. The mechanism of fluorescence enhancement is not understood, but it is usually assumed that adsorption of the sample on the sorbent provides additional nonradiative pathways for loss of the fluorescence excitation energy. These pathways are no longer available when the plate is sprayed with a nonvolatile liquid because the adsorbed solute is transferred to the liquid state. In the liquid phase, other fluorescence-enhancing mechanisms may also be important.

The principal disadvantage of the fluorescence enhancement technique is the spot broadening that occurs while the sample is free to diffuse in the wet

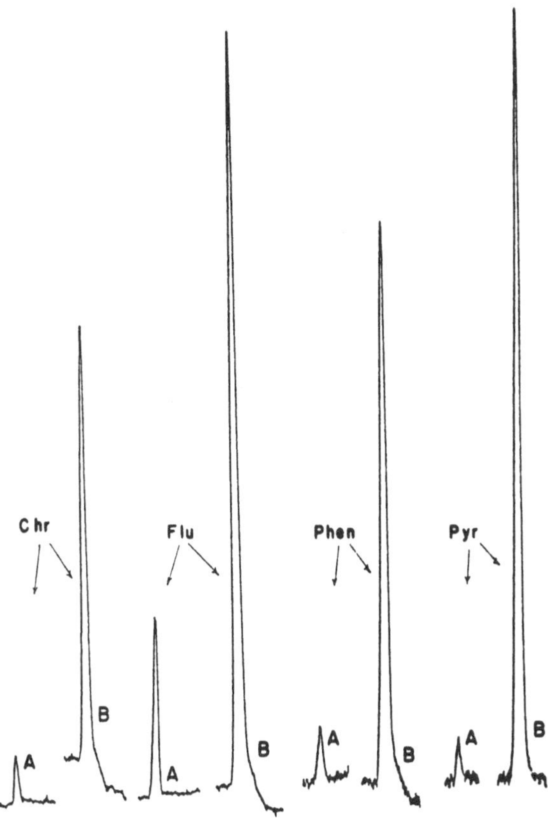

FIG. 1.19. Signal enhancement of some selected polycyclic aromatic hydrocarbons adsorbed onto silica gel. (A) was the normal response and (B) was the response after dipping in a solution of dodecane.

layer. Fomblin Y-Vac, a poly(perfluoroalkyl)ether fluid was found to be particularly effective in minimizing spot broadening of polycyclic aromatic hydrocarbons [33]. A shift in the fluorescence emission maxima to shorter wavelengths may also accompany the application of the fluorescence-enhancing reagent.

RADIOCHROMATOGRAPHY ON THIN-LAYER PLATES

The selection of a method to detect radiolabeled compounds in TLC depends largely on the frequency with which such measurements are made in the laboratory and the general availability of equipment for making radiochemical measurements. Expensive and dedicated radiochemical TLC scanners are commercially available and would be the choice for laboratories making frequent routine measurements of radiochromatograms. Methods employing less sophisticated equipment, or equipment generally available in radiochemical laboratories, would be the choice for those laboratories with only an occasional need for such determinations. The equipment available for thin-layer radiochromatography has been reviewed [35–38].

There are three principal methods for quantifying radioactive substances after separation by TLC. These are autoradiography, liquid scintillation counting, and direct scanning with radiation detectors. Autoradiography requires the direct contact of the TLC layer with a photographic emulsion (e.g., X-ray film). Autoradiography is inexpensive and does not require elaborate apparatus. Quantitative measurements can be made with a photodensitometer to measure the darkening of the developed film. The disadvantages of autoradiography are that the film may require extremely long exposure times, it is difficult to quantify, particularly for wide ranges of radioactivity, and it can show false images as a result of chemical interactions between the TLC layer and the film. For low energy beta emitters (e.g., ^3H), a large part of their energy is absorbed by the chromatographic layer so that only a few electrons reach the photographic emulsion. The sensitivity of autoradiography for low energy beta emitters can be enhanced by incorporating a scintillator into the TLC layer that emits light by absorbing beta particles. This light is absorbed to a lesser extent than the beta particles and can be registered by the film. The scintillator can be applied in a number of ways: by dipping, by spraying, or in a slurry during the preparation of the TLC layer.

Liquid scintillation counting is the most sensitive technique for the quantitative analysis of separated spots on a TLC plate. It represents an indirect method because the adsorbent layer and the sample must be removed from the plate prior to measurement. Automatic equipment that can divide and remove the whole chromatogram as a series of narrow zones is commercially available. The compound to be counted is first eluted from the excised adsorbent, and, after the removal of solvent, the residue is mixed with scintillator solution and determined in the normal manner. Alternatively, the sample can be counted while adsorbed onto the silica gel without elution. To avoid self-quenching, a gelling agent is added to the scintillator cocktail to keep the adsorbent in suspension. Correction factors based on internal or external standards are required for quantitative measurements.

Scintillation counting is probably the most widely used method of quantifying TLC radiochromatograms, partly because of the general availability of scintillation counting equipment in bioanalytical laboratories. The most convenient method, however, is direct scanning with radiation detectors. The necessary equipment is expensive and can be justified only by those laboratories with a high routine workload. Scanning is performed by moving the plate on a motor-driven stage under, over, or between radiation detectors. The most suitable detectors are Geiger–Muller detectors or flow-through proportional counters. The technique is very rapid when the level of radioactivity is high, and the data are provided in a convenient form as a chromatogram on a strip-chart recorder. Its main disadvantage is a lack of sensitivity for detecting small quantities of low energy beta isotopes as a result of self-absorption problems within the adsorbent layer.

CONCLUSIONS

The performance of contemporary HPTLC plates is probably not impaired by scanning densitometry, provided that a well-designed instrument operated under optimum conditions is used. Detection limits in the nanogram range for absorption and the picogram range for fluorescence are adequate for most studies. Contrary to many strongly held beliefs, quantitation in HPTLC is both accurate and precise. As the detection process is essentially static, multiple scanning with different measuring conditions is easily performed to aid in the spectroscopic determination of unresolved sample components and to develop substance-specific response ratios for sample identification.

The use of HPTLC in radiochromatography is hardly established. In many respects it offers the same rewards and challenges discussed for photometric recording. Instrumental evaluation methods using in situ scanning will become the norm and the present practice of using off-line measuring techniques will decline.

ACKNOWLEDGMENT

Work in the authors' laboratory is supported by the United States Environmental Protection Agency and the Michigan Heart Association. Although this research was funded in part by the USEPA under assistance agreement number R-808854-01-0 to C.F. Poole, it has not been subjected to the Agency's required peer and administrative review and, therefore, does not necessarily reflect the view of the Agency and no official endorsement should be inferred.

REFERENCES

1. Zlatkis, A. and Kaiser, R.E.: *HPTLC: High Performance Thin-Layer Chromatography*. Elsevier, Amsterdam, 1977.
2. Bertsch, W., Hara, S., Kaiser, R.E. and Zlatkis, A.: *Instrumental HPTLC*. Huthig, Heidelberg, 1980.
3. Fenimore, D.C. and Davis, C.M.: High performance thin-layer chromatography. *Anal. Chem.* **53**:252A–266A, 1981.
4. Coddens, M.E., Butler, H.T., Schuette, S.A. and Poole, C.F.: Quantitation in high performance thin-layer chromatography. *Liq. Chromatogr. HPLC Magzn.* **1**:282–289, 1983.
5. Poole, C.F. and Schuette, S.A.: *Contemporary Practice of Chromatography*. Elsevier, Amsterdam, 1984, pp. 619–700.
6. Poole, C.F., Coddens, M.E., Butler, H.T., Schuette, S.A., Ho, S.S.J., Khatib, S., Piet, L. and Brown, K.K.: Some quantitative aspects of high performance thin-layer chromatography. *J. Liquid. Chromatogr.*, 1985, in press.

7. Halpaap, H. and Krebs, K.-F.: Thin-layer chromatographic and high-performance thin-layer chromatographic ready-for-use preparations with concentrating zones. *J. Chromatogr.* **142:**823–853, 1977.
8. Fenimore, D.C. and Meyer, C.J.: Contact spotting: a new approach to sample application in thin-layer chromatography. *J. Chromatogr.* **186:**555–561, 1979.
9. Davis, C.M. and Harrington, C.A.: Quantitative determination of chlorpromazine and thioridazine by high-performance thin-layer chromatography. *J. Chromatogr. Sci.* **22:**71–74, 1984.
10. Fenimore, D.C.: Sample application in instrumentalized HPTLC. In *Instrumental HPTLC,* Bertsch, W., Hara, S., Kaiser, R.E. and Zlatkis, A., eds., Huthig, Heidelberg, 1980, pp. 81–95.
11. Pollock, V.: Progress in photometric methods of quantitative evaluation in TLC. *Advan. Chromatogr.* **17:**1–51, 1979.
12. Touchstone, J.C. and Sherma, J.: *Densitometry in thin-layer chromatography.* John Wiley and Sons, New York, 1979.
13. Hurtubise, R.J.: *Solid Surface Luminescence Analysis: Theory, Instrumentation, Applications.* Marcel Dekker, New York, 1981.
14. Ebel, S. and Post, P.: Theory of reflectance measurements in quantitative evaluation in TLC and HPTLC. *J. High Resolut. Chromatogr. Chromatogr. Commun.* **4:**337–342, 1981.
15. Bush, I.E. and Greeley, H.P.: Computer simulation of light transmission through scattering and absorbing chromatographic media. *Anal. Chem.* **56:**91–96, 1984.
16. Coddens, M.E., Khatib, S., Butler, H.T., and Poole, C.F.: Mode selection and optimization of parameters for recording high performance thin-layer chromatograms by transmission measurements. *J. Chromatogr.* **280:**15–22, 1983.
17. Cheng, M.-L. and Poole, C.F.: Minor component tablet analysis by high performance thin-layer chromatography. *J. Chromatogr.* **257:**140–145, 1983.
18. Butler, H.T., Schuette, S.A., Pacholec, F. and Poole, C.F.: Characterization of a scanning densitometer for high performance thin-layer chromatography. *J. Chromatogr.* **261:**55–63, 1983.
19. Butler, H.T. and Poole, C.F.: Optimization of a scanning densitometer for fluorescence detection in high performance thin-layer chromatography. *J. High Resolut. Chromatogr. Chromatogr. Commun.* **6:**77–81,1983.
20. Coddens, M.E. and Poole, C.F.: Sensitivity of scanning densitometers for thin-layer chromatography. *Anal. Chem.* 55:2429–2431, 1983.
21. Coddens, M.E. and Poole, C.F.: Protocol for measuring the sensitivity of slit-scanning densitometers. *Liq. Chromatogr. HPLC Magzn.* **2:**34–36, 1984.
22. Lee, K.Y., Poole, C.F. and Zlatkis, A.: Simultaneous multi-mycotoxin analysis by high performance thin-layer chromatography. *Anal. Chem.* **52:**837–842, 1980.
23. Zhou, L., Poole, C.F., Triska, J. and Zlatkis, A.: Continuous development HPTLC and sequential wavelength scanning for the simultaneous determination of nine clinically important β-blocking drugs. *J. High Resolut. Chromatogr. Chromatogr. Commun.* **3:** 440–446, 1980.
24. Butler, H.T., Coddens, M.E. and Poole, C.F.: Qualitative identification of polycyclic aromatic hydrocarbons by HPTLC and fluorescence scanning densitometry. *J. Chromatogr.* **290:**113–126, 1984.
25. Butler, H.T., Coddens, M.E., Khatib, S., Poole, C.F.: Quantitative determina-

tion of polycyclic aromatic hydrocarbons in environmental samples by high performance thin-layer chromatography and fluorescence scanning densitometry. *J. Chromatogr. Sci.* **23**:320–324, 1985.

26. Poole, C.F., Butler, H.T., Coddens, M.E., Khatib, S. and Vandervennet R.: Comparison of methods for separating polcyclic aromatic hydrocarbons by high performance thin-layer chromatography. *J. Chromatogr.* **302**:149–158, 1984.
27. Lee, K.Y., Nurok, D., Zlatkis, A. and Karmen, A.: Simultaneous determination of antiarrhythmia drugs by high performance thin-layer chromatography. *J. Chromatogr.* **158**:403–410, 1978.
28. Ebel, S., Albert, D. and Schaefer, U.: Calibration in TLC/HPTLC using Michaelis-Menten function. *Chromatographia* **18**:23–27, 1984.
29. Bethke, H., Santi, W. and Frei, R.W.: Data-pair technique, a new approach to quantitative thin-layer chromatography. *J. Chromatogr. Sci.* **12**:392–397, 1974.
30. Butler, H.T. and Poole, C.F.: Two-point calibration method applied to fluorescence scanning densitometry and HPTLC. *J. Chromatogr. Sci.* **21**:385–388, 1983.
31. Funk, W., Kerler, R., Schiller, J. Th., Dammann, V. and Arndt, F.: Prechromatographic derivatization of samples for HPTLC. *J. High Resolut. Chromatogr. Chromatogr. Commun.* **5**:534–538, 1982.
32. Oztunc, A.: Fluorimetric determination of acetaminophen and its dansyl derivative. *Analyst* **107**:585–587, 1982.
33. Ho, S.S.J., Butler, H.T. and Poole, C.F.: Fluorescence enhancement of polycyclic aromatic hydrocarbons separated on silica gel high performance thin-layer chromatography plates. *J. Chromatogr.* **281**:330–339, 1983.
34. Uchiyama, S. and Uchiyama, M.: Thin-layer chromatography-fluorometry of ethoxyquin using Triton X-100. *J. Chromatogr.* **262**:340–345, 1983.
35. Prydz, S.: Summary of the state of the art in radiochromatography. *Anal. Chem.* **45**:2317–2326, 1973.
36. Touchstone, J.C. and Dobbins, M.F.: *Practice of Thin-Layer Chromatography*. John Wiley and Sons, New York, 1978, pp. 275–300.
37. Roberts, T.R.: *Radiochromatography: The Chromatography and Electrophoresis of Radiolabelled Compounds*. Elsevier, Amsterdam, 1978.
38. Filthuth, H.: Radioscanning of TLC. In *Advances in Thin Layer Chromatography: Clinical and Environmental Applications,* Touchstone, J.C., ed., John Wiley and Sons, New York, 1982, pp. 89–124.

CHAPTER 2

Radioanalytical Techniques: ITLC, TLC, Mini-Columns, and Electrophoresis

ALAN P. CARPENTER, JR.

INTRODUCTION

The separation techniques available for use today in quality control and research evaluations of radiopharmaceuticals are the same techniques that have been used in other fields over the past 20 years. Most of the techniques now being used, in fact, were developed before the advent of radiopharmaceutical science itself. Radiopharmaceutical researchers quickly became aware of the value of these analytical techniques, and in the past 5–10 years, separation methods such as TLC (thin-layer chromatography), HPLC (high performance liquid chromatography), and electrophoresis have become relatively common in radiopharmaceutical studies. With the use of these classical analytical methods in the study of radiopharmaceutical preparations, has come a refinement and modification of the instrumentation and methodology for application to trace level analysis of radiopharmaceuticals.

This chapter covers the use of ITLC (instant thin-layer chromatography), TLC, mini-column and electrophoretic separation methods in radiopharmaceutical analyses, with particular emphasis on applications to ^{99m}Tc-based radiopharmaceuticals. A review is given of some, but by no means all, of the literature in these areas, along with a discussion of the use of these techniques in the radiopharmaceutical laboratories at Dupont-NEN Products. It is not my intention to cover radio-HPLC because it is reviewed elsewhere in this volume. However, because HPLC plays such a prominent role in radiochromatographic work today, the discussions in this chapter compare the characteristics of ITLC, TLC, mini-column, and electrophoretic measurement methods with HPLC in order to give the reader an appreciation of the strengths and weaknesses of these other methods relative to HPLC.

ITLC

Introduction

Any rapid thin-layer chromatographic method could be considered as part of ITLC. However, for the purpose of staying within the normal historical usage of the ITLC technique, this discussion is restricted to paper chromatography and, in particular, deals mainly with the use of silica gel-impregnated papers as the chromatographic medium.

ITLC is a technique that has been developed to fill the need for a simple and rapid radiochemical analysis of radiopharmaceuticals in the clinical setting. ITLC is the most commonly used and recommended method for measurement of the radiochemical impurities $^{99m}TcO_4^-$ and colloidal ^{99m}Tc-oxides. Many methods for the analysis of ^{99m}Tc-based radiopharmaceuticals have been published by the USP[1]. It is also widely used in the measurement of free iodide in radioiodinated products [1]. ITLC usually involves the separation and measurement of only one radiochemical species (i.e., free pertechnetate and radiocolloid) at a time. Separate solvent systems are necessary for the measurement of free pertechnetate and radiocolloid in ^{99m}Tc-labeled radiopharmaceuticals. Despite our inability to measure all radiochemical components in one chromatographic system, ITLC is still very popular because of its simplicity, speed, and low cost.

Before the use of silica gel-impregnated papers, radiochemical analyses were commonly performed using standard cellulose filter papers. Bailey [2] reviewed the use of paper chromatography in analyzing radiochemical mixtures. The typical usage of paper chromatographic media at that time was in the separation of inorganic ions, such as for the separation of Pb, Bi, and Po using mobile phases of acid/alcohol mixtures [3,4]. This type of technique has also been successfully applied to the analysis of isotopes of interest to nuclear medicine, such as the paper chromatographic separation of Tl^+ from Tl^{3+} reported by Janardhan and Paul [5]. The use of paper chromatography for inorganic radioisotope separations was important to the later development of techniques specific to the study of ^{99m}Tc-based radiopharmaceuticals. Therefore, it is recommended that the earlier works of Lederer [6], Bailey [2], and Höye [7] be consulted for a more detailed review of the use of paper chromatography in radiochemical analysis.

As a result of the requirement that users of radiopharmaceuticals perform some quality control tests at the site of usage, paper chromatography/ITLC systems have been developed over the past 10–15 years for application to ^{99m}Tc-labeled materials. The radiochemical purity analysis of ^{99m}Tc radiopharmaceuticals was originally performed using paper chromatography strips [8]. This approach to radiopharmaceutical quality control later evolved into the use of silica-impregnated papers in ITLC as demonstrated in the works of Zimmer and Holmes [9], Pauwels and Feitsma [10], and Belkas and Archimandritis [11]. A recent review of ITLC methods for specific ^{99m}Tc-based radiopharmaceuticals has been written by P. J. Robbins

[12]. This SNM monograph is highly recommended to those who are interested or involved in the use of ITLC quality control methods.

Experimental Techniques and Applications

The chromatographic media typically used today for radiochemical purity measurements are Gelman ITLC media, impregnated with silica gel (ITLC-SG) or silicic acid (ITLC-SA), and Whatman #1 chromatography paper. Most of the solvent systems for separation and analysis of current ^{99m}Tc radiopharmaceuticals are included in Robbins' monograph, and are not reviewed in detail here. However, the general methods employed in our laboratories are given for several radiopharmaceuticals in Table 2.1, for reference purposes.

The correct choice of solvent system and the technique used are critical to the accurate and reproducible measurement of $^{99m}TcO_4^-$ and colloidal ^{99m}Tc-oxides in radiopharmaceuticals. For example, it has been found that the use of MEK is preferred over acetone as the developing solvent for the measurement of $^{99m}TcO_4^-$ in ^{99m}Tc-Glucoheptonate and other ^{99m}Tc-labeled products. The data in Table 2.2 compare the use of these solvents for the measurement of $^{99m}TcO_4^-$ in ^{99m}Tc-Glucoheptonate. The acetone system for the ITLC-SG assay of $^{99m}TcO_4^-$ tends to be erratic and biased high (relative to the MEK system). The problem with acetone is most likely due to the fact that the ^{99m}Tc-labeled material is highly water soluble in nature (highly polar), and acetone has a higher dielectric constant than MEK and also, generally, has a higher water content than MEK. Another important factor in the use of ITLC is the way in which the sample is developed. For example, if the solvent level in the developing tank is above the level of the applied sample spot on the chromatographic strip, then the solvent becomes contaminated and the results of the analysis will be erroneous. An important requirement for the measurement of $^{99m}TcO_4^-$ is that the strips should be developed immediately after sample spotting. If the sample spot dries and then the strip is developed, low $^{99m}TcO_4^-$ values may be obtained. This is illustrated for ^{99m}Tc-Glucoheptonate in Table 2.3, where the levels of $^{99m}TcO_4^-$ were measured after known quantities of $^{99m}TcO_4^-$ were added to GLUCOSCAN™ vials (after the complete oxidation of Sn^{2+} by addition of excess $NH_4{}^{99}TcO_4$). The low $^{99m}TcO_4^-$ values obtained by ITLC with sample spot drying apparently are a result of slow migration or some trapping of $^{99m}TcO_4^-$ at the origin of the chromatographic strip.

Validation of Procedures

Although the evaluation of the accuracy of ITLC procedures is a very important part of the method validation, it has not been addressed by most radioanalytical publications to date. The determination of the accuracy of ITLC procedures is not accomplished as easily as in nonradioactive pharmaceutical assay validations where classical standard addition techniques are

TABLE 2.1. General ITLC Systems[a] Used for $^{99m}TcO_4^-$ and ^{99m}Tc-Oxide Measurements

Product	*Free Pertechnetate Assay Method*	*Colloidal ^{99m}Tc Assay Method*
^{99m}Tc-Glucoheptonate	Gelman ITLC-SG strips (1/2 × 2″) methyl ethyl ketone solvent	Gelman ITLC-SG strips (1/2 × 2″) normal saline solvent
^{99m}Tc-Pyrophosphate	Gelman ITLC-SG strips (1/2 × 2″) methyl ethyl ketone solvent	Gelman ITLC-SG strips (1/2 × 2″) normal saline solvent
^{99m}Tc-Medronate (MDP)	Gelman ITLC-SG strips (1/2 × 2″) methyl ethyl ketone solvent	Gelman ITLC-SG strips (1/2 × 2″) normal saline solvent
^{99m}Tc-Disofenin (DisIDA)	Whatman #3 filter paper (1/2 × 2″) soaked in 0.3 *M* sodium bicarbonate/carbonate buffer and dried. Methyl ethyl ketone solvent	Gelman ITLC-SG strips (1/2 × 2″) methanol solvent
^{99m}Tc-MAA (macroaggregated albumin)	Gelman ITLC-SG strips (1/2 × 2″) normal saline solvent	Not applicable
^{99m}Tc-μAA (microaggregated albumin)	Gelman ITLC-SG strips (1/2 × 2″) methyl ethyl ketone solvent	Not applicable

[a] Procedures validated and used by Dupont/NEN Products. Sample is spotted ~1/2″ from bottom of ITLC strip and developed to end of strip. The strip is cut ~3/4″ from the top and the free pertechnetate in the top portion of the strip is calculated as a percentage of the total activity. If colloidal ^{99m}Tc is being measured, then the strip is cut 3/4″ from the bottom and the colloidal ^{99m}Tc is calculated as the percentage of the total activity that is present in the bottom 3/4″ strip.

used. Because $SnCl_2$ is present in excess quantities in most radiopharmaceutical kits, a spike of additional $^{99m}TcO_4^-$ into a labeled kit will not be measurable as $^{99m}TcO_4^-$ once added, since it is reduced and is then present in the "bound" state. In order to overcome this problem, we follow a modified standard addition protocol as outlined in Table 2.4. Basically, this procedure involves the addition of sufficient $^{99m}TcO_4^-$ to completely oxidize the Sn^{2+} present (after the initial $^{99m}TcO_4^-$ labeling), followed by further addition of known quantities of $^{99m}TcO_4^-$. The ITLC measurement of free pertechnetate is carried out using ITLC-SG chromatography strips with MEK as the developing solvent. The results of this type of validation for ^{99m}Tc-DisIDA and

TABLE 2.2. Comparison of Acetone and MEK Mobile Phases for ITLC Assay for $^{99m}TcO_4^-$

^{99m}Tc-Glucoheptonate Sample Set Number	*$^{99m}TcO_4^-$ with[a] Acetone Solvent*	*$^{99m}TcO_4^-$ with[a] MEK Solvent*
1	3.0 ± 1.6	1.4 ± 0.9
2	5.3 ± 3.7	1.3 ± 0.6
3	3.6 ± 2.7	0.7 ± 0.2

Note: All assays using acetone and MEK solvents in ITLC were performed on the same sample vials.
[a] % of total radioactivity as $^{99m}TcO_4^-$.

^{99m}Tc-Glucoheptonate preparations are shown in Table 2.5. As is shown in Table 2.5, the ITLC procedures used for these materials are accurate to within 2% of the actual value of $^{99m}TcO_4^-$ present.

In the case of radiocolloid measurements, the accuracy of the assay method may be determined by the standard addition procedure outlined in Table 2.6. Examples of the validation of the radiocolloid determination are shown in Table 2.7 for ^{99m}Tc-DisIDA and ^{99m}Tc-Glucoheptonate. These data demonstrate that the radiocolloid assay (outlined in Table 2.1) is accurate to within ~2% of the true value.

TABLE 2.3. $^{99m}TcO_4^-$ Measurements by ITLC of ^{99m}Tc-Glucoheptonate with Development of Wet vs. Dried Sample Spots

Sample	*Theoretical % $^{99m}TcO_4^-$*	*Measured % $^{99m}TcO_4^-$ Wet Sample Spot Development[a]*	*Measured % $^{99m}TcO_4^-$ Dried Sample Spot Development[a]*
Control, ^{99m}Tc-glucoheptonate	≤1%	1.0, 0.7	0.4, 0.6
$^{99m}TcO_4^-$ added to glucoheptonate only (no $SnCl_2$)	100%	99, 99	31, 34
4.4% $^{99m}TcO_4^-$ added to* ^{99m}Tc-Glucoheptonate after addition of excess $^{99}TcO_4^-$	4.4	5.3, 5.1	2.1, 2.2
33% $^{99m}TcO_4^-$ added to* ^{99m}Tc-Glucoheptonate after addition of excess $^{99}TcO_4^-$	33	34, 34	20, 21

[a] Duplicate assays of same vial.
* *Procedure:* Reconstitute kit with 75 mCi $^{99m}TcO_4^-$/5 ml saline, then add 300–400 μg $NH_4{}^{99}TcO_4$, mix and add $^{99m}TcO_4$ spike at level noted, and perform ITLC $^{99m}TcO_4^-$ assay.

TABLE 2.4. Protocol for Evaluation of Accuracy of ITLC $^{99m}TcO_4^-$ Measurements

1. Reconstitute several kits with the recommended levels of $^{99m}TcO_4^-$ in saline.
2. Perform ITLC assays of $^{99m}TcO_4^-$.
3. Add to each kit 1 to 2 times the amount of $^{99m}TcO_4^-$ required to completely oxidize the $SnCl_2$ in the vial.
4. Mix well.
5. Add a known quantity of $^{99m}TcO_4^-$ to each of the above kits, at levels ranging from 1–100% of the original amount of $^{99m}TcO_4^-$ used in reconstitution of the kit.
6. Measure $^{99m}TcO_4^-$ in each spiked kit by the ITLC method.

Advantages and Disadvantages of ITLC

For the purpose of measuring free, bound, and colloidal ^{99m}Tc species, ITLC does have significant advantages over other more sophisticated analytical procedures. These are: the low cost of equipment and supplies necessary to do the measurements, the relative simplicity of the technique, and the great speed of the analysis, which allows for a rapid quality check of the radiopharmaceutical prior to its use in humans. The most significant limitation of the rapid paper chromatography/ITLC methods is the fact that only single radiochemical components (i.e., $^{99m}TcO_4^-$ or colloidal ^{99m}Tc-oxides) are normally measured per analysis, and very limited information is obtained regarding the "bound" ^{99m}Tc species of the radiopharmaceutical. These limita-

TABLE 2.5. Validation of ITLC Procedure for Accuracy of $^{99m}TcO_4^-$ Measurement[a]

^{99m}Tc-DisIDA	
% $^{99m}TcO_4^-$ Added	*$^{99m}TcO_4^-$ Measured*[b]
0.0	1.5
5.3	5.2 (4)
9.3	7.6
13.0	14.0
20.9	22.9
^{99m}Tc-Glucoheptonate	
% $^{99m}TcO_4^-$ Added	*% $^{99m}TcO_4^-$ Measured*[b]
0.0	1.5
5.4	6.1
9.3	11.8
17.8	17.8

[a] See Table 2.4 for general procedure used.
[b] All values are averages of two measurements, except where noted.

TABLE 2.6. Protocol for Evaluation of Accuracy of ITLC ^{99m}Tc-Colloid Measurement

1. Reconstitute kits with recommended levels of $^{99m}TcO_4^-$ in saline.
2. Perform ITLC assay for radiocolloid.
3. Prepare Stannous Liver Colloid kit (Medi+Physics), according to manufacturer's specifications.
4. Add known quantities of ^{99m}Tc-colloid of ~0–25% of original ^{99m}Tc in labeled kit.
5. Perform ITLC assay for radiocolloid.

tions can be partially overcome by scanning, as demonstrated in the works of Robbins [12] and Jovanovic [13] or by cutting and counting the chromatographic strip as done by Bish et al. [14] However, the inherently low resolution of labeled species obtained in ITLC restricts this radioanalytical measurement principally to the assay of colloidal ^{99m}Tc, $^{99m}TcO_4^-$, and "bound" ^{99m}Tc.

RADIO-TLC

Introduction

As a result of the limitations of ITLC techniques, many researchers have used the higher resolution separation method of high performance liquid chromatography (HPLC) for the study of the radiochemical status of ra-

TABLE 2.7. Validation of ITLC Procedure for Accuracy of ^{99m}Tc-Colloid Measurement[a]

% ^{99m}Tc-Colloid Added	*% ^{99m}Tc-Colloid Measured*[b]
^{99m}Tc-DisIDA	
0.0	1.0
3.5	4.7
8.6	10.1
13.8	14.0
15.3	16.5
19.0	21.3
^{99m}Tc-Glucoheptonate	
% ^{99m}Tc-Colloid Added	*% ^{99m}Tc-Colloid Measured*[b]
0.0	<1.0
2.0	4.0
5.0	5.8
9.0	11.0

[a] See Table 2.6 for general procedures used.
[b] All values are averages of two measurements.

diopharmaceuticals, good examples of which are the works of Pinkerton et al. [15], Srivastava et al. [16], Nunn [17], Loberg et al. [18], and Fritzberg and Lewis [19]. Although HPLC is the radiochromatographic technique with the best resolution, it is relatively costly, fairly complex in nature, and does not always allow for the complete analysis of the sample (i.e., some radiolabeled compounds are trapped by, or adsorbed to, the chromatographic column and are not eluted or detected). Despite being a very powerful research tool, HPLC is not an ideal technique for the quality control testing of radiopharmaceuticals. There is a need for the application of a technique with intermediate resolution, such as TLC, in radiochemical analyses. TLC separations on silica or synthetic polymer stationary phases supported on glass plates, aluminum foil, or cellulose acetate sheets, allow for good resolution of ^{99m}Tc-labeled radiochemical components while maintaining some simplicity and cost advantages over HPLC. Radioanalytical measurements using TLC separations along with the use of a radiochromatographic scanner for quantitation of individual components offer a practical compromise between the high resolution/high cost characteristics of HPLC and the low resolution/low cost characteristics of ITLC [20].

Background

Although the concept of thin-layer chromatography was described by Izmailov and Shraiber as early as 1938 [21], the development and refinement of the experimental methodology in TLC was principally due to the work of Stahl [22]. The state of TLC development up to the mid-1960s is well reviewed in Stahl's textbook [23]. The developments in TLC since that time have been essentially an extension of Stahl's original work.

The most significant changes in TLC over the past 15–20 years have come with the introduction of chemically bonded silica gel stationary phases for TLC. The broad range of stationary phases now available in TLC have made this technique more generally applicable and competitive with HPLC. TLC adsorbents commercially available today include silica, chemically bonded phases on silica, alumina, charcoal, cellulose, and ion exchange resins [24,25]. In addition, the advent of high performance thin-layer chromatography (HPTLC) [26] has led to the development of TLC plates with adsorbents of smaller and more uniform particle sizes (*Note:* The specific area of HPTLC is covered by C. Poole in Chapter 1 and is not discussed here).

Very little work has been published on radio-TLC of radiopharmaceuticals. Höye [7] reviewed the use of silica gel TLC on some early radiopharmaceuticals such as ^{131}I-Iodohippurate and ^{131}I-Rose Bengal. More recently, Jovanovic et al. [13], have described a method using silica gel foils, however, this was only used for the measurement of free, bound, and colloidal ^{99m}Tc in ^{99m}Tc-HIDA.

In this section on radio-TLC, the equipment and procedures used for analysis of ^{99m}Tc-labeled radiopharmaceuticals by TLC are described, along with a few examples of TLC separations of ^{99m}Tc-labeled materials. For the reader who seeks a more fundamental understanding of the basic processes occurring in TLC, the theoretical treatment of Guichon et al. [27], is highly recommended. For the purposes of this article, some of the most important relationships in TLC are outlined in Table 2.8.

Experimental Techniques and Applications

The equipment and techniques used in radio-TLC are of paramount importance. The heart of the TLC separation process is the TLC plate. The choice of the TLC adsorbent is the most important decision to be made in developing a useful separation system. Some of the various types of commercial adsorbents for TLC are shown in Table 2.9. For the bulk of the work we have done on ^{99m}Tc radiopharmaceuticals, reversed phase TLC media have been employed. This is principally because the radiolabeled samples are usually in aqueous solution and are miscible with all common reversed phase developing solvents. In addition, reverse phase separations have been found to offer a greater resolution and range of separation than normal phase chromatographic systems [28]. The choice of the TLC solvent (i.e., mobile phase) to be used is also important to the successful development of a TLC

TABLE 2.8. Definition of Important Parameters in TLC Separations

R_f =	Retardation factor, calculated as d_a/d_m, where d_a = the distance traveled by the analyte, and d_m = the total distance traveled by the mobile phase (i.e., the developing distance).
K' =	Capacity factor, calculated as $(1 - R_f)/R_f$.
N =	Plate number, calculated as $16(d_a/w_a)^2$, where d_a if as described above, and w_a is the base-width of the analyte spot.
R =	Resolution, calculated as $2(d_2 - d_1)/(w_1 + w_2)$, where d_1, and d_2 are the distances moved by components 1 and 2, respectively, and w_1 and w_2 are the corresponding base-widths of the sample spots.
Results:	The resolution of two sample spots of about the same width, w, may also be expressed as: $R = \left(\frac{\sqrt{N}}{4}\right)\left(\frac{\Delta R_f}{R_f}\right)$ The resolution obtainable in TLC is dependent on both the relative capacity factor difference of the components being separated, and on the plate characteristics (the N term).

Recommendations:

1. TLC plates should be chosen with uniform stationary phase coatings and particle sizes. Smaller particle size coatings will give greater resolution with all other factors being equal.
2. The relative capacity factor differences between two components being separated ($\Delta R_f/R_f$) should be optimized by the correct choice of developing solvent.

TABLE 2.9. Some Stationary Phases Available for Use in TLC[a]

Type	*Stationary Phase*	*Commercial Source*
Adsorption	Silica gel	Whatman, E. Merck, Analtech
Adsorption	Alumina	E. Merck, J. T. Baker, Analtech
Reversed phase	C-18 bonded phase	Whatman, Analtech, E. Merck
Reversed phase	C-8 bonded phase	Whatman, E. Merck
Normal phase	Cellulose	Analtech, J.T. Baker, E. Merck
Normal phase	Polyamine phase	J.T. Baker, E. Merck
Ion exchange	Cation exchange resin	Brinkmann
Ion exchange	Anion exchange resin	Analtech, E. Merck, Brinkmann

[a] This list is not intended to be a complete compilation of the types of phases available, or of the suppliers of these materials, but rather to illustrate the variety of TLC stationary phases available for use.

separation system. Some of the commonly used developing solvents for various types of TLC stationary phases are listed in Table 2.10. In general, the optimal resolution is achieved by choosing the mobile phase solvent to give an R_f of ~0.3–0.6 for the component of interest. The minimum TLC plate development time and development distance required for separation is achieved for R_f's of ~0.35–0.50 for the analyte as shown in the work of Guichon et al. [27].

The developing tank used in radio-TLC can be the standard type for housing 20 × 20 cm TLC plates, the micro TLC tank for microscopic slide TLC plates, or the horizontal, closed TLC developing system for 10 cm × 10 cm TLC plates. Upon development of the TLC plate, the plate must be scanned on a suitable radiochromatographic scanner. The TLC scanning equipment used in our work has been discussed in detail [29]. Basically, as shown in Figure 2.1, the TLC scanner system includes a variable speed flatbed scanner table, a lead-shielded NaI (Tl) detector linked to a scaler/

TABLE 2.10. Commonly Used Mobile Phases in TLC Separations

Type of Chromatographic Phase	*Commonly Used Mobile Phases*
Adsorption	Hexane, methylene chloride, chloroform, acetone, methanol, isopropanol Normally used as binary, ternary, or quarternary mixtures
Normal phase partition	Same solvents as those usable with adsorptive chromatographic media
Reversed phase partition	Aqueous buffers of phosphate or acetates in binary or ternary mixtures with methanol, acetonitrile, THF, or other water-miscible polar organic solvent
Ion exchange resin	Aqueous buffers of relatively high salt contents (e.g., NaCl or other alkali halide salt in phosphate buffers)

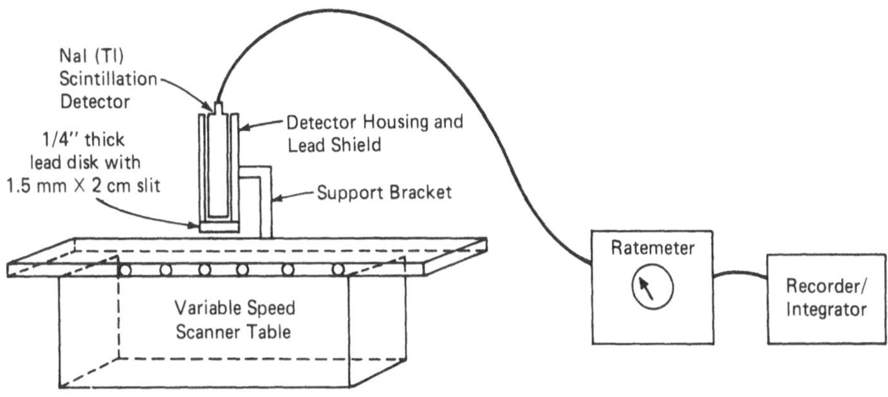

FIG. 2.1. Block diagram of gamma TLC scanner.

ratemeter, which is hooked to an electronic integrator capable of filtering, plotting, and integrating the relatively noisy radiodecay signal.

The general procedure followed in the TLC analysis of ^{99m}Tc-radiopharmaceuticals is outlined in Table 2.11. It is important that the sample spot volume be minimized in order to limit the resulting band width of the sample components after the separation has taken place. For a kit in the range of ≥20 mCi/ml ^{99m}Tc, the spotted volume does not need to be any greater than 1 μl using our current TLC scanning system. Accurate integration of sample component peaks is obtained on this TLC scanning system in the range of ~1–100 μCi per component. Below 1 μCi, accurate integration is not possible because of the background noise, and above 100 μCi the detector response becomes nonlinear.

In reverse phase TLC using a 10 cm developing distance, the separation time is about 45–50 minutes for a 50/50 methanol/dilute aqueous buffer system. The developing time varies with the viscosity of the solvent being used. This means that reversed phase TLC is more rapidly performed when a greater percentage of organic solvent is used in the mobile phase.

Before the radio scan of the sample TLC plate is performed, it must be established that the instrumentation is working properly. This is done by a routine check of the TLC scanner using a "phantom" of ^{57}Co-impregnated epoxy strips embedded in a plexiglass block that was constructed at Dupont/NEN Products. The epoxy strips contain known quantities of ^{57}Co, and are at known distances apart in the plexiglass block so that both an accuracy and resolution check of the scanner can be made using this device. The phantom used and the scan obtained of this phantom are shown in Figure 2.2. Once the scanner has been shown to be working properly, the sample TLC plates are scanned.

The ability to do the simultaneous measurement of several labeled ^{99m}Tc species in radiopharmaceutical preparations is one of the main advantages of using a radio-TLC procedure in place of an ITLC assay method. An example

TABLE 2.11. General Procedure for Radio-TLC Analysis of ^{99m}Tc-Radiopharmaceuticals

1. Mark the TLC plates with pencil at the sample application line and at the intended final solvent development point (e.g., draw lines at 2 and 12 cm from one edge of the plate for a sample spot at 2 cm from the end of the plate and a 10 cm development distance for the sample).
2. Add to the TLC tank the mobile phase solvent (up to a level of ~1 cm from the bottom of the tank).
3. Cover the TLC tank to allow for equilibration of mobile phase vapors within the tank.
4. Reconstitute the radiopharmaceutical kit with $^{99m}TcO_4^-$ in saline (following the manufacturer's recommended procedures).
5. Draw up 0.2–2.0 μl (~20–100 μCi) of the sample into a 1- or 10-μl glass syringe (a blunt teflon-coated needle on a glass syringe is preferable). Apply the sample to the TLC plate at the origin line marked in pencil. Approximately 0.2–0.4-μl aliquots should be applied at a time with partial drying by using a gentle nitrogen stream over the sample spot between applications. If possible, do not allow the sample spot to become >2–3 mm i.d. Also, do not completely dry the sample spot before developing the plate.
6. Uncover the TLC tank. Put the TLC plate (with the sample spot end down) in the tank, taking care not to contact the sample spot with the developing solvent. Replace the cover on the tank. Wait until the developing solvent reaches the upper pencil line. Remove the TLC plate and dry briefly.
7. Scan the TLC plate on an appropriate scanning device (see [20, 29] for details).

of this is illustrated by the study of the radiochemical purity of $^{99m}Tc(dmpe)_2Cl_2^+$ (dmpe = bis-dimethylphosphinoethane), which was evaluated as a potential myocardial imaging agent, and has been described by Deutsch et al. [30]. An example TLC separation of $^{99m}Tc(dmpe)_2Cl_2^+$ from other radiochemical species is shown in Figure 2.3. This type of chromatographic separation allows one to assess the radiochemical purity from the integrated activity trace, and allows the optimization of reaction parameters for forming the desired ^{99m}Tc species. This TLC analysis requires a knowledge of the R_f value of the ^{99m}Tc species of interest. In order to determine this, the analogous ^{99}Tc complex, which had been previously characterized by elemental analysis, mass spectrometry, and and X-ray crystallography, was run under the same conditions and found to have an identical R_f to the major ^{99m}Tc complex seen in the γ detector scan of the TLC plate. The separation and scan time for this radio-TLC procedure is about one hour per run. If there is more than one plate being developed at a time, then the analysis time per sample is lowered considerably.

As noted by Guichon et al. [27], the optimum development distance is dependent upon the relative retention of the compound being analyzed from the other components of the sample. The separation of $^{99m}Tc(dmpe)_2Cl_2^+$

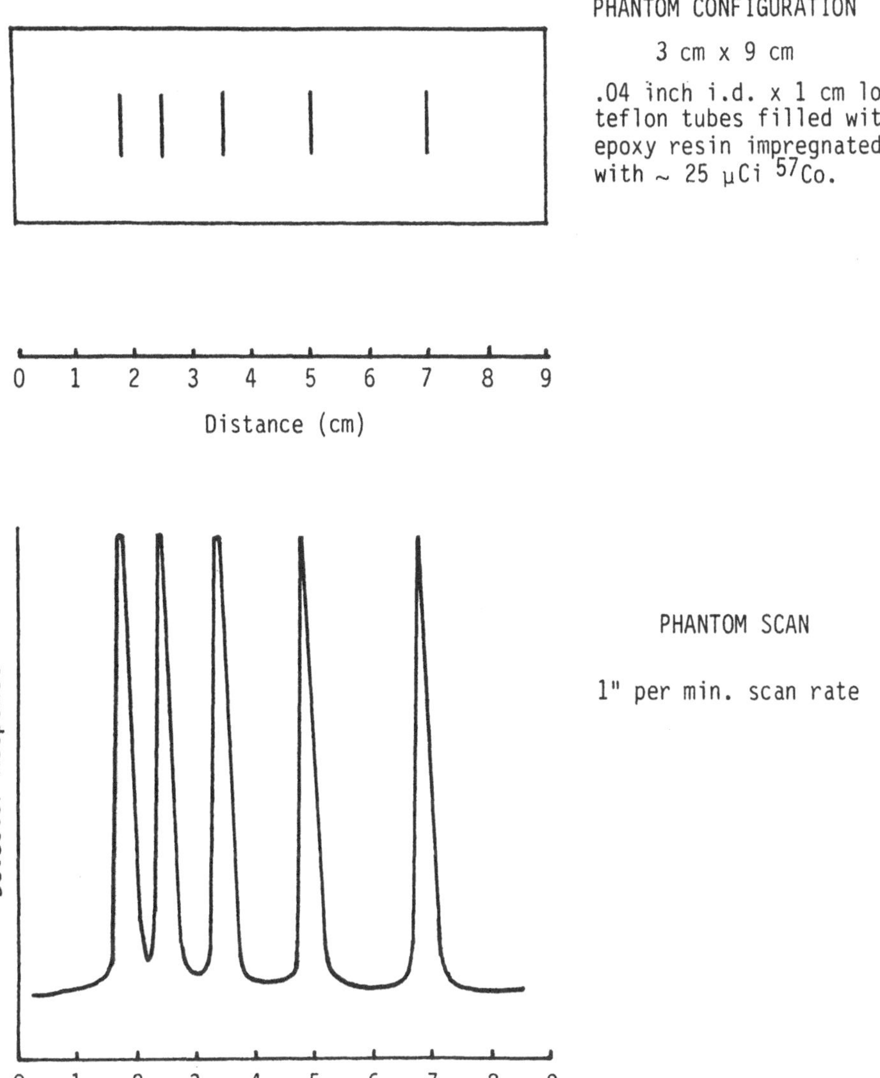

FIG. 2.2. Phantom design and example calibration scan.

from radiochemical impurities in the sample is very good for a development distance of 10 cm, however, even shorter plate lengths may be used with adequate resolution for quantitation of the ^{99m}Tc complex of interest. This is depicted in Figure 2.4 for the separation of a $^{99m}Tc(dmpe)_2Cl_2^+$ sample containing higher levels of radiochemical impurities using ~6 cm microscope slide TLC plates. In this case, some resolution is lost using the shorter TLC

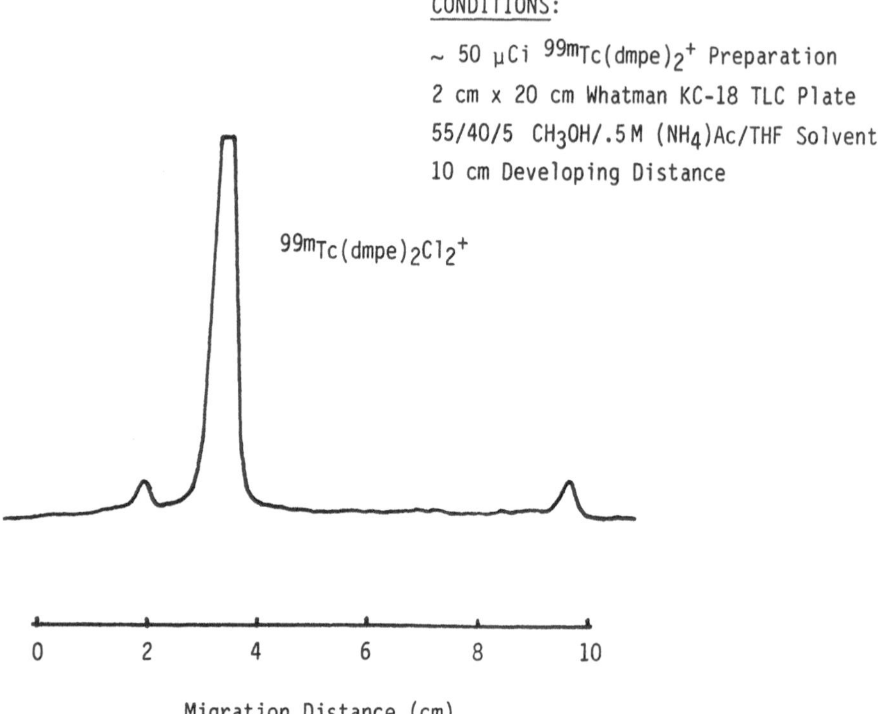

FIG. 2.3. Radio-TLC of $^{99m}Tc(dmpe)_2Cl_{2+}$ preparation.

plates, but the ^{99m}Tc complex of interest is measurable with a significant reduction in analysis time (~15 minutes). As this illustrates, considerable flexibility exists in the choice of the TLC conditions for various types of analyses. In all cases, there is a trade-off to be made between the separation time and the resolution achieved. The best choice is dependent on the type of analysis sought, and must be decided upon on a case-by-case basis.

Radio-TLC procedures have a broad range of applicability to radiopharmaceutical analysis. In Figure 2.5, is shown the reverse phase TLC separation of ^{99m}Tc hexakis(isopropyl), a compound described by Davison et al. [31]. The radio-TLC profile of the $^{99m}Tc(CNC_3H_7)_6^+$ preparation is seen to be about 95% radiochemically pure based on the radio-TLC analysis.

In Figure 2.6, are shown the reversed phase radio-TLC profiles obtained for $^{99m}TcO_4^-$ and colloidal ^{99m}Tc. The colloidal ^{99m}Tc has an R_f of 0.0, and the $^{99m}TcO_4^-$ has an R_f of 0.9 under these conditions (same conditions as in Figure 2.3). This shows that it is possible to detect colloidal ^{99m}Tc and $^{99m}TcO_4^-$ at the extremes of the TLC scan.

Validation of Radio-TLC Procedures

As noted earlier for ITLC procedures, the validation of the accuracy and reproducibility of radioanalytical methods has not received much attention in the past. In our studies of the radio-TLC analysis of $^{99m}Tc(dmpe)_2Cl_2^+$, the assay validation was performed by preparation of a kit of radiochemically pure $^{99m}Tc(dmpe)_2Cl_2^+$ followed by spiking known quantities of this into a kit prepared with a low percentage of $^{99m}Tc(dmpe)_2Cl_2^+$ complex. The TLC analysis was then performed at each spiked level prepared. A plot of the measured vs. expected percentage of ^{99m}Tc as $^{99m}Tc(dmpe)_2Cl_2^+$ is shown in Figure 2.7. The linear regression fit to the data indicates that the assay is

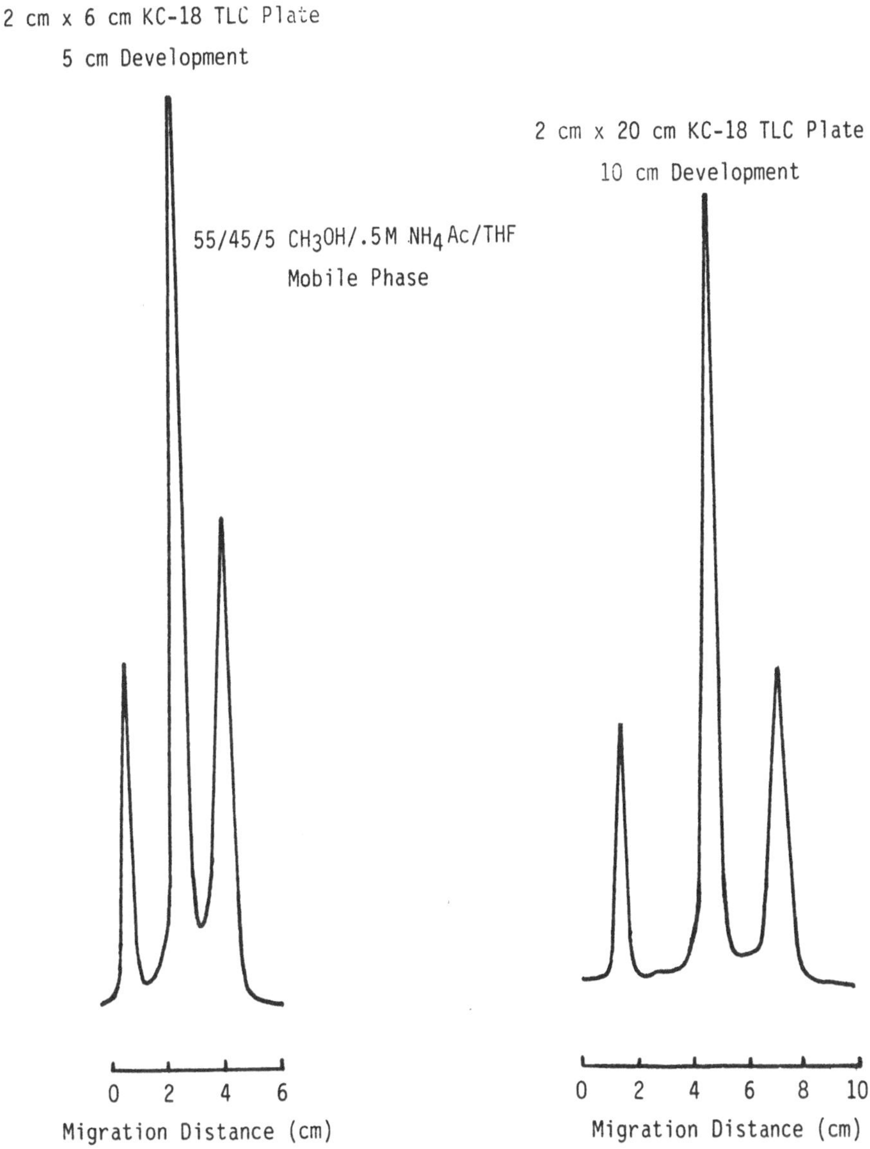

FIG. 2.4. Comparison of short and long TLC plates in separation of an impure $^{99m}Tc(dmpe)_2Cl_{2}+$ preparation.

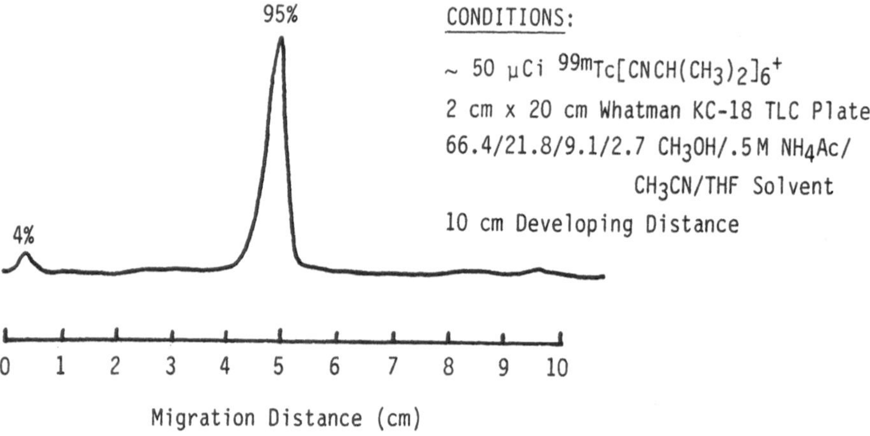

FIG. 2.5. Radio-TLC of $^{99m}Tc[CN\ CH(CH_3)_2]_6^+$.

accurate to within ±1% of the theoretical value, with a standard error of ~3% and a correlation coefficient of 0.97 for the fit. This type of assay validation should be considered for all routine radioanalytical procedures performed in order to ensure the reliability of the procedures followed.

Advantages and Disadvantages of Radio-TLC

Radio-TLC is slightly more complex and expensive in the equipment and techniques used when compared with ITLC. However, TLC provides greater resolution of the labeled species present than ITLC. The separation time and the resolution achievable by radio-TLC are easily controllable by proper choice of the mobile phase and the TLC plate. Radio-TLC is capable of the simultaneous separation of ^{99m}Tc-colloid, $^{99m}TcO_4^-$, and the various forms of "bound" ^{99m}Tc present in the radiopharmaceutical. These attributes suggest that radio-TLC with the properly chosen stationary and mobile phases can be used as a replacement of, or supplement to, ITLC methods now in use.

Radio-TLC has a lower intrinsic resolution compared with HPLC. In addition, the ability to do gradient elutions in HPLC further enhances the resolution advantage of HPLC over TLC. This makes HPLC the radioanalytical method-of-choice in cases where sample resolution needs to be optimized. However, the future application of gradient TLC equipment now commercially available [26] may partially offset this resolution advantage of HPLC.

The major advantages of radio-TLC over HPLC are its relative simplicity, lower cost, and ability to detect 100% of the sample radioactivity. In HPLC studies of $^{99m}Tc(dmpe)_2Cl_2^+$ and ^{99m}Tc-DisIDA, we have found that as much as 5–10% of the sample activity is not recovered from the chromatographic

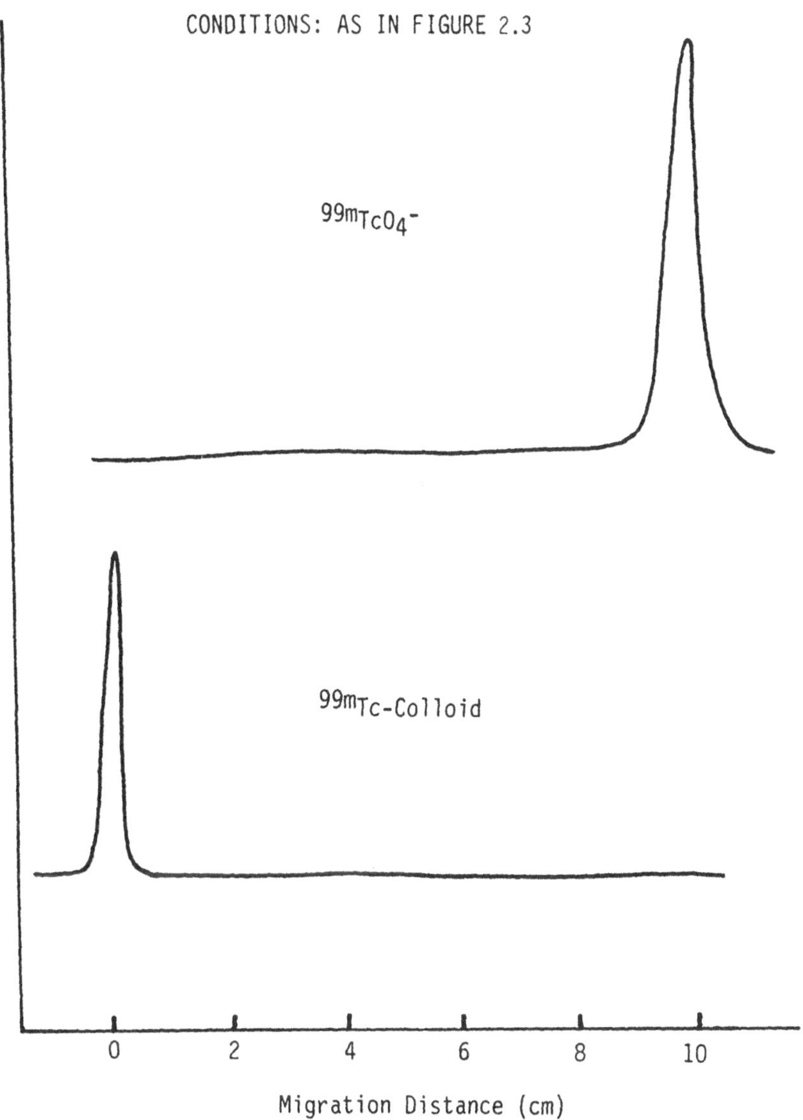

FIG. 2.6. Reverse phase TLC analysis of $^{99m}TcO_4^-$ and ^{99m}Tc-radiocolloid.

column due to adsorption or trapping of some of the radiolabeled species (see Table 2.12). At least a portion of this unrecovered material is radiocolloid, which is trapped at the head of the HPLC column. In contrast, radiocolloid is detectable on the TLC plate with an $R_f = 0.0$. This makes radio-TLC a slightly better quantitative procedure for radiochemical analysis when compared with HPLC due to the 100% sample detectability achieved by TLC.

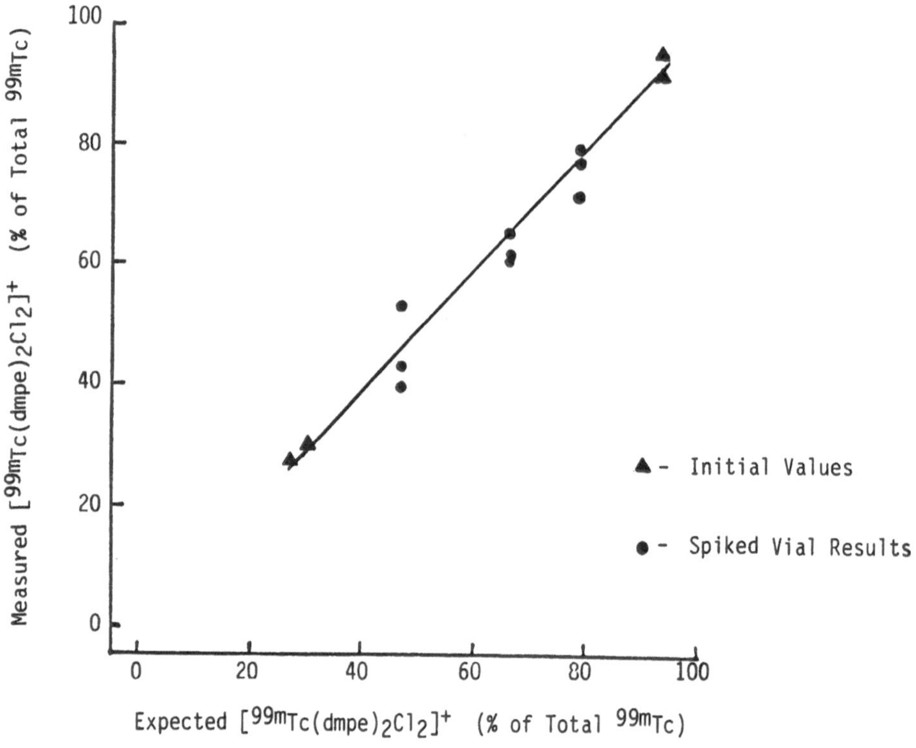

FIG. 2.7. Measured vs. expected % $^{99m}Tc(dmpe)_2Cl_{2+}$.

MINI-COLUMNS

Introduction

The utility of mini-column assays of radiochemical impurities in radiopharmaceuticals is based upon the speed and simplicity of the technique. The concept of rapid column assays has been described by Colombetti et al. [32]. In their work, columns of alumina were used for the separation of colloidal and bound ^{99m}Tc from $^{99m}TcO_4^-$. The $^{99m}TcO_4^-$ was selectively eluted from the alumina column with 0.9% saline solution, while the colloidal and bound ^{99m}Tc complexes remained on the column. Very good correlations were obtained between ITLC and rapid column assays of $^{99m}TcO_4^-$ for the several products tested. The assay time reported in this work was 3–5 minutes per sample, not including the time for column preparation. Zimmer et al. [33] have used miniaturized gel chromatography columns for the separation and analysis of $^{99m}TcO_4^-$ and colloidal ^{99m}Tc in ^{99m}Tc-labeled HIDA analogs. The agreement between ITLC and miniaturized column assays of $^{99m}TcO_4^-$ and collidal ^{99m}Tc-oxides was very good, and the analysis time was reported as less than 4 minutes per sample.

TABLE 2.12. Percentage Recovery of Injected Activity from HPLC[a]

Sample	*% Recovery*	*Ave. % Recovery*
$^{99m}Tc(dmpe)_2Cl_2^+$	91.6	
	89.3	
	98.0	
	91.6	
	96.3	95.1
	99.1	
	90.4	
	91.0	
	102.0	
	100.0	
	95.4	
	96.4	
^{99m}Tc-DisIDA	94.3	
	96.0	
	93.4	
	95.9	95.4
	101.3	
	96.8	
	89.9	
	95.2	

[a] *HPLC system for* $^{99m}Tc(dmpe)_2Cl_2^+$
4.6 mm × 10 cm Lichrosorb RP-8 column, 0 → 90%, 20 min linear methanol gradient in 0.05 *M* $(NH_4)_2SO_4$, 2 ml/min flow rate.
HPLC system for ^{99m}*Tc-DisIDA*
4.6 mm × 2.5 cm Zorbax C-18 BP-ODS column, 30 → 70%, 15 min linear acetonitrile gradient in 0.02 *M* pH 6.5 K_2HPO_4, 1 ml/min flow rate.

Despite these encouraging reports of rapid mini-column assays of common radiochemical impurities in ^{99m}Tc-based radiopharmaceuticals, this approach to quality control of radiopharmaceuticals has not been widely adopted. This may be due to the lack of commercial sources of prepacked mini-columns until recently.

Experimental Techniques and Applications

In work done in our laboratories, prepacked mini-columns from commercial suppliers such as Waters Associates and Analytichem, Inc., have been used. This has simplified the use of mini-columns beyond that reported in the literature owing to the elimination of the need to prepack the columns.

The dead volume of mini-columns is extremely low (<0.5 ml) and, consequently, the elution of the sample can be carried out with very small volumes ($\leq$ 4 ml). Octadecyl silane bonded phases have been used in our labs for the mini-column measurement of the hydrophilic (polar) ^{99m}Tc components of $^{99m}Tc(dmpe)_2Cl_2^+$ preparations. The total measurement time is ~2–3 minutes by this method. In this assay, the polar labeled radiochemical impurities, with high R_f values in the TLC separation system, are eluted from the mini-column with a normal saline rinse. Conceptually, the TLC and mini-column assays are compared in Figure 2.8. Under these conditions, free pertechnetate is unretained on the column, and any $^{99m}TcO_4^-$ present is ~100% recovered in the eluent as shown in Table 2.13. All activity eluted from the C-18 mini-column is a sum of the labeled polar ^{99m}Tc species and any free pertechnetate that may be present. When comparing the TLC-measured values for the percentage of ^{99m}Tc complexes of R_f's > 0.5 with the mini-column-measured percentage of activity in the normal saline eluent, a very good agreement is seen (see Table 2.14). This same type of reversed phase proce-

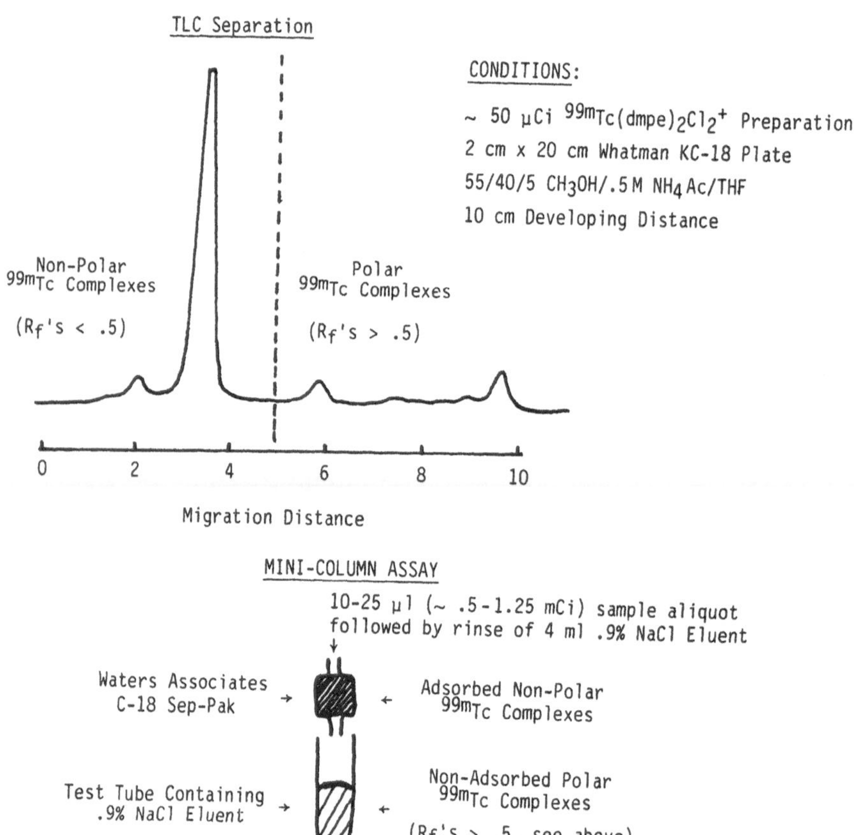

FIG. 2.8. TLC and mini-column separations of an impure $^{99m}Tc(dmpe)_2Cl_2+$ preparation.

TABLE 2.13. Recovery of $^{99m}TcO_4^-$ from C-18 Reverse Phase Mini-Column[a] Eluted with 4 ml 0.9% NaCl

Individual % Recovery Measurements	*Ave. Recovery ± σ*
99.6	
99.6	
99.7	99.6 ± 0.1
99.6	
99.7	
99.6	

[a] C-18 Sep-Pak, Waters Associates.

dure can be applied to the measurement of polar ^{99m}Tc radiochemical impurities in any product where the desired labeled species are retained on the column and the radiochemical impurities are eluted in saline or other polar solvent.

Although only adsorption and reverse phase partition stationary phases have been proven to be of value to date, there are many other types of chemically bonded phase mini-column materials available. This makes the extension of this concept into the area of ion-exchange and normal phase mini-column assays of specific ^{99m}Tc complexes likely in the future.

TABLE 2.14. Comparison of Radio-TLC[a] and Mini-Column[b] Measurements of the Percentage of ^{99m}Tc, as $^{99m}Tc(dmpe)_2Cl_2^+$ at Various Levels of Radiochemical Purity

Sample	*Radio-TLC Assay*	*Mini-Column Assay*
1	65.2, 61.9, 61.6	63.1, 63.2, 63.7
	(62.9 ± 2.0)	(63.3 ± 0.3)
2	69.8, 69.3, 69.4	73.3, 73.9, 73.6
	(69.5 ± 0.3)	(73.6 ± 0.3)
3	82.4, 83.3, 83.1	80.7, 80.9, 81.1
	(82.9 ± 0.6)	(80.9 ± 0.2)
4	85.9, 87.5, 87.0	85.7, 85.5, 85.1
	(86.8 ± 0.8)	(85.4 ± 0.3)
5	91.1, 91.5, 90.5	89.2, 88.8, 88.6
	(91.0 ± 0.5)	(88.9 ± 0.3)

[a] Whatman KC-18 TLC plates used with mobile phase of 55/40/5 Methanol/0.5 *M* ammonium acetate/THF.

[b] C-18 Sep-Paks, Waters Associates; 4 ml saline elution of 25-μl sample applied to column.

Advantages and Disadvantages of Mini-Columns

The obvious advantages of the mini-column approach to radiochemical purity analyses are the great speed and simplicity of the method. The only instrumentation required for mini-column assays is a radioactivity measurement device. This technique is even faster than ITLC. However, ITLC and mini-columns may not necessarily give the same information on a product. The choice of the type of mini-column stationary phase and the eluting solvent used will determine the type of information obtained. For example, the use of alumina mini-columns gives good $^{99m}TcO_4^-$ measurements for many products as illustrated by Colombetti's work [32]. The use of reversed phase mini-columns allows for the measurement of $^{99m}TcO_4^-$ plus other unretained or polar ^{99m}Tc complexes present in the radiopharmaceutical preparation. This is not necessarily a disadvantage of using reversed phase mini-columns, however, because other ^{99m}Tc impurities in addition to $^{99m}TcO_4^-$ can be responsible for the degradation of scintigraphic image quality. The ultimate goal of any radioanalytical test procedure, including mini-column assays, should be to selectively measure the radiochemical species responsible for lowering the quality of the images obtained with use of the radiopharmaceutical. In order to ensure that the correct radiochemical species are being measured in the mini-column assay procedure, a high resolution assay method must be used as the "referee" method, as has been demonstrated in the data showing the equivalency of radio-TLC and mini-column assays of polar ^{99m}Tc impurities in $^{99m}Tc(dmpe)_2Cl_2^+$ preparations (Table 2.14). There is sufficient merit in the high speed and low cost of mini-column assays to warrant future research in this area for the development of routine mini-column quality control measurements on radiopharmaceuticals.

ELECTROPHORESIS

Introduction

The use of electrophoresis in the separation of inorganic ions has been fairly extensive. Electrophoretic separations were commonly carried out on paper media during the late 1950s and early 1960s. The papers of Lederer [6], and Bailey and Yaffe [34], and Gross [35] are good references to the types of studies being formed at that time. Much of the early work on paper "ionophoresis" has been summarized by Blasius and Preetz [36]. The application of electrophoresis to radioisotope studies began very early because the properties of radiotracers provided a convenient means of detecting the separated bands. The review of Bailey [2] covers many of these studies of radioisotopes. Some of this type of work done in the 1960s also included studies on materials of radiopharmaceutical interest. For example, Bailey and Yaffe [34] described an electrophoretic system for the separation of

$^{99m}TcO_4^-$ and $^{99}MoO_4^{2-}$. In 1967, Höye published a method for the separation of ^{131}I-labeled human serum albumin from $^{131}I^-$ and $^{131}IO_3^-$ [7].

Despite these significant early efforts, the use of electrophoresis as a separation method in radiopharmaceutical studies has not gained much popularity. This can be mainly attributed to the fact that electrophoresis, on paper, does not give a very reproducible separation; it is highly dependent on the technique and procedures being followed. In addition, paper as a chromatographic medium is not ideal because it can interact in an adsorptive or ion exchange manner with labeled species, which leads to broad or tailing bands for many metal complexes (Bailey [2]). The resolution of electrophoresis is also relatively poor compared with chromatographic methods such as HPLC and TLC. This is due to the significant diffusive and electroosmotic band spreading that occurs during separation. This has been partially overcome by working at extremely high voltage gradients (up to 200 V/cm) on paper strips 100 cm long, which has allowed the electrophoretic run time to be routinely less than 30 minutes. However, the use of high voltage gradients leads to problems of uneven heating along the paper strip and this causes band spreading and problems of irreproducible separations from run to run.

HPLC, TLC, mini-columns, and ITLC analysis methods available today are more convenient and reproducible than electrophoresis. Despite this, electrophoresis continues to be used as a research tool in radiopharmaceutical studies. The reason for this is that electrophoresis gives information on the charge state of ^{99m}Tc complexes that can not be easily obtained by other techniques. From the standpoint of determining whether a specific radiolabeled complex is anionic, cationic, or neutral, electrophoresis is the best technique currently available.

Paper electrophoresis has been used successfully for the determination of the relative charge on some ^{99m}Tc complexes [18,37,38]. Marzilli et al. [38] have applied electrophoresis in an interesting manner to the determination of the number of ligands bound to technetium. Besides these few reports, very little work has been done on the electrophoresis of ^{99m}Tc complexes.

Experimental Techniques and Applications

A very rigorous treatment of the theory behind electromigration and electrophoresis is given in the text edited by Deyl [39]. The most important relationships, for the purpose of this discussion, are outlined in Table 2.15. From Table 15, it is seen that the electromigration rate is directly related to the charge and inversely related to the ionic radius of the complex being separated. Also, this means that in the ideal case, two species will have a relative migration rate that is a simple function of their relative charges and ionic radii. This ideal case is not often achieved, however, The major nonideal behavior is that variable interactions occur between the paper electropho-

TABLE 2.15. Migration Rate Equations for Electrophoretic Separations

$$dx/dt = \left(\frac{e}{6\pi\eta r}\right) E$$

Where: dx/dx = rate of migration of charged species
e = ionic charge of component
r = ionic radii of charged component
E = field gradient (volts/cm)
η = viscosity of the medium

For two charged species being separated under the same conditions, the relative migration rate:

$$\frac{dx_1/dt}{dx_2/dt} = \frac{r_2 e_1}{r_1 e_2}$$

retic medium and the complexes being separated. This type of interaction is particularly severe for cationic materials [2] because of the presence of negatively charged groups on the cellulose. In addition, if lipophilic complexes are being studied, it has been our experience that adsorption to the paper becomes a major problem, in some cases leading to no migration at all for charged lipophilic complexes. For this reason, we have recently been performing electrophoresis using polyacrylamide gels in place of paper.

The instrumentation used in our electrophoresis work is pictured in Figure 2.9. It consists of a flatbed electrophoresis chamber, which is cooled by a circulating ethylene glycol solution held at ~3°C, and a power supply that can be operated in either constant voltage, current, or power modes. The experimental techniques used in our electrophoresis studies are outlined in

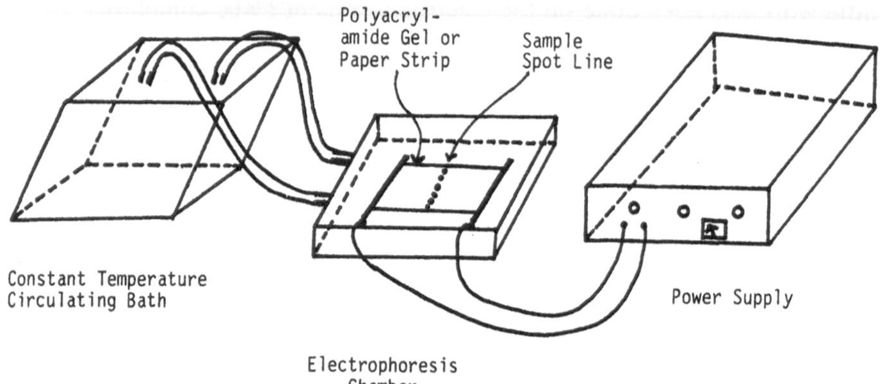

FIG. 2.9. Apparatus for electrophoresis studies.

TABLE 2.16. General Procedure for Paper Electrophoresis Studies

1. Cut Whatman #1 filter paper into 2 × 25 cm strips.
2. Prepare 10 m*M* sodium bicarbonate solution.
3. Wet paper strips with bicarbonate and insert each end in the electrolyte reservoir (i.e., bicarbonate).
4. Load 1–3 μl of sample in the center of strip.
5. Perform electrophoresis at: Voltage = 400 V
Current = 25 mA
Power = ~100 watts

   Run time = 20–30 minutes
6. Scan the strips with an appropriate TLC scanner from the cathode to the anode end of the strip to determine migration of sample components.

Table 2.16, for the paper medium, and in Table 2.17, for the polyacrylamide gel system.

For the purpose of determining the charge state of nonlipophilic materials, such as pertechnetate, paper or gel electrophoresis media may be used as shown in Figure 2.10. However, electrophoresis on a more lipophilic cationic material, polyacrylamide gels are a superior medium relative to paper. As shown in Figure 2.11, the $^{99m}Tc(dmpe)_2Cl_2^+$ complex migration toward the cathode using paper media is associated with significant band spreading and perhaps decomposition as well. In contrast, the gel electrophoresis separation clearly reflects the cationic nature of the complex, gives a symmetric,

TABLE 2.17. General Procedure for Slab Gel Electrophoresis Studies

1. Take Ampholine Polyacrylamide PAG PLATE pH 3.5–9.5 (obtained from LKB, Cat. #1804-101).
2. Place gel on a flatbed electrophoresis unit (which is hooked up to a circulating water bath set at 3°C).
3. Place the paper wicks (wetted with electrolyte) along the edge of gel next to the electrodes. (*Note:* The wick should be 0.5 cm wide; the length is determined by the length of the gel.)
4. Insert the platinum electrode assembly to contact the wicks.
5. Load samples (~1–3 μl) along the centerline of the gel, approximately 2 cm apart.
6. Perform electrophoresis at: Voltage = 400 V
Current = 25 mA
Power = 10 watts

   Run time = 20 minutes
7. Cut gel strips in ~2 × 11 cm segments.
8. Scan the gel strips with an appropriate TLC scanner from the cathode to the anode ends of the gel to determine sample mobility.

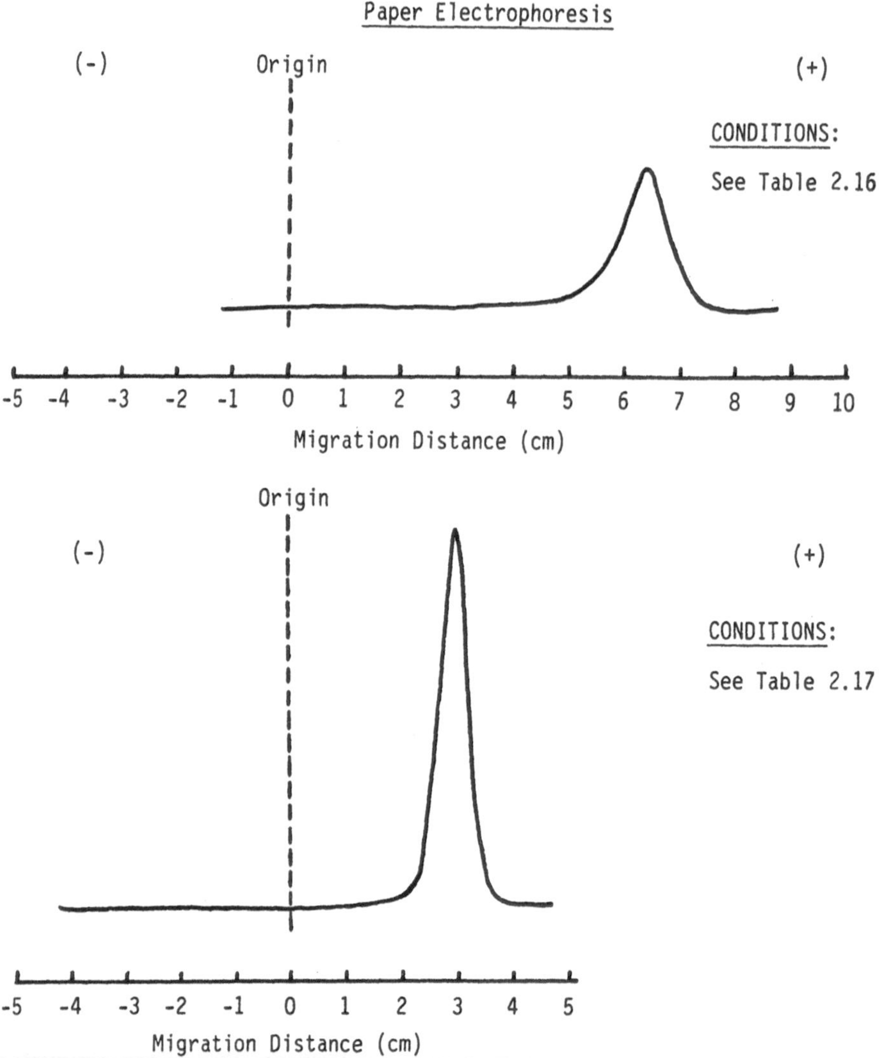

FIG. 2.10. Electrophoresis separations and polyacrylamide gel electrophoresis of $^{99m}TcO_{4^-}$.

narrow, sample band and also gives charge information on the other minor impurities in the sample preparation.

In the quantitation of individual radiochemical species separated by gel electrophoresis, the same chromatographic scanner apparatus used for scanning TLC plates is employed. An example of the integrated electrophoresis scan is shown in Figure 2.12 for $^{99m}Tc(dmpe)_2Cl_2^+$ along with a TLC profile obtained on this same type of preparation. Good qualitative and quantitative agreement is obtained on the measurement of the percentages of each radiochemical component between the two methods. From the electrophoresis scan, the cationic, anionic, and neutral ^{99m}Tc complexes present can be equated with specific TLC bands.

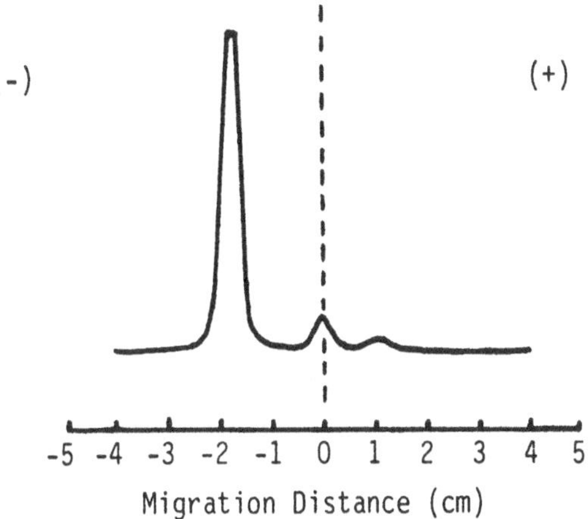

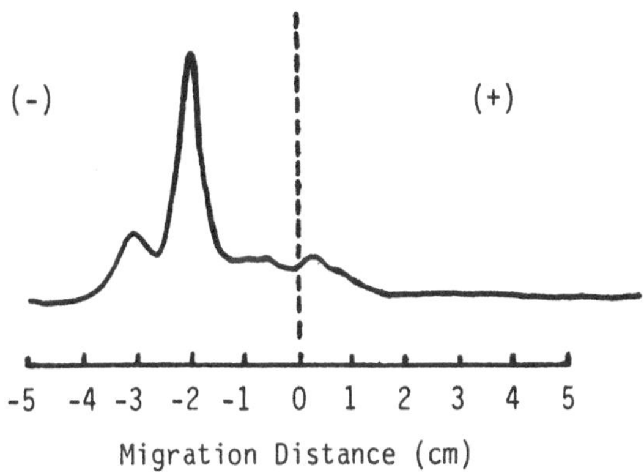

FIG. 2.11. Polyacrylamide gel and paper electrophoresis separations of ^{99m}Tc $(dmpe)_2Cl_2^+$ preparations.

A similar example is shown in Figure 2.13 for a ^{99m}Tc-DisIDA electrophoresis scan along with an HPLC profile of ^{99m}Tc-DisIDA. Good agreement is obtained for the percentages of the radiochemical species detected by electrophoretic analysis with the HPLC data. The overall negative charge of ^{99m}Tc-DisIDA seen here is in agreement with previously published data on ^{99m}Tc-HIDA [37].

Advantages and Disadvantages of Electrophoresis

Electrophoresis as applied to radiopharmaceutical separations has some significant limitations when paper electrophoretic media are used. Paper electrophoresis works well when separating charged, hydrophilic radiolabeled

Polyacrylamide Gel Electrophoresis Separation

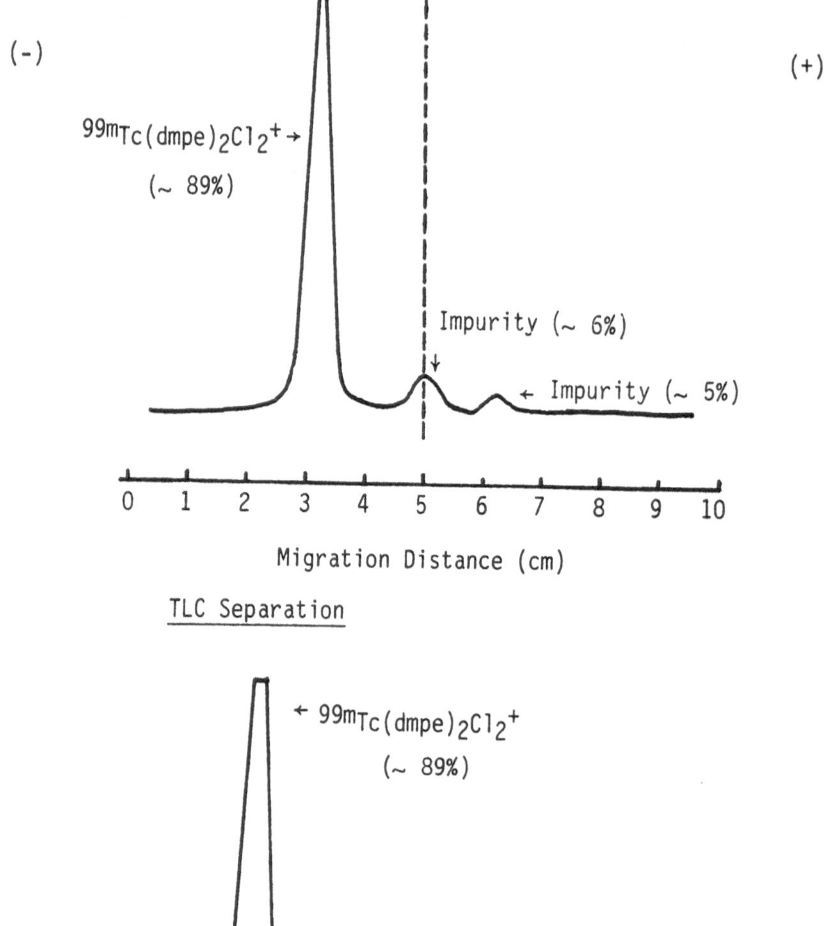

FIG. 2.12. Electrophoresis vs. TLC separations of $^{99m}Tc(dmpe)_2Cl_2^+$ preparation.

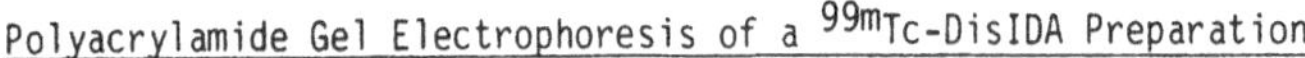

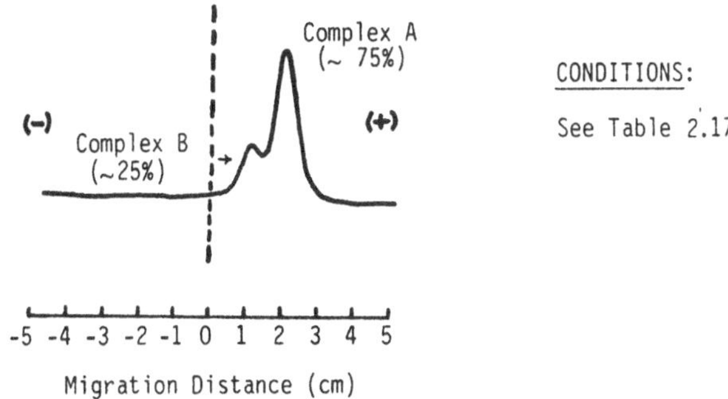

Radio-HPLC of a ^{99m}Tc-DisIDA Preparation

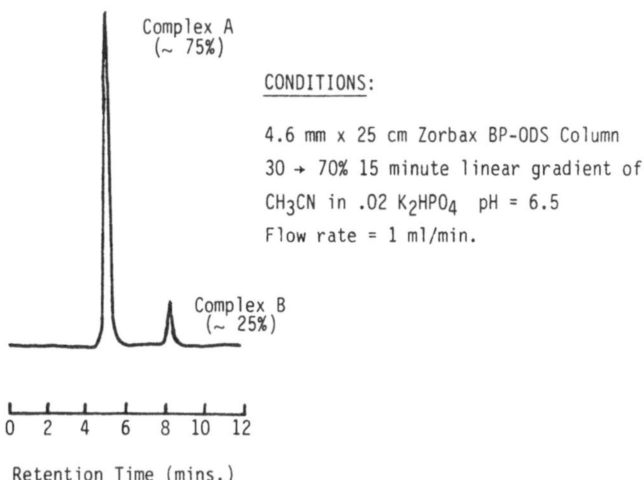

FIG. 2.13. Electrophoresis vs. HPLC profiles of ^{99m}Tc-DisIDA preparations.

materials, but it does not work well on highly lipophilic complexes. The use of polyacrylamide gels in place of paper extends the utility of the electrophoretic technique further because adsorptive effects are diminished. This technique is very informative regarding the charge state on radiolabeled materials in radiopharmaceuticals. It has the capability of being used as a quantitative method for the assay of radiochemical purity, although it is not as convenient as many of the chromatographic methods employed for this purpose.

The use of electrophoresis will remain important in research studies on new, unknown ^{99m}Tc-labeled complexes. However, the slow separation speed and poor reproducibility of electrophoresis will most likely prevent its adoption in routine quality control studies.

CONCLUSIONS

The separation methods discussed here all have application to specific aspects of the analysis of radiopharmaceuticals. The choice of the separation method used in any radioanalytical study must be based primarily on the type of information sought, and secondarily, on the convenience, cost, and difficulty of performing the analysis.

HPLC is the method-of-choice for radioanalytical studies requiring the highest resolution possible. ITLC or mini-column assays are the best methods available for the rapid analysis of specific types of radiochemical components such as free pertechnetate or colloidal technetium. TLC, which offers moderate resolution and good speed, is a practical compromise between HPLC and ITLC/mini-column methods. TLC allows for 100% detection of the labeled species in the sample, which is an advantage over HPLC in cases where the greatest quantitative accuracy is needed. Electrophoresis is the only technique commonly used for the assessment of the charge state of radiopharmaceutical components. The use of electrophoresis, however, appears to be limited to research only, as a result of its poor reproducibility and speed characteristics.

FUTURE CONSIDERATIONS

Recent developments in high performance thin-layer chromatography and microbore HPLC techniques have demonstrated the advantages of these techniques in high speed separations of pharmaceuticals. The extension of these procedures to radiopharmaceutical analyses in a research environment may take place over the next few years.

With the judicious choice of analysis conditions, radio-TLC and mini-column procedures for the quality control study of radiopharmaceuticals may become more popular in the clinical laboratory over the near future, particularly as radiopharmaceutical scientists become more aware of the need to assay for other radiochemical impurities, in addition to $^{99m}TcO_4^-$ and colloidal ^{99m}Tc.

Electrophoresis, as it is currently being employed, will continue to be limited by problems associated with the adsorption of large lipophilic radiopharmaceuticals to the electrophoretic medium. The use of open tubular (*supportless*) methods, such as capillary isotachophoresis, may be required for future electrophoretic studies of radiopharmaceuticals.

REFERENCES

1. *U.S. Pharmacopeia,* USP XX. USP Convention Inc., Rockville, 1980, pp. 410–411.
2. Bailey, R.A.: *Paper Chromatographic and Electromigration Techniques in Radiochemistry,* NAS-NS #3106. U.S. Atomic Energy Commission, Oak Ridge, Tenn., 1962.

3. Dickey E.E.: Separation of radium D, E and F by paper chromatography. *J. Chem. Ed.* **30**:525, 1953.
4. Lima, F.W.: Experiments in paper partition chromatography with tagged material. *J. Chem. Ed.* **31**:153, 1954.
5. Janardhan, P.B. and Paul, A.: Separation of tin and thallium valences by paper chromatography. *Separation Science* **2**(5):597, 1967.
6. Lederer, M.: *Progress de la Chromatographie, Troisieme Partie, Chromatographie sur Papier des Radioelements.* Herman et Cie, Paris, 1956.
7. Höye, A.: Determination of radiochemical purity of some radiochemicals and pharmaceuticals by paper chromatography, thin layer chromatography and high voltage electrophoresis. *J. Chromatogr.* **28**:379, 1967.
8. Eckelman, W.C. and Richard, P.: analytical pitfalls with ^{99m}Tc-labeled compounds. *J. Nucl. Med.* **13**:202, 1972.
9. Zimmer, A.M. and Holmes, R.A.: A precise chromatography system for specific technetium-99m radiopharmaceuticals. *Nuklearmedezin* **14**:192, 1975.
10. Pauwels, E.K.J. and Feitsma, R.I.J.: Radiochemical quality control of ^{99m}Tc-labeled radiopharmaceuticals. *Eur. J. Nucl. Med.* **2**:97, 1977.
11. Belkas, E.P. and Archimandritis, S.: Quality control of colloid and particulate ^{99m}Tc-labeled radiopharmaceuticals. *Eur. J. Nucl. Med.* **4**:375, 1979.
12. Robbins, P.J.: *Chromatography of Technetium-99m Radiopharmaceuticals—a Practical Guide.* Society of Nuclear Medicine, Inc., New York, 1984.
13. Jovanovic, J., Konstantinovska, D. and Memedovic, T.: Determination of radiochemical purity and stability of ^{99m}Tc-diethyl HIDA. *Eur. J. Nucl. Med.* **6**:375, 1981.
14. Bish, R.E., Silverstein, D. and Bebe, J.J.: Radiopharmaceutical-radiochromatography quality assurance with a radioisotope calibrator. *J. Radioanal. Chem.* **57**:565, 1980.
15. Pinkerton, T., Ferguson, D., Deutsch, E., Heineman, W.R. and Libson, K.: Invivo distributions of some component fractions of Tc($NaBH_4$)-HEDP mixtures separated by anion exchange high performance liquid chromatography. *Int. J. Appl. Radiat. Isot.* **33**:907, 1982.
16. Srivastava, S.C., Bandyopadhyay, D., Meinken, G. and Richards, P.: Characterization of Tc-99m bone agents (MDP, EHDP) by reverse phase and ion exchange high performance liquid chromatography. 28th Society of Nuclear Medicine Meeting Abstract, 1981.
17. Nunn, A.: Correlation between the reverse-phase capacity factors for Tc-99m phenylcarbamoylmethyliminodiacetic acids and their renal elimination. *J. Chromatogr.* **255**:91, 1983.
18. Loberg, M.D., Nunn, A.D. and Porter, D.W.: Development of hepatobiliary imaging agents. In *Nuclear Medicine Annual,* Freeman, L.M. and Weissman, H.S., eds., Raven Press Publications, New York, 1981, pp. 1–33.
19. Fritzberg, A.R. and Lewis, D.: HPLC analysis of Tc-99m iminodiacetate hepatobiliary agents and a question of multiple peaks: concise communication. *J. Nucl. Med.* **21**:1180, 1980.
20. Jansholt, A.L., Krohn, K.A., Vera, D.R., and Hines, H.H.: Design and validation of a radiochromatogram scanner with analog and digital output. *J. Nucl. Med. Tech.* **8**:222, 1980.
21. Izmailov, N.A. and Shraiber, M.S.: The application of analysis by drop chromatography to pharmacy. *Farmezia* **3**:1, 1938.

22. Stahl, E.: Dünnschicht-Chromatographie. *Pharmazie* **11**:633, 1956.
23. Stahl, E., ed. *Thin Layer Chromatography: A Laboratory Handbook*. Academic Press, New York, 1965.
24. Issaq, H.J.: Modifications of adsorbent, sample and solvent in thin layer chromatography. *J. Liquid Chromatogr.* **3**:1423, 1980.
25. Issaq, H.J.: Recent developments in thin layer chromatography—II. *J. Liq. Chrom.* **4**:955, 1981.
26. Zlatkis, A. and Kaiser, R.E., eds. *HPTLC*. Elsevier, Amsterdam, 1977.
27. Guichon, G., Bressolle, F. and Siouffi, A.: Studies of the performances of thin layer chromatography: IV—optimization of experimental conditions. *J. Chromatogr. Sci.* **17**:368, 1979.
28. Snyder, L.R. and Kirkland, J.J.: *Introduction to Modern Liquid Chromatography*. John Wiley and Sons, New York, 1979, pp. 301–316.
29. Carpenter, A.P. Jr., Mysliwy, T.J., Sivakoff, S.I., Doshi, M.N., Wolshin, E.M., Matthew, M.A., Schnaper, G.H., Subramanyam, V. and Camin, L.: Characterization and use of a gamma radiation TLC scanner for radiochemical purity measurements of radiopharmaceutical kit preparations of $^{99m}Tc(dmpe)_2Cl_2^+$. *J. Radioanal. Chem.*, in press.
30. Nishiyama, H., Deutsch, E., Adolph, R.J., Sodd, V.J., Libson, K., Saenger, E.L., Gerson, M.C., Gabel, M., Lukes, S.J., Vanderheyden, J.L., Fortman, D.L., Scholz, K.L., Grossman, L.W. and Williams, C.C.: Basal kinetics studies of Tc-99m DMPE as a myocardial imaging agent in the dog. *J. Nucl. Med.* **23**:1093, 1982.
31. Davison, A., Jones, A. and Abrams, M.: Investigations on a new class of technetium cations. 29th Society of Nuclear Medicine Meeting Abstract, 1982.
32. Colombetti, L.G., Pinsky, S. and Moerlien, S.: A rapid column chromatographic determination of free ^{99m}Tc in labeled radiopharmaceuticals. *Radiochem. Radioanal. Letters* **20**:77, 1974.
33. Zimmer, A.M., Majewski, W. and Spies, S.M.: Rapid miniaturized chromatography for Tc-99m IDA agents: comparison with gel chromatography. *Eur. J. Nucl. Med.* **7**:88, 1982.
34. Bailey, R.A. and Yaffe, L.: The separation of some inorganic ions by high-voltage electromigration in paper. *Can. J. Chem.* **38**:1871, 1960.
35. Gross, D.: High-voltage paper electrophoresis of inorganic cations. *J. Chromatogr.* **10**:221, 1963.
36. Blasius, E. and Preetz, W.: Application of paper ionophoresis and electrochromatography to the study of metal complexes in solution. In *Chromatographic Reviews,* vol. 6, Lederer, M., ed., Elsevier, Amsterdam, 1964, pp. 191–213.
37. Burns, D., Marzilli, L. Sowa, D., Bauma, D. and Wagner, H.N. Jr.: Relationship between molecular structure and biliary excretion of technetium-99m HIDA and HIDA analogs. 24th Society of Nuclear Medicine Abstract, 1977.
38. Marzilli, L., Worley, P. and Burns, H.D.: A new electrophoretic method for determining ligand: technetium stoichiometry in carrier free ^{99m}Tc radiopharmaceuticals. *J. Nucl. Med.* **20**:871, 1979.
39. Deyl, Z., ed.: *Electrophoresis: A Survey of Techniques and Applications*. Part A, Techniques. *J. Chrom.* library, vol. 18, Elsevier, Amsterdam, 1979.

CHAPTER 3

Radio-Thin-Layer Chromatogram Imaging Systems—Performance and Design

SHERYL J. HAYS

INTRODUCTION

Thin-layer chromatography (TLC) is a valuable tool for the radiochemist and radiopharmacist. It helps in following the progress of radiochemical reactions and in identifying and determining the purity of radiolabeled compounds. In the past, thin-layer radiochromatograms have been most commonly analyzed by autoradiography, by radiochromatogram scanners (i.e., Packard 720 scanner), and by scraping or cutting of plates followed by scintillation counting. These methods are time consuming and all demonstrate one or more limitations. Autoradiography is generally slow, has poor sensitivity, and cannot provide accurate, quantitative results. Radiochromatogram scanners have poor sensitivity and often are not equipped to provide quantitative analysis. Cutting or scraping of plates followed by scintillation counting has excellent sensitivity and quantitative accuracy, but the process is time consuming and spatial resolution is often compromised [1].

If radioisotope thin-layer chromatography is to remain competitive with such techniques as HPLC, improvements are needed for quantitating radiolabeled compounds on the TLC plate. This need has been filled by the development of radiochromatogram imaging proportional counters. These instruments have been available since 1981, and two companies, Berthold from West Germany and Bioscan from Washington D.C., manufacture and market these systems. These instruments retain most of the advantages of the previously mentioned analytical techniques while eliminating many of the shortcomings. They provide excellent quantitation of thin-layer chromatograms with average analysis time of less than ten minutes. These systems also demonstrate good sensitivity, with resolution surpassed only by autoradiography. The purpose of this chapter is to further describe and compare the design and performance of these imaging systems.

INSTRUMENTATION

TLC imaging systems count the whole thin-layer plate simultaneously by utilizing a proportional detector positioned directly over the chromatogram as shown in Figure 3.1. The chamber over the plate is filled with P-10 gas. This is a combination of 90% argon and 10% methane and is readily available from most gas suppliers. A high voltage anode wire runs the length of the gas-containing chamber. When radiation such as gamma rays or beta particles interacts with the gas, the gas becomes ionized. This produces free electrons, which are accelerated toward the anode wire. The electrons then enter the high electric field near the wire surface and are accelerated to sufficient energy to produce further ionization of the gas and a resultant pulse on the anode wire of sufficient magnitude to detect with electronics. The imaging capability of the counters is produced by providing an electronic signal that changes as the position of the incident radiation varies. A preamplifier is attached at both ends of the detector. When a pulse of electrons is collected at the resistive anode, the current is divided and part of the pulse flows toward each end. The ratio of the two parts of the pulse is determined by the amount of resistance between the original collection point and each of the preamplifiers. The two amplified pulses are further converted to a digital output on a pulse height scale and the quotient shown is computed to give a numerical result for the position. A digital image is built up in the computer memory by adding one count to the total, which is stored in the memory location corresponding to a specific location along the anode [1]. As the counts are collected, the radiochromatogram is simultaneously viewed on the visual display monitor. Upon completion of the scan, the chromatogram can be stored and hard copies can be printed [1–3].

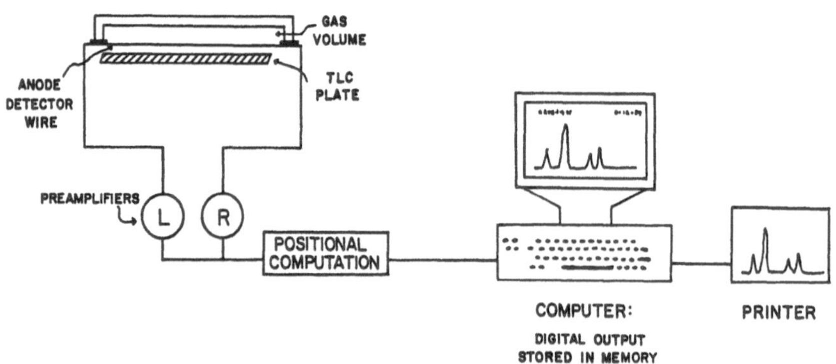

FIG. 3.1. Diagram of radio-TLC imaging system.

RESOLUTION AND EFFICIENCY

A primary limitation on the resolution of the TLC imaging system results from the finite 5-ml depth of the detector. In general, the spread will be approximately comparable to the detector depth for high energy betas and gammas. This spread will decrease as the energy of the beta decreases [1]. Figure 3.2 shows a scan of a TLC plate spotted with 1) tritium, (β^- emitter, 18 Kev), 2) carbon-14 (β^- emitter, 155 Kev), 3) iodine-125 (Auger electrons, 30 Kev), 4) iodine-131 (β^- emitter, 608 Kev) and flourine-18 (β^+ emitter, 635 Kev). The isotopes were spotted with a micropipette and the spots had intrinsic diameters of approximately 2 mm. It is clear from the chromatogram that the best resolution is obtained with beta-emitting tritium, while the higher energy-producing isotopes, iodine-131 and fluorine-18, produce slightly broadened peaks. This broadening can be reduced by the use of aperture plates, which can be placed over the detector head to provide mechanical collimation. Bioscan's aperture plates are easily snapped in place and are, therefore, more convenient to use than Berthold's screw type. In a further effort to improve resolution of the higher energy-emitting isotopes, Berthold offers a second detector head, termed a P-32 head. This detector, which was designed for use with the more energetic phosphorus-32 isotope, simply moves the detector anode wire closer to the TLC plate. It

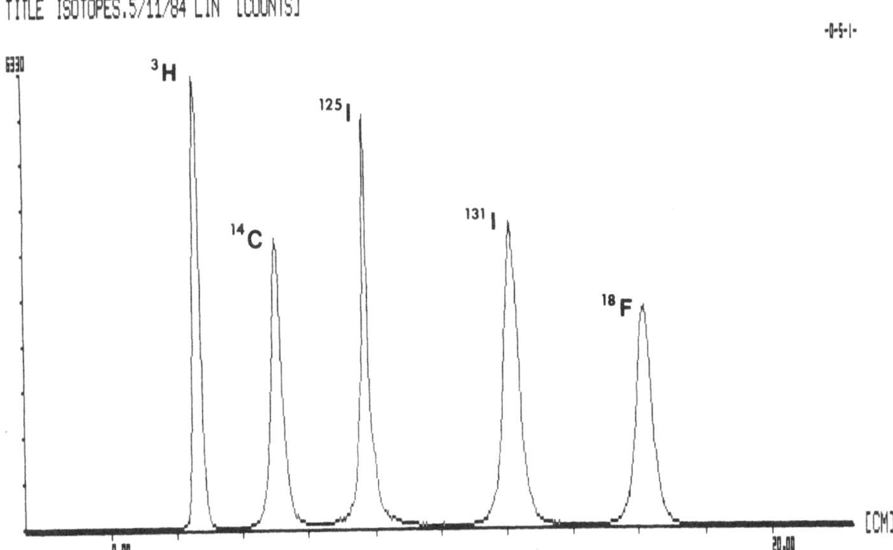

FIG. 3.2. Five isotopes were spotted (1–2 mm diameter) and scanned with a Berthold 2832 imaging system on gain setting 1. In general, the spread will be approximately comparable to the detector depth for high energy betas and gammas. This spread will decrease as the energy of the beta decreases.

TABLE 3.1. Counting Efficiencies of Common Nuclides from TLC Plates of 200 Thickness

Nuclide	*Emitter*	E(*max*)	*CPM/DPM*
^{3}H	Beta	18 Kev	0.5–3%
^{125}I	Auger Gamma	30 Kev	2–5%
^{14}C	Beta	155 Kev	<20%
^{35}S	Beta	200 Kev	<20%
^{131}I	Beta	608 Kev	<40%
^{18}F	Positron	635 Kev	<40%
^{11}C	Positron	970 Kev	<40%
^{32}P	Beta	1700 Kev	<40%
^{99m}Tc	Gamma	140 Kev	5%
^{123}I	Gamma	159 Kev	5%

[a] Plates were scanned with a Berthold LB 283 imaging systems.

then becomes necessary to wrap the plate with a thin polymeric sheet to avoid contamination problems. This option is feasible for laboratories that exclusively utilize higher energy isotopes, but may present limitations for those researchers who also make use of lower energy isotopes, such as tritium.

Detector efficiency is also correlated with the energy of the emitted beta particles. As the energy of the betas increases, efficiency also increases. Counting efficiency of pure gamma emitters such as technetium-99m and iodine-123, is reduced compared with most beta-emitting isotopes. This is a result of the reduced interactions of gamma radiation with the detector gas compared with the more massive beta particles [4]. Another major limitation on efficiency is the self-absorption of low energy betas by the thin-layer absorbent itself. Also, because the detector head is positioned above the plate, downward and lateral emissions from the plate cannot be detected. Consequently, a detector efficiency of 50% is all that is theoretically possible. Table 3.1 shows commonly used isotopes, their energies, and counting efficiencies.

PERFORMANCE AND OPERATION OF THE TLC IMAGING SYSTEMS

The Berthold and Bioscan instruments come with many standard operating features as well as several optional features. The Berthold is sold with an Apple IIe computer, while the Bioscan 200 imaging scanner can be interfaced to either an Apple IIe or a Hewlett-Packard 85B computer. Both instruments provide three types of sample changers: 1) a manual changer, 2) an automatic plate changer with a 20 × 20 cm sample area, and 3) an auto-

TABLE 3.2 Software Features Available for the Berthold LB 2832 and the Bioscan 200 Imaging Scanners

Software Features	*Berthold LB 2832*	*Bioscan 200*
Menu-driven program for control, display, and data analysis	Yes	Yes
Real-time data display	Yes	Yes
Automatic and manual peak integration	Yes	Yes
Background subtraction	Yes	Yes
One-to-one scale graphics (for easy location of the radiolabeled compound on the plate)	Yes	Yes
Reduced-scale graphics	Yes	Yes
Storage and recall of data on disk	Yes	Yes
Multi-user custom setup files	Yes	Yes

matic plate changer that is large enough to accommodate at least three 20 × 20 cm plates (approximately 60 counting lanes). The automatic sample changers will advance from one counting lane to the next after each analysis without operator attention. These instruments have a number of software features that can be seen in Table 3.2. In 1984, the cost of the entire Berthold or Bioscan system, including detector, sample changer, computer interface, computer, software, and printer was between $15,000 and $28,800, depending on model and options selected. While this sum may seem staggering at first glance, these instruments are comparably priced with other major laboratory equipment, such as scintillation counters and microprocessor-controlled HPLC systems. In addition, the prices of several models of TLC-imaging systems have been reduced in the past twelve months, primarily owing to competition between Bioscan and Berthold. If this trend continues, these systems will be within the financial means of more academic radiochemical laboratories.

The greatest gain from using a TLC imaging system is derived from its ease of use and ability to quantitate peaks. Figure 3.3 illustrates a chromatogram with several operator-selected regions of interest. The peak printout shows the beginning, end, and center of each region, the total counts, counts per minute, and the percentage of total. This information was obtained in less than two minutes of scanning time with high reproducibility.

CONCLUSION

Proportional imaging scanners for analysis of radiochemical thin-layer chromatograms have been available since 1981. While the initial cost of these systems is high, they demonstrate advantages in sensitivity, resolution, and quantitation over conventional methods of TLC radiochromatographic analysis. These instruments are convenient, fast to use and represent important,

5C-243
Plate = 0
Lane = 0

Date: 2/29/84 Start time: Accum. time: 00:06:20

Total counts (0.0 - 20.0 cm) = 2937.0, CPM = 463.74
Electronic Resolution Setting = N

PEAK #	START (cm)	STOP (cm)	CENTROID	TOTAL CNTS	CPM	% of TOTAL
1	0.5	3.0	1.9	100.0	15.79	3.4
2	4.1	6.9	5.6	89.0	14.05	3.0
3	7.1	10.2	8.9	362.0	57.16	12.3
4	10.3	11.9	11.2	998.0	157.58	34.0
5	12.5	14.2	13.2	1187.0	187.42	40.4
6	14.6	15.8	15.2	65.0	10.26	2.2

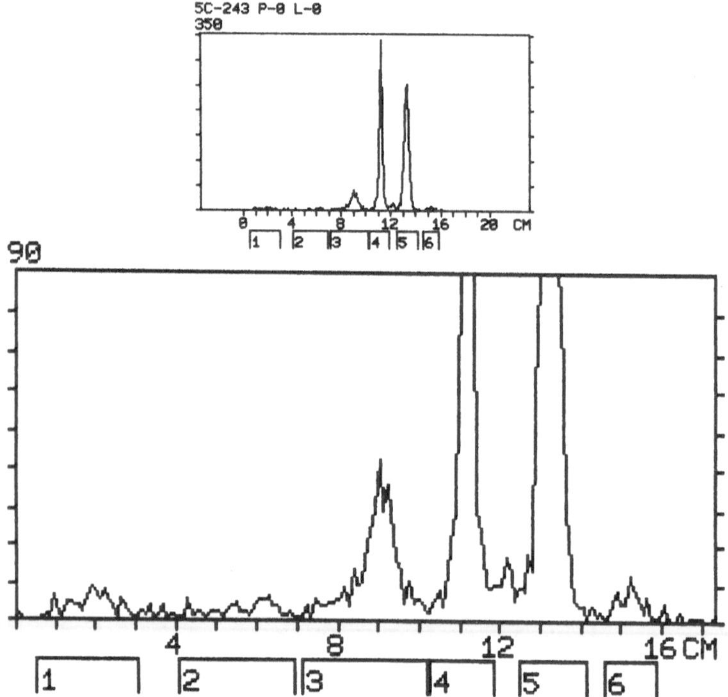

FIG. 3.3. This chromatogram was scanned with a Bioscan 200 imaging system interfaced with an Apple IIe computer. In this scan, regions of interest have been operator defined and the peaks quantitated by CPM and the percentage of total.

state-of-the-art systems for radiochemical laboratories and nuclear medicine pharmacies.

ACKNOWLEDGMENTS

I would like to thank Drs. Richard Ehrenkaufer and Thomas Mangner of the University of Michigan and Dr. Seth Shulman of Bioscan for their assistance and cooperation.

REFERENCES

1. Shulman, S.D.: A review of radiochromatogram analysis instrumentation. *J. Liquid Chromatogr.* **6**:35–53, 1983.
2. Filthuth, H.: State of the art in TLC scanning. In *Synthesis and Applications of Isotopically Labelled Compounds: Proceedings of an International Symposium*, Duncan, W.P. and Susan, A.B., eds. Elsevier, Amsterdam, pp. 447–452, 1983.
3. Filthuth, H.: Linear analyzer improves detection in radio-TLC tests. *Indust. Res. Devel.* **6**:140–145, 1981.
4. Wang, C.H., Willis, D.L. and Loveland, W.D. Characteristics of ionizing radiation in *Radiotracer Methodology in the Biological, Environmental and Physical Sciences*. Prentice-Hall, Englewood Cliffs, pp. 72–74, 1975.

CHAPTER 4

Detection of Radiochromatograms and Electropherograms with Position-Sensitive Wire Chambers

HEINZ FILTHUTH

INTRODUCTION

New developments in the detection technique of radiochromatograms and electropherograms were presented for the first time at the Symposium of Chemical and Environmental Applications of Quantitative Thin Layer Chromatography in Philadelphia, 1980 [1,2]. Since then, the methods we have described have been refined and applied in practice in many laboratories and considerable experience has been gained. First, we provide a short description of the new methods, then show examples of their application and results.

What Do We Want To Detect?

Figure 4.1 illustrates the situation in which two radiolabeled compounds are separated by chromatography in a thin-layer plate. Beta particles are emitted isotropically from the surface above the compounds. The problem is to detect the position of the compounds and the amount of radioactivity they contain. The classical method for the detection of materials in radiochromatography is *autoradiography*. This technique involves the exposure of thin-layer plates or electrophoresis gels to conventional X-ray film. After several days—or even several weeks—one can identify the radioactive regions on the film. Using photometric methods, one can evaluate the relative intensities of the radioactivity associated with the various spots.

One obvious limitation in this method is the small dynamic range of the X-ray film. A far more precise method is the *scintillation counting* technique.

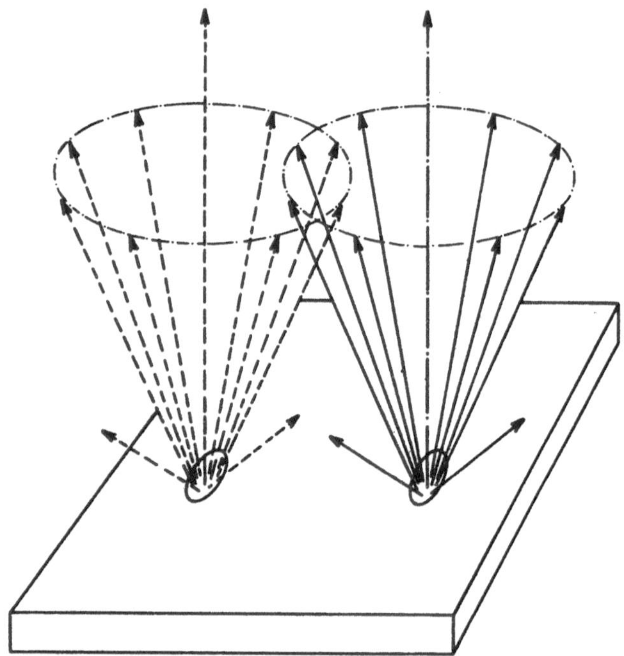

FIG. 4.1. Beta emission of two separated compounds from TLC plate.

This method requires the removal of the identified radioactive regions from the plates, gels, or paper. Then the substances of interest must be removed from the carrier materials and introduced into the scintillation liquid. The method is time consuming; it demands extensive labor; and it presents the additional problem of disposal of poisonous radioactive waste.

About 1965 the *TLC scanner* was introduced. This instrument replaced the autoradiographic technique in many applications that required a spatial resolution of only a few millimeters. Because this counter has a small entrance window (approximately 2 × 20 mm), the scanning time for the surface of the TLC plate was reduced to several hours, or a day, depending on the radioactivity in the plate.

What Would Be the Ideal Detector?

Two methods can be regarded as ideal:

1. To use an infinitely *thin detector*, attached directly to the emitting surface, to overcome the problem of detecting a divergent particle beam of isotropically emitted beta particles.
2. To design a detector that can "see" the particle trajectories in space and trace them down to the origin of emission. We call this a *three-dimensional detector*.

Autoradiography makes use of a very thin detector, the photoemulsion. But this technique is not quantitative and is especially time consuming.

The second approach, a three-dimensional detector, is applied mostly in particle physics at the accelerator laboratories. Cloud chambers, bubble chambers, spark- and multiwire proportional chambers, and time projection chambers are the powerful tools in particle physics. One can detect particle trajectories in space to a very high precision and can determine their origin of emission to a precision of about 100 μm. Unfortunately, these experimental setups at the accelerator laboratories must be considered very big, complicated, and very expensive, when measured on the scale of normal laboratories. We have chosen, therefore, the approximation to method (1), developing a very thin (flat) detector. Our aim was to design a detector that would be relatively simple and provide inexpensive measurement of radiochromatograms and electropherograms. We have called our instrumentation the *TLC Linear Analyzer*.

THE TLC LINEAR ANALYZER

For the detection of one-dimensional radiochromatograms, we use the technique involving position-sensitive counters. These devices are widely used in particle physics. The linear analyzer counts the whole chromatogram trace simultaneously. It is approximately 100 times faster and, therefore, more sensitive than the old type of TLC scanner.

The basic principle of the linear analyzer, which appears in Figure 4.2, is described in the following discussion. The heart of the linear analyzer is a position-sensitive proportional counter, shown in Figure 4.3. When the instrument is operated, beta particles and gamma rays emitted from the chromatogram enter the counter from the bottom through a 250 × 15 mm open entrance window. This configuration permits the detection of low energy beta particles, such as those emitted from ^{3}H and ^{125}I. Gamma rays, i.e., from ^{99m}Tc and ^{123}I, are also detected being converted to electrons by photo- and compton effects inside the counter. The counting wire is stretched along the center line of the counter. Since the detector operates as a proportional counter, the electron avalanche caused by gas amplification of the primary ionization is limited to locations directly above those at which particles are emitted from the TLC plate. The height of the count pulses is proportional to the electron avalanches; the time of each pulse occurrence is proportional to the position of the incident particles. The pulses are induced into a delay line, running the full length of the counting chamber, and they pass along the delay line in both directions. The arrivals of pulses at either end of the delay line are recorded electronically. The time difference between the arrivals of two associated pulses is a direct measure of the location of an incident particle. After preamplification, the pulses are converted into the fast output pulses, *start* and *stop*. The stop-start time difference is digitized in the time

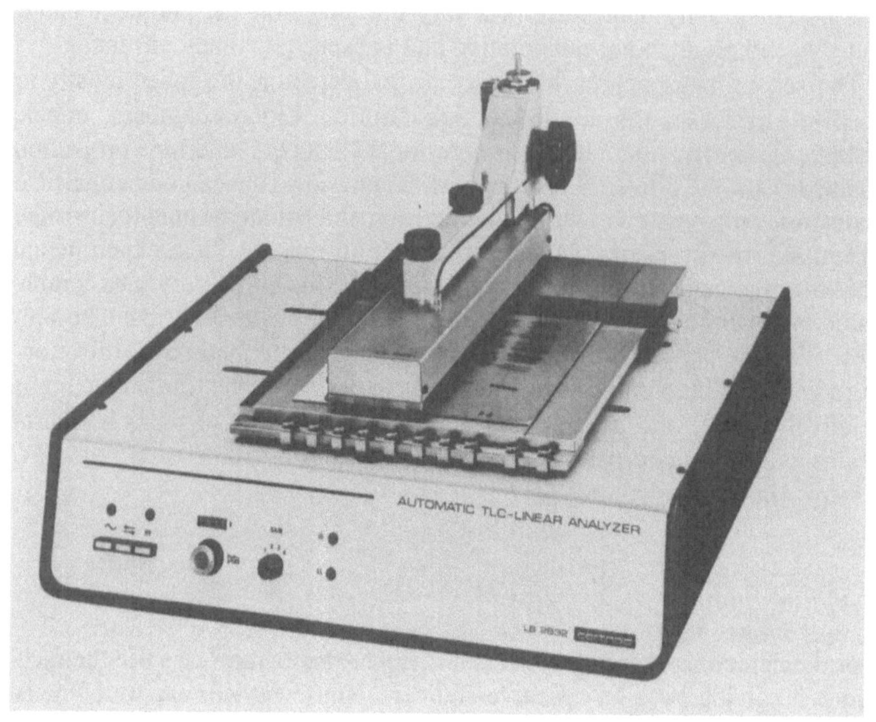

FIG. 4.2. LB 283 linear analyzer.

digital converter. The digitized signal then goes to a data acquisition system. The time difference is

$$T_{(\text{stop})} - T_{(\text{start})} = 2T_1 \quad (1)$$

where T_1 is the time required for the stop pulse to run from the position, X_1, to zero (see Figure 4.3). The instrument can measure time differences from 10^{-9} to 10^{-6} sec. This corresponds to positions from 0.25 to 250 mm on the wire.

The inherent resolution of the counter is 0.25 mm. This is the resolution for a nondivergent particle beam entering the counter perpendicular to the wire. In practice, the beta sources from TLC plates radiate more or less isotropically.

To obtain a good spatial resolution, the TLC plate is positioned as close as possible to the entrance window. One can control the accepted opening angle, θ, of the emitted beta particles electronically. No mechanical collimator is needed. A small θ means high resolution but some loss in detected particles. The signal-to-noise ratio does not change from a small to a large θ. With this arrangement, we obtain for ^{14}C sources a resolution of 1 mm, for ^{3}H sources, 0.5 mm.

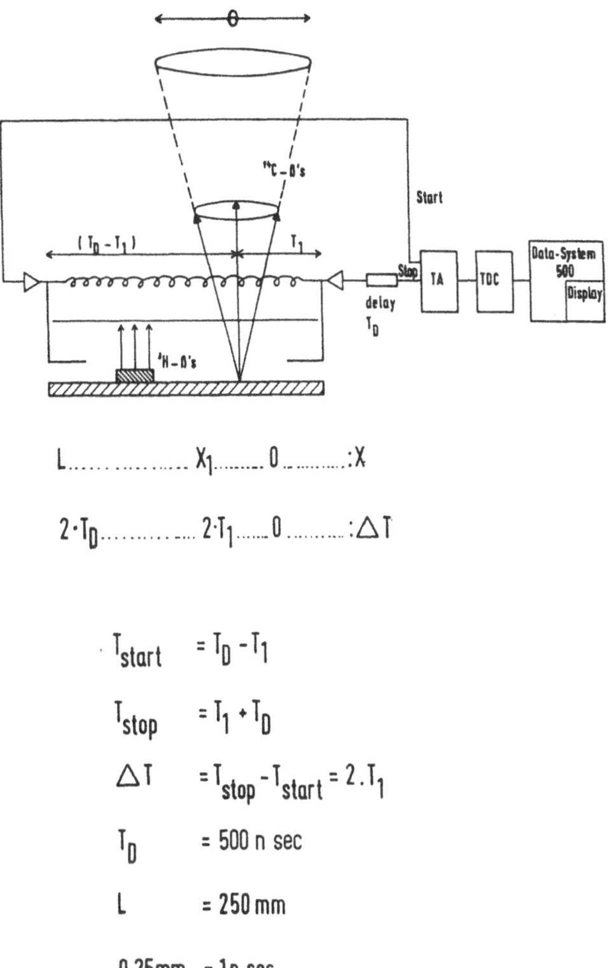

FIG. 4.3. Principle of position-sensitive proportional counter with delay line readout.

DATA ACQUISITION SYSTEM

To "read" the detected signals, the linear analyzer has to be connected to a data acquisition system. One can use a normal multichannel pulse-height analyzer or an analog digital converter connected to a computer system. We have recently developed the data acquisition system LB 500 incorporating the well-known Apple II computer and the system LB 511 with the Berthold computer. With the developed software one can measure automatically any number of chromatogram tracks. During measurements there is a live display of the data on the CRT screen. When the measurement of any track is completed, the instrument stores the data in memory, then sends a signal to position the next preselected track below the diaphragm of the counter and

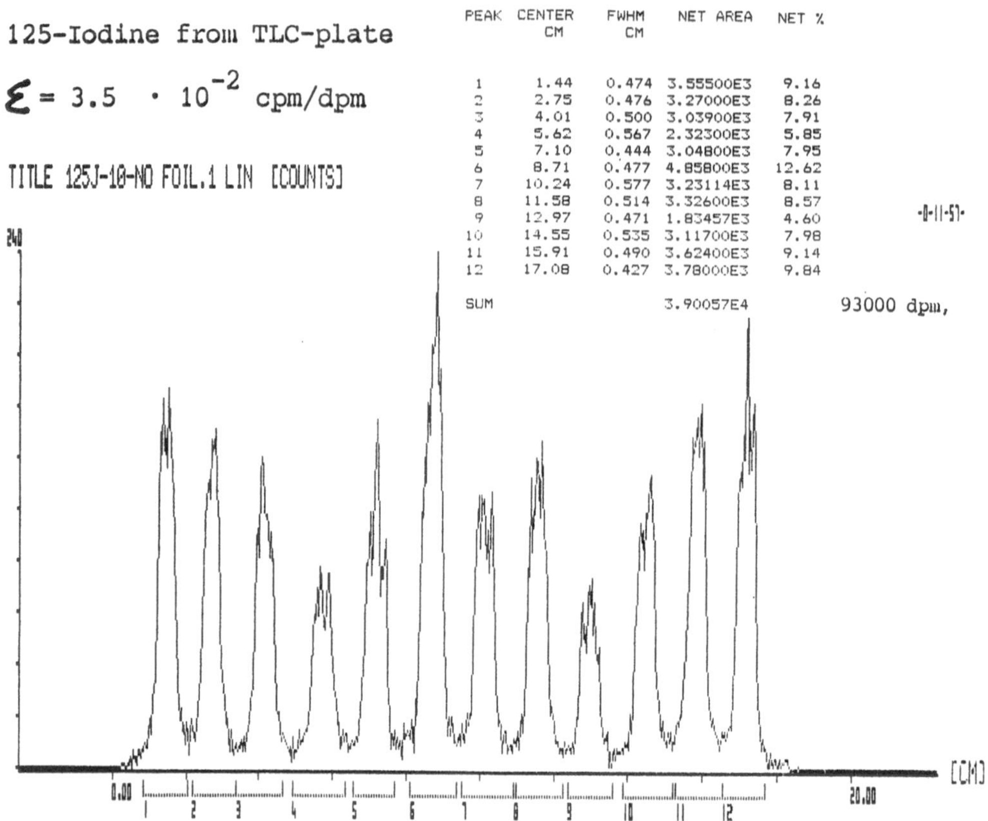

PEAK	CENTER CM	FWHM CM	NET AREA	NET %
1	1.44	0.474	3.55500E3	9.16
2	2.75	0.476	3.27000E3	8.26
3	4.01	0.500	3.03900E3	7.91
4	5.62	0.567	2.32300E3	5.85
5	7.10	0.444	3.04800E3	7.95
6	8.71	0.477	4.85800E3	12.62
7	10.24	0.577	3.23114E3	8.11
8	11.58	0.514	3.32600E3	8.57
9	12.97	0.471	1.83457E3	4.60
10	14.55	0.535	3.11700E3	7.98
11	15.91	0.490	3.62400E3	9.14
12	17.08	0.427	3.78000E3	9.84
SUM			3.90057E4	

FIG. 4.4. Iodine-125 from TLC plate with hard copy report.

initiates a new measurement. After data acquisition has been completed for any track, or for several tracks, the data are automatically analyzed. The program finds all the peaks, subtracts background, and gives a full report of the data analysis on a hard copy, including peak position, width, counts, percentage areas, and a graphics plot of the measured chromatogram, the regions of interest being labeled as shown in Figure 4.4.

MULTIPLATE DETECTOR

To allow the automatic measurement of several chromatogram plates, we have enlarged our present linear analyzer, which is shown in Figure 4.5. This new instrument is capable of measuring up to four TLC plates, 20 × 20 cm, fully automatically. The plates are stationary, fixed on the measuring table, which is approximately 1 m long. They are pressed on two borders against two thin steel bands from below. Under computer control, the counter glides on the surface of these steel bands, always keeping a fixed distance of 0.3 mm from the plate surface. With the computer program, any number of chromatogram tracks to be measured can be preselected, i.e., their position, the measuring time, or number of counts.

PERFORMANCE OF THE LINEAR ANALYZER

Detector Heads

Three different types of detector heads have been developed: N, normal; HS, high sensitivity; and HR, high resolution. They have similar characteristics, except that the HS detector has two to three times the sensitivity for ^{3}H and ^{125}I detection as the N detector. The HR detector has position resolution that is twice as good for ^{14}C and beta emitters of similar energy or higher and is twice as sensitive for ^{14}C detection relative to the N detector. It is especially applicable for ^{32}P and ^{131}I detection and essential for the detection of gamma emitters, like ^{99m}Tc and ^{123}I.

Open Entrance Window

The open entrance window is 250 × 15 mm. For measurements of radioisotopes other than ^{3}H and ^{125}I, the window can be closed by a thin foil, 0.15 mg/cm^2. (In this case, the resolution resulting from multiple scattering of the beta particles is slightly worse.)

Background Radiation

Background radiation is 50 cpm/250 mm = 0.20 cpm/mm.

Sensitivity

The detection efficiency for low energy beta particles is practically 100%. But because of their absorption in the thin layer, only a small fraction enters the counter. We have for a thin layer of 20 mg/cm^2 determined the following efficiencies:

^{3}H: 2.0% cpm/dpm (250 dpm detected in 20 min) (HS head)
^{14}C: 3 to 15% depending on "electronic collimation" (50 dpm detected in 20 min)
^{32}P: up to 40% cpm/dpm (HR head)
^{99m}Tc: 5% cpm/dpm (HR head)

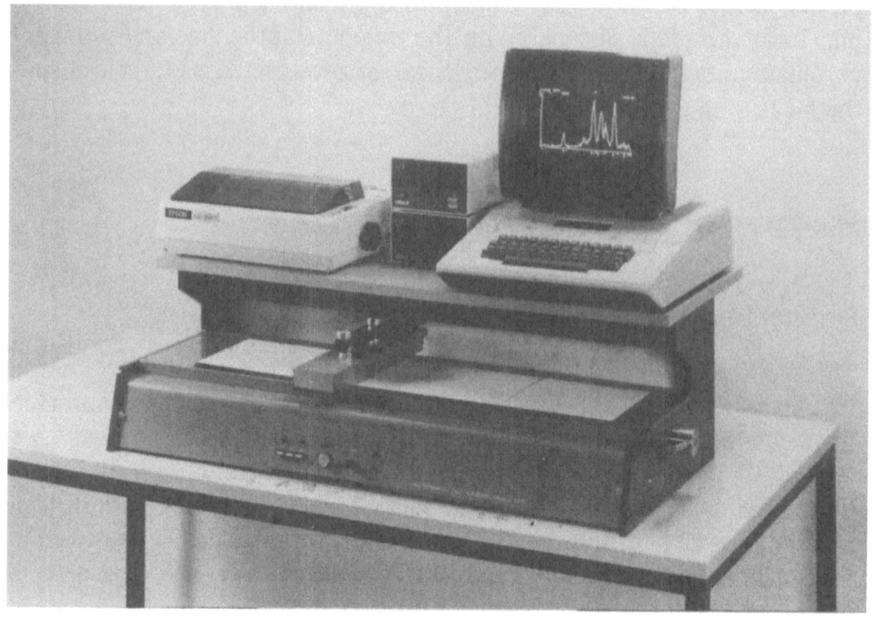

FIG. 4.5. Multiplate linear analyzer with data system LB 500.

Resolution

The resolution obtained for three isotopes is as follows:

^3H: 0.5 mm (for a source covered with ½-nm plastic strip) N head
^{14}C: 1.0 mm (for a source covered with 1-nm plastic strip) HR head
^{32}P: 1.0 mm (for 2 strips on glass plate) HR head

QUANTITATIVE MEASUREMENTS

The present system permits quantitative analysis of beta sources. The technique avoids the limitations inherent in quantitation using a scintillation counter. For example, results for analysis of material labeled by ^3H and ^{14}C show that, within about 10%, the measured counts per minute are directly proportional to the amount of radioactivity introduced into the respective thin-layer gels, as shown in Figures 4.6 and 4.7. For the technique to work well, the chromatogram should be uniformly distributed in depth within the gel. In this respect, very thin layers result in more precise quantitative measurements. For a quantitative measurement, one has to consider several important factors, which influence the result and certainly limit the precision. Only beta particles penetrating the surface of the thin layer or gel are detected, as shown in Figure 4.8. A particle originating at $Z = Z_0$ does not follow a straight line to the surface, $Z = O$. It collides with the atoms of the thin layer and therefore deviates from its original direction by the angle θ.

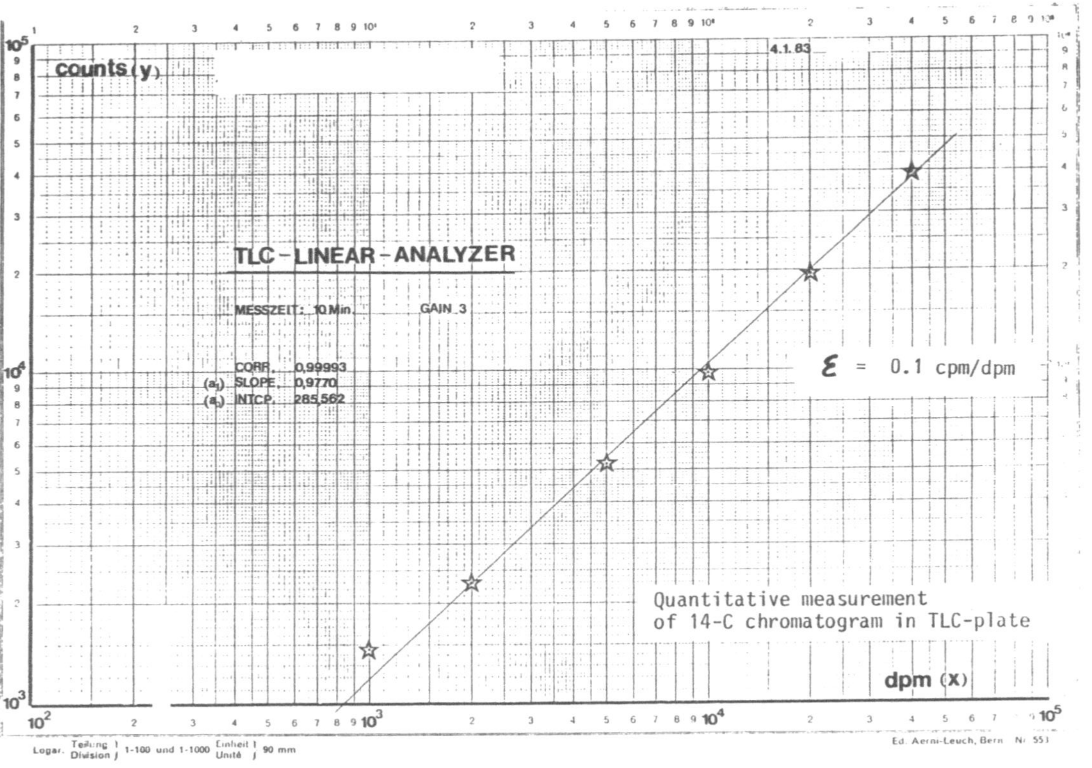

FIG. 4.6. Sensitivity for ^{14}C. Quantitative measurement of ^{14}C chromatogram in TLC plate.

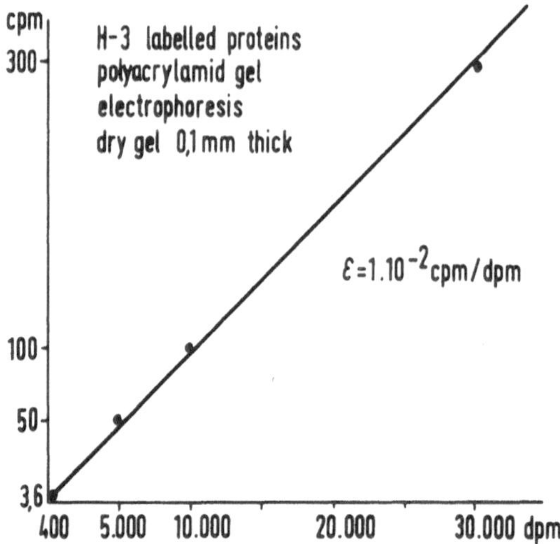

FIG. 4.7. Sensitivity for ^{3}H. Quantitative measurement of ^{3}H labeled proteins separated by electrophoresis.

The mean multiple scattering angle is

$$\langle\theta\rangle = 15 \cdot (Z_0/Z_{rad})^{1/2}/p \cdot v \tag{3}$$

p is the momentum, v the velocity of the particle, pv in MeV. Z_{rad} is the radiation length characteristic for the material being traversed. (The energy loss by radiation of an electron traversing the material of thickness Z_0 is $\Delta E/E = 1 - \exp[-Z_0/Z_{rad}]$.) For a beta particle of 100 keV energy traversing a 100-μm layer, the angle is about 30°.

The multiple scattering of the beta particle therefore limits the spatial resolution. The beta particle loses energy before reaching the surface, therefore, one detects beta particles only from a certain layer of thickness Rmax, corresponding to its maximum energy in the beta decay of the nuclide. For example ^{3}H has Rmax = 0.55 mg/cm^2 = 8-μm layer of thin-layer plate, and already after $R(\frac{1}{2})$ = 0.06 mg/cm^2 = 0.8 μm, half of the emitted ^{3}H beta particles are absorbed and do not reach the surface. Therefore, introducing a certain activity N (dpm) into the thin layer, only $\Sigma_1 \cdot N$ gets to the surface and $\Sigma_1 \cdot \Sigma_2 N$ (dpm) are detected. Here Σ_1 is the absorption factor and Σ_2 is the detection efficiency of the detector, which is between 50 and 100%.

From a 200-μm TLC plate, we can detect under the very best conditions

$$\sum_1 = 8\ \mu/200\ \mu\text{m} \cdot 2\pi/4\pi = 2\% \tag{4}$$

of the ^{3}H decays, assuming that the ^{3}H labeled compounds have a uniform distribution in depth of the thin layer and that the density of the thin layer does not vary over the whole volume (X,Y,Z). To deduce from the measured counts (cpm) the final result (dpm), we should know the distribution of the

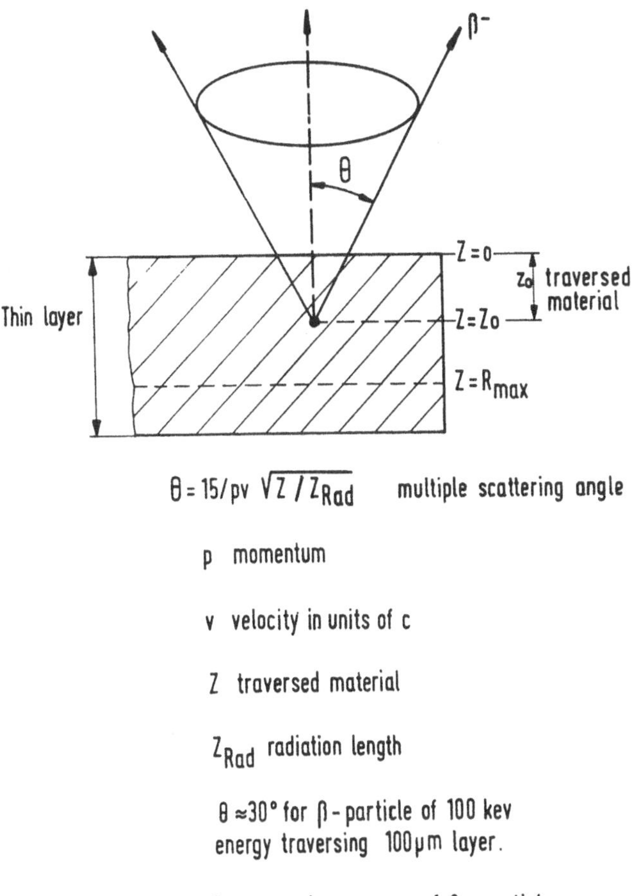

FIG. 4.8. Emission of beta particles from thin-layer plate. θ Is the multiple scattering angle; R_{max} is the maximum range of the beta particle.

compound in X, Y, and Z. This can be measured in running a chromatogram with known radioactive compounds in parallel to the unknown chromatogram. To reduce the effect of a variable depth distribution, one should use very thin layers, 100 μm thick and less, and take all the necessary precautions to generate the chromatogram.

Another important factor to deduce the final result, dpm of the compound, from the number of cpm is the correct *data analysis*. To integrate the number of counts related to the separated compound, one has to define its position and its contours. This is fairly simple in the case of a single peak without background or a uniform, nonrising, linear background. One can integrate between the two border lines and subtract the background measured in one or two sections near the peak. All other cases are more complicated, and the simple method I have indicated is not sufficient to obtain a precise result.

If we knew the mathematical analytical expression for the distribution of a chromatogram or electropherogram, we could analyze the data with standard χ^2-fitting methods. We have done this assuming that the peak has a gaussian shape and the background can be approximated by a polynomial up to second order. This means the chromatogram distribution can be described by the expression

$$N(X) = \sum_{o}^{n} a_n X^n + \sum_{i} h_i \cdot \exp[-(X - X_i)^2/2\sigma_i^2] \tag{5}$$

where X_i is the position of the peak, h_i the height of the peak, and $\Gamma_i = 2.35 \cdot \sigma_i$ the full width at half height. The area of the peak, A_i is

$$A_i = \sqrt{2\pi} \cdot h_i \cdot \sigma_i \approx h_i \Gamma_i . \tag{6}$$

The χ^2-fit finds all the parameters, a_n, h_i, σ_i, X_i, and allows then the correct calculation of the interesting quantities. An example of such a fit was already shown in Figure 4.4.

The analytical method has the advantage of defining precisely the position, boundaries, and area of a compound. It also is capable of identifying two overlapping compounds even if they are not clearly separated visually, as shown in Figure 4.9. Instead of assuming a Gaussian peak, one can of course approximate the shape of the peak by a different mathematical expression.

The definition of background varies with the experimenter and the experiment. Background in our case is the radiation not originating from the pure and, therefore, separated compound. The background is then composed of signals from cosmic rays (1 cpm/cm^2), environmental radiation, and signals from unseparated, overlapping compounds in the chromatogram. As long as one clearly defines the numbers extracted from a measurement (chromatogram), there cannot be any confusion. Considering all the factors I have described, it is certainly possible to achieve a precise measurement with small, known, systematic errors; the statistical error can be reduced to practically any desired level.

EXPERIENCE WITH THE LINEAR ANALYZER

Originally we have seen the application of the linear analyzer in detecting TLC chromatograms. In the meantime, the application has been expanded. We can detect radiolabeled electropherograms in gels and radioactive distributions in organic tissues, and we can detect special nuclides like ^{125}I, ^{99m}Tc, ^{32}P.

The following isotopes have been detected successfully:

Beta emitters: ^{3}H, ^{125}I, ^{131}I, ^{14}C, ^{35}S, ^{33}P, ^{32}P
Gamma emitters: ^{99m}Tc, ^{123}I, ^{59}Fe, ^{75}Se.

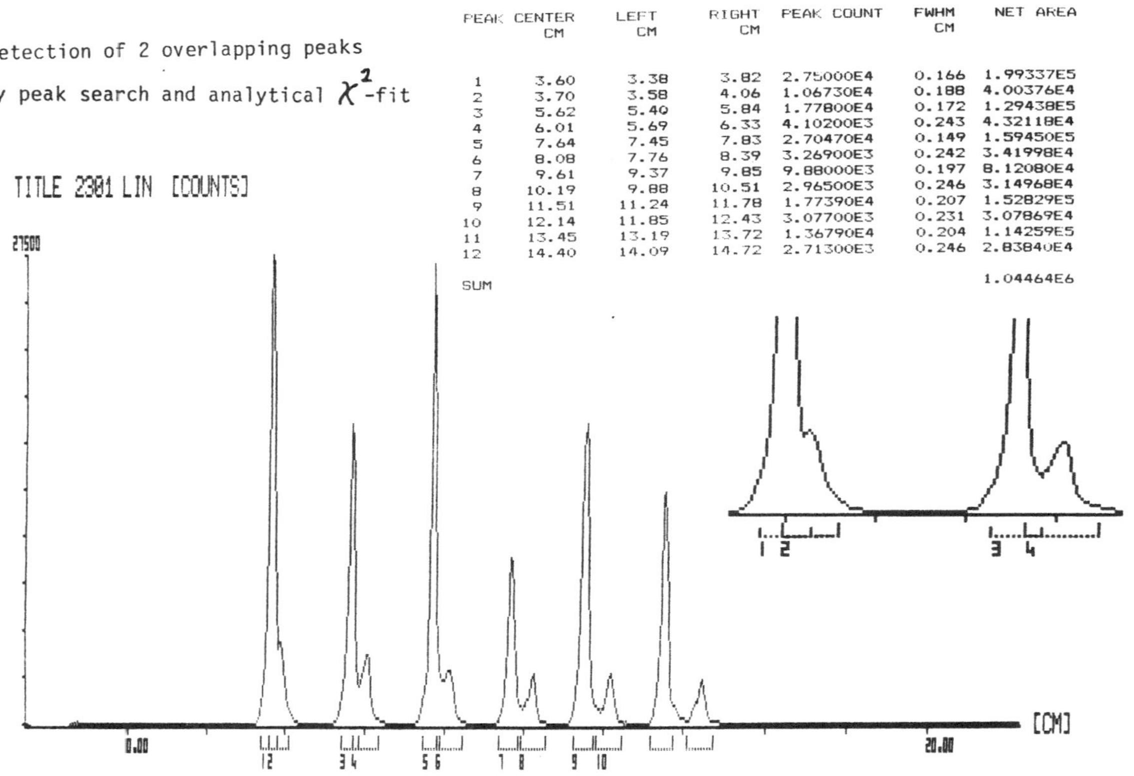

PEAK	CENTER CM	LEFT CM	RIGHT CM	PEAK COUNT	FWHM CM	NET AREA
1	3.60	3.38	3.82	2.75000E4	0.166	1.99337E5
2	3.70	3.58	4.06	1.06730E4	0.188	4.00376E4
3	5.62	5.40	5.84	1.77800E4	0.172	1.29438E5
4	6.01	5.69	6.33	4.10200E3	0.243	4.32118E4
5	7.64	7.45	7.83	2.70470E4	0.149	1.59450E5
6	8.08	7.76	8.39	3.26900E3	0.242	3.41998E4
7	9.61	9.37	9.85	9.88000E3	0.197	8.12080E4
8	10.19	9.88	10.51	2.96500E3	0.246	3.14968E4
9	11.51	11.24	11.78	1.77390E4	0.207	1.52829E5
10	12.14	11.85	12.43	3.07700E3	0.231	3.07869E4
11	13.45	13.19	13.72	1.36790E4	0.204	1.14259E5
12	14.40	14.09	14.72	2.71300E3	0.246	2.83840E4
SUM						1.04464E6

FIG. 4.9. Detection of two overlapping peaks by peak search an analytical χ^2-fit.

The performance of the linear analyzer in measuring the most common nuclides in TLC chromatography has already been summarized [1,2]. We have improved our measuring technique, thereby obtaining better spatial resolution and increasing the sensitivity.

Table 4.1 summarizes the detection efficiencies of the measured isotopes. In the case of ^{125}I, we detect the Auger electrons and not the gamma rays. The spatial resolution for ^{125}I detection is about the same as for ^{3}H, i.e., better than 1 mm. The detection efficiency from a 400-μm thin-layer plate is 1.2%, and from a 200-μm plate is 2.5%.

Because the high energy of the ^{32}P beta particles, E (max) = 1.7 Mev, it is difficult to detect them with good spatial resolution. The energetic beta particles produce secondary electrons in the counter wall, in the delay line, and in the thin-layer plate or gel, therefore, broadening the primary spots originating from ^{32}P. Multiple scattering of the beta particles in the gel layer also contributes to the broadening. We have reduced this broadening effect by designing the very thin HR detector. The same applies for the detection of gamma-ray emitters, such as ^{99m}Tc and ^{123}I.

In the case of ^{32}P, clear separation down to 3 mm can be seen, as in Figure 4.10. Extrapolating these data, a separation down to 1 mm is possible, the peak-to-valley ratio being about 3:2. If the experiment demands a higher resolution than is obtainable with ^{32}P labeled compounds, we recommend using ^{33}P, which has a maximum beta energy of 200 keV. For ^{33}P, similar spatial resolutions to ^{14}C can be obtained.

TABLE 4.1. Total Detection Efficiency, $\Sigma_1^a \cdot \Sigma_2^b$, from Thin-Layer Plate of 200-μm Thickness with LB 283

Nuclide	*Emitter*	E(*max*)	*Half-life*	*cpm/dpm*
^{3}H	beta	18 keV	12 yr	2%
^{125}I	gamma-beta (Auger)	30 keV	60 d	2.5
^{131}I	beta-gamma	608 keV 723 keV	8 d	up to 20%
^{14}C	beta	155 keV	5736 yr	up to 20%
^{35}S	beta	167 keV	87.5 d	up to 20%
^{32}P	beta	1.7 MeV	14.3 d	up to 40%
^{33}P	beta	200 keV	25.3 d	up to 20%
^{99m}Tc	gamma	140 keV	6 hr	5%
^{123}I	gamma	159 keV	13 hr	5%
^{201}Tl	gamma	75 keV	76 hr	~5%
^{51}Cr	gamma	321 keV	27.7 d	>5%
^{58}Co	gamma	811 keV	70.8 d	>5%
^{59}Fe	gamma	1.10 MeV	44.6 d	>5%

[a] Σ_1 is the fraction of beta particles penetrating the surface of TLC plate.
[b] Σ_2 is the detection efficiency with LB 283: 50% for ^{3}H, 90% for ^{14}C, ^{32}P.

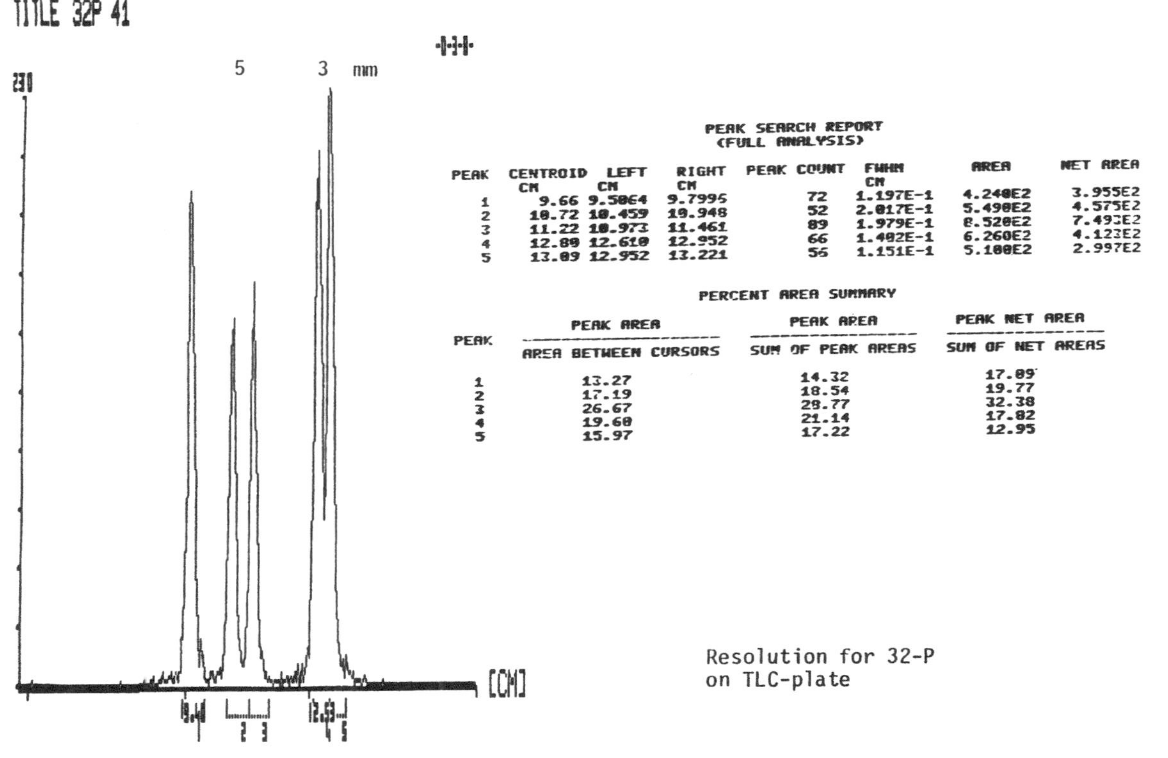

PEAK SEARCH REPORT
(FULL ANALYSIS)

PEAK	CENTROID CM	LEFT CM	RIGHT CM	PEAK COUNT	FWHM CM	AREA	NET AREA
1	9.66	9.5064	9.7995	72	1.197E-1	4.240E2	3.955E2
2	10.72	10.459	10.948	52	2.017E-1	5.490E2	4.575E2
3	11.22	10.973	11.461	89	1.979E-1	8.520E2	7.493E2
4	12.89	12.610	12.952	66	1.402E-1	6.260E2	4.123E2
5	13.09	12.952	13.221	56	1.151E-1	5.100E2	2.997E2

PERCENT AREA SUMMARY

PEAK	PEAK AREA / AREA BETWEEN CURSORS	PEAK AREA / SUM OF PEAK AREAS	PEAK NET AREA / SUM OF NET AREAS
1	13.27	14.32	17.09
2	17.19	18.54	19.77
3	26.67	28.77	32.38
4	19.60	21.14	17.82
5	15.97	17.22	12.95

FIG. 4.10. (a) Resolution for ^{32}P on TLC plate.

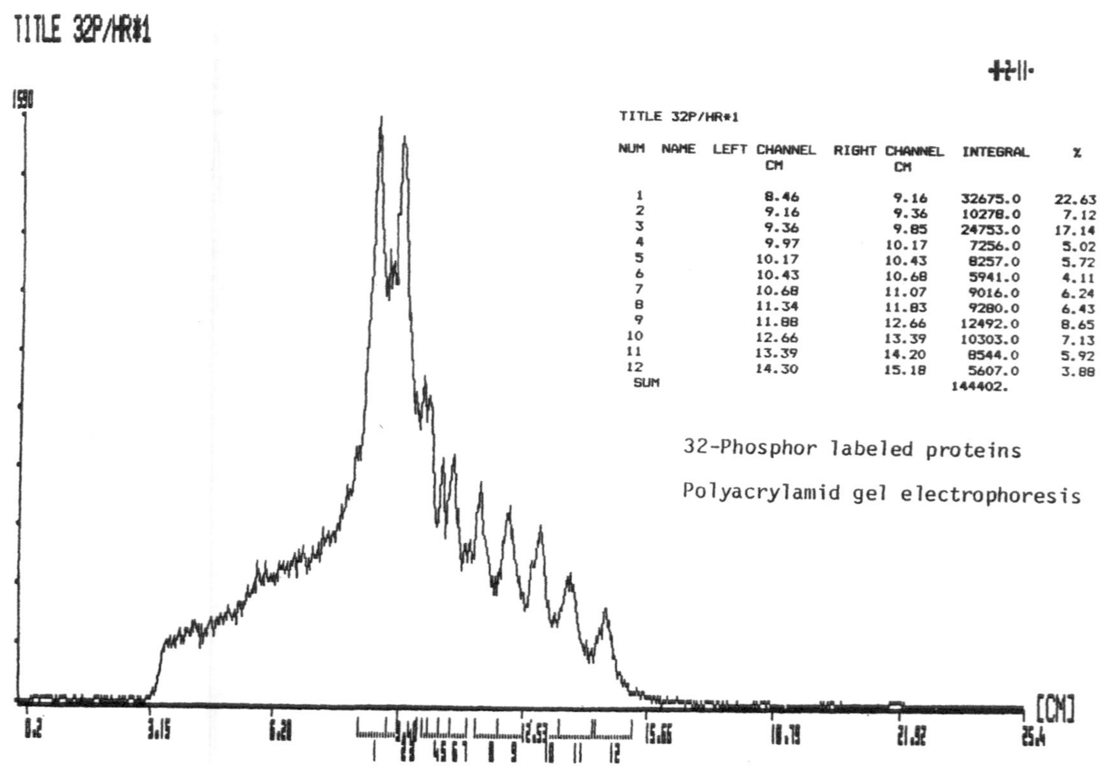

TITLE 32P/HR#1

NUM	NAME	LEFT CHANNEL CM	RIGHT CHANNEL CM	INTEGRAL	%
1		8.46	9.16	32675.0	22.63
2		9.16	9.36	10278.0	7.12
3		9.36	9.85	24753.0	17.14
4		9.97	10.17	7256.0	5.02
5		10.17	10.43	8257.0	5.72
6		10.43	10.68	5941.0	4.11
7		10.68	11.07	9016.0	6.24
8		11.34	11.83	9280.0	6.43
9		11.88	12.66	12492.0	8.65
10		12.66	13.39	10303.0	7.13
11		13.39	14.20	8544.0	5.92
12		14.30	15.18	5607.0	3.88
SUM				144402.	

FIG. 4.10. (b) Detection of ^{32}P-labeled proteins. Polyacrylamid gel electrophoresis.

^{131}I is a fairly high beta energy nuclide, $E(\max) = 608$ keV. A spatial resolution down to 1 to 2 mm can be expected, as for ^{32}P.

Figure 4.11 shows the results for ^{99m}Tc being a pure gamma-ray emitter of 140 keV energy. The two peaks of 2.5 mm distance are clearly separated with a peak-to-valley ratio of 3:1. For 1 mm separation, a peak-to-valley ratio of approximately 3:2 is expected. The total detection efficiency here is 5%.

^{123}I is a gamma emitter of similar gamma energy as ^{99m}Tc. We get comparable results as for ^{99m}Tc.

The detection of very low activity, 100 dpm, 60 dpm, and 10 dpm of ^{14}C is shown in Fig. 4.12.

Very low activity, 2 cpm of ^{14}C, can be detected very well in the presence of high activity, 2000 cpm, as shown in Fig. 4.13.

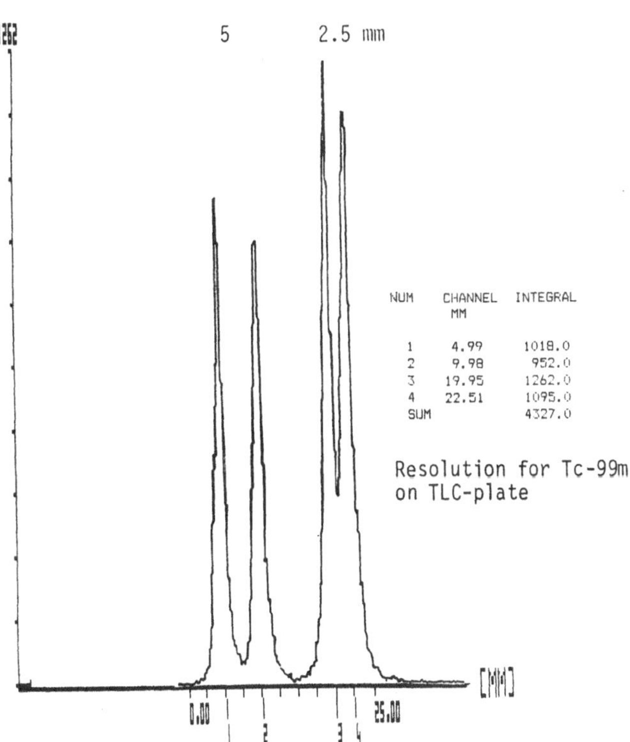

FIG. 4.11. Resolution for ^{99m}Tc on TLC plate.

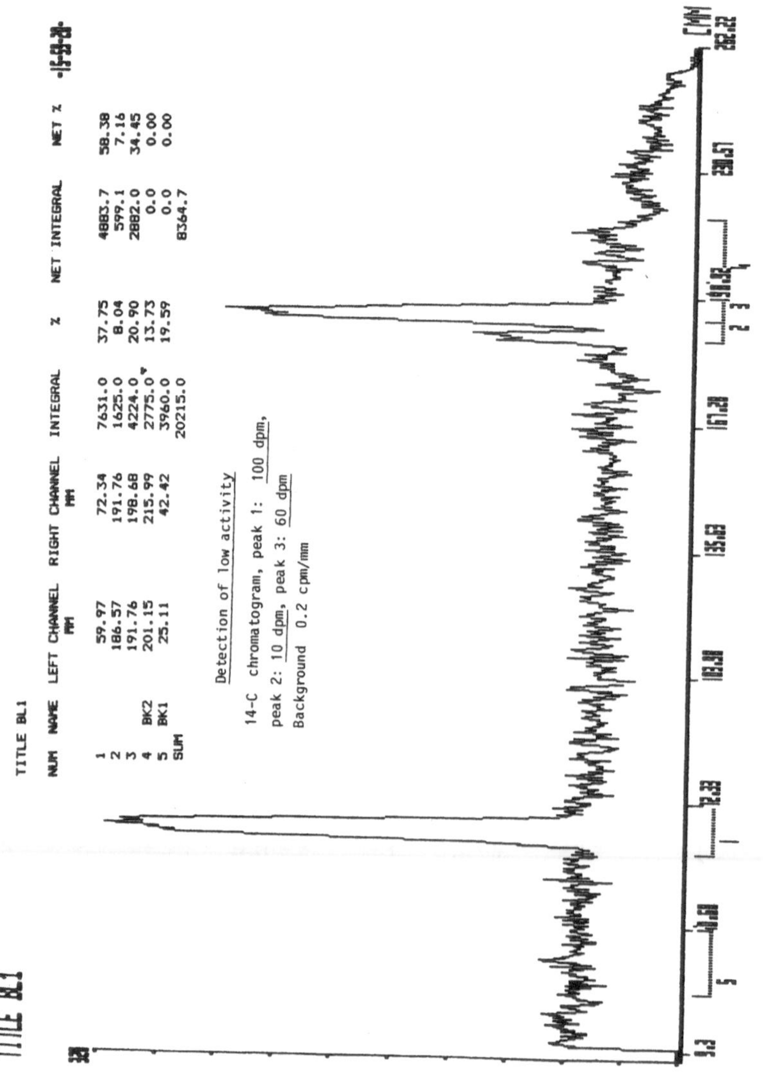

FIG. 4.12. Detection of low activity ^{14}C chromatogram. Peak 1: 100 dpm; peak 2: 10 dpm; peak 3: 60 dpm, background 0.2 cpm/mm.

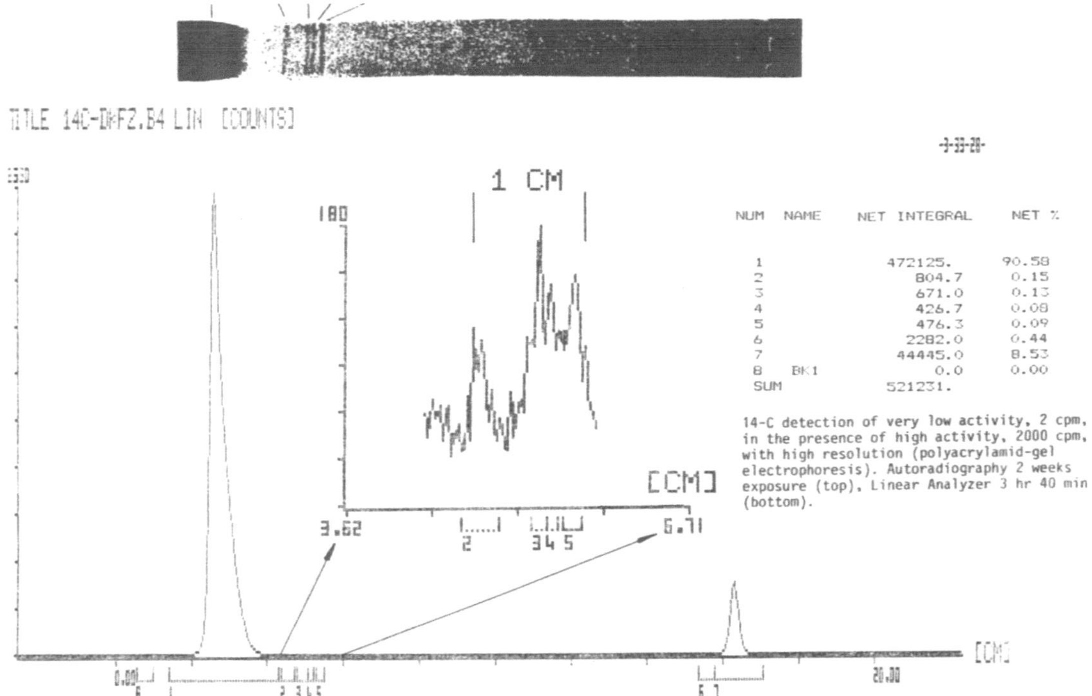

FIG. 4.13. ^{14}C detection of very low activity, 2 cpm, in the presence of high activity, 2000 cpm, with high resolution (polyacrylamid gel electrophoresis). Autoradiography 2 weeks exposure (top), linear analyzer 3 hr 40 min (bottom).

DETERMINATION OF THE RADIOCHEMICAL PURITY OF RADIOPHARMACEUTICALS

Radiopharmaceutical gamma emitters that are frequently used in clinical treatment were examined as to their radiochemical purity by paper and thin-layer chromatography or electrophoresis, respectively. It is known that radiochemical impurities may result in an unnecessary exposure of the patients to be examined. For this, the activity distribution in the developed radiochromatograms was assessed by four different measuring methods (TLC linear analyzer, TLC scanner with NaI(Tl) detector, TLC scanner with gas flow counter and gamma sample counter). The most significant results from detailed investigations of Hammermaier, et al. [5], comparing different measuring methods for various radiopharmaceuticals (or the different nuclides used for labeling radiopharmaceuticals) are compiled in Table 4.2.

As shown by the analysis in Table 4.2, only the TLC linear analyzer and the gamma sample counter (measurement of chromatograms or electropherograms cut into strips) are generally suitable methods for determining the radiochemical purity of radiochemicals.

SUMMARY

New developments in the detection technique of radiochromatography and electrophoresis, i.e., the application of multiwire chambers and the appropriate data analysis, are described. For the detection of *one-dimensional* radiochromatograms we use the technique involving position-sensitive counters. Our instrument, the Linear Analyzer, counts the whole chromatogram trace simultaneously. It is about 100 times faster and therefore more sensitive than the TLC scanner. The TLC scanner only detects that part of

TABLE 4.2. Comparison of Different Measuring Methods for Purity Determination of Radiopharmaceuticals

Nuclide	*Measuring Method*	*TLC Linear Analyzer*	*TLC Scanner with NaI(Tl) Detector*	*TLC Scanner with Gas Counter*	*Gamma Sample Counter*
^{58}Co		+	−	0	+
^{51}Cr		+	−	+	+
^{59}Fe		+	−	0	+
^{131}I		+	−	+	+
^{99m}Tc		+	−	+	+

+ Good suitability of measuring method.
0 Limited suitability.
− Not suitable.

the chromatogram that is directly below its slit diaphragm, which is typically 2 mm wide.

Results detecting the most frequently used isotopes are presented, as ^{3}H, ^{14}C, ^{32}P, ^{33}P, ^{35}S, ^{125}I, ^{131}I, and ^{99m}Tc. High spatial resolution down to 0.5 mm for ^{3}H and 1 mm for ^{14}C can be obtained. Results of quantitative measurements and their limitations are discussed.

REFERENCES

1. Filthuth, H.: Linear analyzer improves detection in radio-TLC tests, *Indust. Res. & Dev.*, June 1981: p. 140–145.
2. Filthuth, H.: State of the Art in scanning TLC. Detection of radiochromatograms and electropherograms with position sensitive wire chambers. In *Synthesis and Applications of Isotopically Labeled Compounds. Proceedings of an International Symposium,* Kansas City, June 6–11, 1982, W.P. Duncan and A.B. Susan (eds.), 447–452, Elsevier Publishing Company, Amsterdam, 1983.
3. Filthuth, H.: Radioscanning of TLC. In *Advances in Thin Layer Chromatography,* chapter 7, Touchstone, J.C. (ed.), John Wiley & Sons, New York, 1982.
4. Kruppa, J.: Localization and direct quantitation of ^{3}H-labeled proteins and RNAs in slab gels by a new detection system. *Biochem. and Biophys. Res. Com.* **113**(2):703–709, 1983.
5. Hammermaier, A., Reich, E. and Bögl, W.: Vergleichende Untersuchung von verschiedenen Me*β*-verfahren an Radiochromatogrammen zur Beurteilung der Radiochemischen Reinheit von zehn Radiopharmaka. ISH-Heft 36, 1984. Institut für Strahlenhygiene des Bundesgesundheitsamtes Neuherberg/München.
6. Filthuth, H.: New Methods in Analyzing Radiochromatograms and Electropherograms. In *Techniques and Applications of Thin Layer Chromatography,* chapter 10. Touchstone, J.C. (ed.), John Wiley & Sons, New York, 1985.

High Pressure Liquid Chromatography

CHAPTER 5

Components for the Design of a Radio-HPLC System

ADRIAN D. NUNN and ALAN R. FRITZBERG

INTRODUCTION

High pressure or high performance liquid chromatography (HPLC) has made a large impact on the analysis of compounds since its beginning in the late 1950s and throughout its continuous development to the present. HPLC consists of a number of components that contrast it to ordinary liquid chromatography: reusable columns, precise injection system, high pressure pumps that drive solvents through a relatively impermeable column at precise flow rates, column packings consisting of small particles designed to provide high separation efficiencies, and detectors that provide sensitive continuous monitoring of column effluent. The coupling of radioactivity detectors to such systems makes it possible to apply HPLC to radiochemistry and hence to radiopharmaceuticals.

This chapter describes applications of HPLC to radiopharmaceutical analysis with emphasis on the equipment used. It is based on the combined experience and personal preferences of more than a decade of radiopharmaceutical HPLC done by the authors.

A number of books are available on HPLC [1–8]. The reader is encouraged to consult them for detailed theoretical background material on the subject.

Abbreviations: high pressure liquid chromatography, HPLC; pounds per square inch, psi; solvent strength values, s; acetonitrile, ACN; tetrahydrofuran, THF; polytetrafluoroethylene, PTFE; ultraviolet/visible, UV/VIS: no-carrier-added, nca; Geiger–Müller, GM; multichannel analyzer, MCA; radiochemical purity, RCP; analog-to-digital converter, ADC

COLUMNS AND PACKINGS

The heart of the separation potential of an HPLC system obviously lies in the packing material used. In general, HPLC packings utilize small particles based on a rigid silica matrix that can withstand high pressures of several thousand pounds per square inch (psi). The movement toward small particles is based on the inverse relationship between particle size and plate count or separation efficiency. The number of plates per meter for identical columns made with 3-, 5-, or 10-μm particles decreases [2]. Unfortunately, as particle size decreases, the pressure drop along the column increases so that these same columns would have a 15,000, 8000, or 3000 psi/m pressure drop, respectively, hence the technical demands for high pressure pumps. Several of the reference books on HPLC include large tables of column packing descriptions, including manufacturers of the materials [9,10]. Also described are the technologies involved in packing preparation and derivatization.

HPLC uses a variety of derivatized supports (bonded phases) in addition to most of the types of adsorbents common to gravity flow low pressure liquid chromatography. Thus, much of the experience obtained in low pressure liquid chromatography or thin-layer chromatography can be applied to normal phase HPLC.

The following sections briefly describe popular adsorbents, typical solvent conditions used with them, and considerations involved in their applications.

Normal Phase

The routine manufacture of small silica particles with closely controlled size has allowed the transfer of silica gel-type gravity or low pressure normal phase chromatography to high resolution silica-based HPLC. At particle sizes of 5 μm, efficiencies of up to 20,000 theoretical plates per 25 cm column can be realized [11]. The use of underivatized silica together with solvents ranging from nonpolar pentane to increasingly polar chorinated solvents, ethers, and alcohols allows the well-known elutropic series for silica and alumina to be used as a guide to solvent selection. Interactions between the polar groups of the compounds to be separated and the stationary phase are primarily responsible for retention and selectivity. Normal phase chromatography is particularly effective for separations between isomers and between different polar groups or classes of compounds.

Normal phase chromatography is also carried out on silica to which polar organic molecules such as glycols, amino-, dimethylamino-, diamino-, and cyano groups have been attached. The resulting bonded stationary phases are relatively deactivated compared with untreated silica gels. Advantages include a more uniform surface for interactions, lower chemical reactivity, and more rapid equilibration with solvents, which is useful when gradient elution is carried out. Separations are often comparable to silica gel, but carried out under milder conditions [12]. A detailed description of solvent choice development based on thin-layer chromatographic behavior using different strategically chosen solvents has been described [13]. The systematic use of pentane or hexane and combinations of ether, chloroform, and methylene chloride as representatives of different solvent classes has been described [14]. The mechanisms of interaction differ between these solvents so their resolving power may differ. The solvent optimization procedure is carried out with the HPLC system, and extrapolation is not required.

Reversed Phase

Reversed-phase HPLC refers to chromatographic conditions in which the stationary phase is of low polarity, and solvent elution power increases as the solvent system employed changes from water to mixtures that comprise a high or complete percentage of less polar solvents such as methanol, acetonitrile, or tetrahydrofuran. Polar compounds are weakly retained and are rapidly eluted by largely aqueous solvent systems; nonpolar compounds are more strongly retained and require increased proportions of organic solvent for elution.

The most common type of packing for reversed-phase HPLC is silica gel to which nonpolar or alkyl groups are covalently attached. Detailed descriptions of the processes used for derivatizing the silica are available and are not discussed in detail here [15,16]. The most popular linkage is based on the siloxane group (Si-O-Si-C), which is stable through the pH range 2 to 8.5. Reagents and conditions have been developed to provide a maximum of derivatization or coverage and a minimum of remaining silanol groups. Residual silanol groups may interact with the eluates in a normal phase sense and may increase tailing of peaks. Some companies have introduced reversed-phase columns based on nonpolar supports, e.g., polystyrene, which do not have residual silanol groups.

The most popular bonded phase packing is a C_{18} or octadecyl silane type, but a variety of other carbon chain lengths are available. In general, selectivity increases rapidly between one and five carbons and then less rapidly between eight and eighteen carbons for polar organic compounds. For hydrocarbon solutes, selectivity increases dramatically from low numbers to eighteen carbons [17]. Retention also increases with carbon load on the silica gel, up to about 15% (w/w). Sample capacity increases with increasing chain length with a rough doubling between four and eight carbon ligates [18]. Sample capacity also increases with temperature (lowered viscosity, improved interaction with bonded phase) and fraction of organic solvent in the eluant.

Retention on medium or long chain alkyl bonded phase silica columns tends to be predominantly reversed phase with highly aqueous solvents. As water content decreases, residual silanols, which typically are a significant fraction of the original number, begin to exert an interactive effect with the solutes, and mixed modes of retention may be seen. At high organic solvent levels, silanol effects may be significant [19]. The point at which reaction with silanol groups becomes significant, and the extent of the reaction, depends mainly upon the bonded phase. Some manufacturers go to considerable lengths to reduce this reaction, for instance, by endcapping residual silanol groups on a C_{18} column with trimethylsilane groups. A detailed description of the solvophobic theory as a basis for understanding retention in reversed-phase chromatography is available [19].

Solvent selection strategies range from beginning with methanol or acetonitrile mixed with an aqueous phase and increasing or decreasing the proportions of the aqueous phase depending on the desired retention, to using seven fixed combinations of water and three organic solvents (acetonitrile, methanol, and tetrahydrofuran), and then extrapolating [14]. These organic solvents are miscible with water in all proportions and have different modes of interaction with solutes [20]. Thus, selectivity may differ from one solvent combination to another, even though mixtures of these solvents and water may have been adjusted to give identical retention for a model compound.

If organic solvents have no specific reaction with a series of solutes, their eluting power or solvent strength can be determined experimentally [21]. Changes from one composition to another of equal strength can be related by the following expression [21]:

$$o_c = o_b \frac{(S_b)}{(S_c)} \tag{1}$$

where o represents volume percent of solvents, S represents solvent strength values, and b and c refer to solvents of interest. If a starting mobile phase is 40% acetonitrile, a similar solvent strength would require 28% tetrahydrofuran as given by the expression

$$o_{\mathrm{THF}} = o_{\mathrm{ACN}} \frac{(S_{\mathrm{ACN}})}{(S_{\mathrm{THF}})} = 40 \frac{(3.1)}{(4.4)} = 28.2\% \tag{2}$$

In terms of dominant types of interactions with solutes, methanol is a proton donor, tetrahydrofuran a proton acceptor, and acetonitrile strongly dipolar. Whenever solutes interact with solvents via one of these types of interactions, that solvent will have increased elution power. On this basis, the order of elution of a mixture of compounds may change even though the solvent "strength" is held constant.

The values of S used above are those of Snyder et al. [21]. They obtained similar solvent strength values of 3.0 and 3.1 for methanol and acetonitrile. Other studies comparing solvent effects disagree with these values. One study using a 10-μm C-8-bonded (LiChrosorb RP-8) column used water–methanol (50 : 50) and water–acetonitrile (60 : 40) as mixtures to give identical retention times for benzene [22]. The other study used water–methanol

(37 : 63) and water–acetonitrile (48 : 52) and gave comparable retentions for naphthalene [14]. If a value of 3.0 is assumed for methanol, these mixtures give 3.7 and 3.6 for acetonitrile, respectively. In these same studies, tetrahydrofuran values were 4.0 and 4.8, respectively, based on 3.0 for methanol and mixture comparisons.

In practical terms, S values can be used to pick alternate organic modifiers when specific solvent/solute interactions are known or suspected. They are also useful when alternate organic modifiers are needed because of incompatibility with the aqueous phase, e.g., when high organic concentrations cause salting out of buffer ions.

Charged species are in general much less lypophilic than the corresponding uncharged species, so the retention of an ionizable species is very dependent upon the degree of ionization. The two extremes are the wholly charged or wholly uncharged conditions. If the reversed-phase HPLC separation is performed near the pK_a of the ionizable group, changes in pH, temperature, percent organic modifier or counter ions can greatly affect retention. A potential source of problems in this regard is the use of high volume injections in which the buffering of the mixture to be analyzed is greater than the HPLC solvent buffer capacity.

Columns made with small particles can cause high pressures if the flow rate is high. We have found that longer column life results from not allowing pressures to exceed 3000 psi. This means typical maximum flow rates of about 1.5 ml/min for 5-μm particle columns of 4.6 $\times$ 250 mm size. The actual maximum reasonable flow rate depends on what solvents are used. Methanol and ethanol mixtures with water have high viscosity values near the 50% organic solvent level. In order to minimize pressures, pump limits should be set, especially during alcohol–water gradient runs.

Ion Exchange and Ion Pairing

The presence of ionizable groups in the mixtures to be analyzed provides another basis for separation. The well-known use of ion exchange columns using sulfonate (strong anion), carboxylate (weak anion), quaternary amine (strong cation), or amine (weak cation) groups bound to supports by an organic linking group has also been carried over to HPLC in the form of small particle supports. Such packings have had important applications in the separations of metal cations, but often good separations of amino acids, peptides, proteins, and inorganic complexes are also possible. Discussions of eluant conditions are provided in detail with regard to ionic strength and mobile phase anions for anion–solute elution and cations for cation–solute elution [23].

Ion exchange HPLC packings suffer from disadvantages of decreased stability, particularly at high pH, and decreased capacity compared with other bonded phase packings. Some are also prone to volume changes on changing eluants, which can ruin the column. Because chloride ions attack stainless steel, they are not recommended as mobile phase anions. Some of the disadvantages of the ion exchange HPLC columns are avoided with ion pair chromatography using alkyl bonded phase packings. In effect, the ion exchange column is formed in situ.

Ion pair chromatography is simple in principle. Lipophilic cations (quaternary amines such as tetrabutyl ammonium) or anions (alkyl sulfonates) are added to the mobile phase of a reversed-phase system. Solutes of the opposite charge form a lipophilic ion pair that is retained on the column. This can be written as:

$$(\text{butyl})_4\text{N}^+ + \text{RCO}_2^- \rightleftharpoons \text{RCO}_2^-(\text{butyl})_4\text{N}^+ \text{ ion pair} \tag{3}$$

An equilibrium relationship that describes interaction with, or retention by, the bonded organic phase of the column then is given by:

$$\text{Retention} = \frac{[\text{RCO}_2^-(\text{butyl})_4\text{N}^+]_{\text{org}}}{[\text{RCO}_2^-]_{\text{aq}}[\text{butyl})_4\text{N}^+]_{\text{aq}}} \tag{4}$$

where aq and org refers to aqueous and organic packing phase, respectively. Theories of retention range from uptake of ion pairs by the bonded organic phase of the packing to adsorption of the counter ion on the bonded phase and retention by usual ion exchange principles. A range of behavior probably results, depending on the chain length of the counter ion and conditions involved.

Important variables peculiar to the selectivity of ion pair chromatography are (a) pH, (b) concentration of counter ion (usually 0.005 *M* or less), (c) fraction of organic solvent in the mobile phase, (d) competing ions, and (e) length of counter ion alkyl group. The first four affect ion pairing, the last variable affects the lipophilicity, and thus the retention of the resulting ion pair.

In practice, ion pair conditions convert a reversed-phase system into an ion exchange type of system allowing an interchange between two methods with a single column. A concern with its use in radiochemical analysis, however, is that trace amounts of counter ion in the system (injector, column) can still markedly affect the behavior of the trace level concentrations found in radiochemical analyses. Thus, using a column for different methods (reversed phase/ion pairing) may take extended washing and is not recommended. Because many biochemicals contain ionizable groups, and inorganic complexes often are charged species, ion pair chromatography on reversed-phase columns can be a powerful tool for the radiopharmaceutical chemist.

COMPONENTS OF A RADIO-HPLC SYSTEM

There are five major parts to an HPLC system: (a) pump(s), (b) injector, (c) column, (d) detector(s), and (e) controller/integrator.

Pumps

All pumps should be capable of delivering a relatively pulse-free supply of solvent at 1–5 ml/min and at pressures up to 5000 psi. Most manufacturers meet these specifications, and so the important factors to consider when choosing pumps are the reliability, compatibility, and the intended use.

Reliability

Reliability is by far the most important factor to consider. Flow rate reproducibility must be of the order of 1% or less. Accuracy (delivered flow vs. indicated flow) is not so critical and will vary depending on how the pump adjusts for the compressibility of the solvent. Normally the only replaceable items in a pump are the seals in the pump head. With careful use, they should last six months to a year. The actual figure varies from manufacturer to manufacturer, and some have better reputations than others. The ease with which the seals can be replaced should also be considered. The check values should last at least a year, but they can be fouled by organic impurities in the solvents and inorganic salts can take their toll.

HPLC pumps need constant attention, which is not to say that they are delicate, but that they are precision instruments working under strenuous conditions.

Particulates in the mobile phase can damage the high pressure seals, so all mobile phases should be filtered through a 0.2-μm filter before use. Filters made of Nylon 66 are the most versatile, as they can be used with most organic solvents. However, it is not normally the organic solvents that are the problem, especially if prefiltered-distilled-in-glass solvents are used. Most of the particulate material comes from the buffer salts. Phosphate salts are particularly bad because they throw precipitates on standing, probably because of the formation of insoluble metal phosphates. An added advantage of filtering premixed solvent is that degassing is achieved at the same time. The lines leading from the (filtered) solvent reservoirs should start with a stainless steel frit, which removes particulates that are introduced after filtering. Glass solvent reservoirs should be recognized as sources of particulates, especially if glass stoppers are used at any time, because microscopic chips of glass are easily produced by mechanical or thermal insult. High purity water must be used, otherwise organic impurities from the water can foul the column. The water should have a conductivity of >10 mhos. Plastic solvent reservoirs can also be a source of organic contaminants.

Pumps should not be allowed to run dry. Most systems have low pressure limits designed to stop the pumps if this occurs.

Most of the wetted surfaces in an HPLC system are stainless steel, which is resistant to most agents; however, halogen ions and acetic acid will corrode it quite rapidly. These ions should be used at the lowest concentration possible and should not be allowed to stand in the system. Buffers containing any inorganic ions should not be allowed to stand in HPLC systems because crystallization can occur. If this happens in the pump head, the piston chamber can be irreparably scored when the pump is turned on. Elsewhere it can clog frits or columns, causing high back pressures. In a similar manner, precipitation of inorganic salts can be caused by changing to an incompatible organic modifier or percentage of modifier before flushing the system of the aqueous phase. The stainless steel surfaces should be protected by passivation with about 15% nitric acid for half an hour every month or so. This is done by replacing the columns with tubing and using the nitric acid as eluant. Do not pass nitric acid through a column because the packing will be destroyed.

Compatibility

Compatibility with other equipment is important as systems are inevitably upgraded and modified. You might want to split a two-pump system into two single-pump systems or vice versa. Manufacturers do not always ensure that existing pumps are compatible with new controllers, for instance, and the controllers are invariably incompatible with other manufacturers' pumps. Pump controllers that may also be designed to interface with a company's own mass detector may not be able to interface with a radioactivity detector. It is not necessary or necessarily desirable to buy a complete system from one manufacturer. A good pump manufacturer is not necessarily a good detector manufacturer.

Intended Use

The intended use for the pumps will influence the type of pump you buy. If any developmental work is planned, a two-solvent system is a minimum requirement. This allows you to cover a range of organic modifiers without making up fresh solvent each time. (The relative proportions are changed by the delivery system.) Caution must be exercised to ensure that on-line mixing of mobile phases does not extend beyond the miscibility of the solvents being used. Phase separation or salting out of buffer components can occur and normally results in a rapid rise in operating pressure, which will damage the system if pressure limits are absent. There are systems that have one motor, but can still deliver different solvent systems by using valves or multiple pump heads. They may be more accurate, but are not as versatile. For instance, a two-pump, two-solvent system can be split into two one-pump, premixed solvent systems. With multiple solvent delivery capabilities, gradients can be used. They are best avoided because they reduce sample throughput and can cause increased column degradation, although they are sometimes the best way of achieving the desired separation. If many routine separations are performed, it is cost efficient to have dedicated systems. These need only consist of a pump, injector, and column. The detectors can be common to many systems, connected by switching valves. In such a system, it is not necessary to have a $5000 pump working in isocratic mode. We routinely use $900 pumps with minimum pulse damping. A discussion of basic pump types and their advantages and disadvantages is provided by Snyder and Kirkland [2].

Dissolved gases in the solvents can be a problem if they come out of solution in the system. This normally happens at a point where there is a sudden drop in pressure, e.g., at the column outlet. The bubbles so formed cause very sharp "peaks" to be detected by most mass detectors. Degassing is not normally necessary if sufficient back pressure is maintained in the detector. This can be achieved with a length of narrow-bore tubing. The pressure limit of most mass detectors is about 300 psi, but some have limits as high as 4000 psi.

Injectors

An HPLC injector usually consists of two loops in parallel between the pump and the column—a sample loop and a bypass loop. The solvent flow is switched between the two paths by the injector. The sample loop can be isolated from the eluant flow, filled at low pressure, and reintroduced into the flow without interrupting the flow. The sample size can be changed by partially filling the loop or by completely filling loops of different sizes. Injectors function under the same conditions as pumps and need the same care. Damage can occur because particulate material is introduced by a pre-column (sometimes called a saturator column), which is in the eluant flow between the pump and the injector, or from the sample itself.

Manual

The Rheodyne 7125 injector is probably the one used most commonly. It can be operated manually or by a pneumatic actuator, the 7126. It is original equipment on many systems offered for normal, as opposed to microbore, HPLC. There are alternatives (e.g., Valco, Altex), and some manufacturers of HPLC systems provide their own (Waters). Many injectors require specific needles to be used on the injection syringe. If a bevelled needle is used instead of a square-ended needle with a 7125 injector, the stator plate can be ruined. Electro-tapered square-ended needles leak and give irreproducible partial-loop injections with the 7125 series.

Automatic

An automatic injector such as the 7126 is required if unattended operation is desired. These injectors are controlled by microprocessors or computers that control the air pressure by means of solenoid valves.

Column Hardware

There are many different configurations for the columns and couplings. Shorter columns are becoming the norm, a common configuration is 150 × 4 mm. Microbore columns (diameters <1 mm) and very short columns have been introduced, but have not taken over the market. They have the advantage of fast separations with small volumes of solvents, but suffer from low loading ability both in terms of mass and sample volume. This means that detector sensitivity must be high. At first glance, this seems perfect for radio-HPLC, but it must not be forgotten that nonradioactive components in the sample can saturate such columns at very low radioactive concentrations that are too small to detect accurately. The column-end fittings and associated couplings are not standard throughout the industry. Most columns can be ordered with the fittings of your choice, so it is best to choose one configuration and stick with it. The same goes for the frits in the ends of the columns. Staying with one type of fitting prevents the problem of trying to fit a zero dead volume fitting into a low dead volume fitting, or of having, for instance, a male fitting on the column and also on the connecting tubing. Fittings and connecting tubing should have a minimum swept volume between the injector and the detectors, otherwise band spreading will occur with a subsequent loss of resolution. Stainless steel tubing is essential between the pumps and the analytical column, because of the high pressures involved. It is not essential after the analytical column because pressures are typically <300 psi, and small-bore PTFE tubing can withstand these pressures. However, stainless steel is recommended, because it does have greater intrinsic stopping power than PTFE and so gives better radiation protection.

Many people recommend using pre-columns and guard columns to protect the analytical column. A silica-packed pre-column (or saturator column) is a small column put between the pumps and the injector. It is used to saturate the mobile phase with silica. This prevents dissolution of a silica-based analytical column. As the sample does not go through the pre-column, it does not have to be packed to high standards nor need it be made from uniform particles, so it is relatively inexpensive. A pre-column is not necessary or desirable with nonsilica-based analytical column packings.

A guard column is put between the injector and the analytical column and is made from the same packing material as the analytical column. It is designed to remove strongly retained material from the sample before it gets to the analytical column and so prevents a change in properties of the analytical column resulting from absorption of strongly retained material. It must be packed to the same standard as the analytical column and is frequently destroyed, but its small size makes it relatively inexpensive. There are some disadvantages to guard columns. The main one is that the additional couplings they require can add more void volume to the system, which causes band spreading and poorer separations. Secondly, as the main column size has been reduced (because of smaller particles, which give better separations), the size of the guard column has approached that of the main column. As a result, a significant part of the separation is done by the guard column. In practice, we find that it is better to run without guard columns and accept the faster degradation of the analytical column. We do use a silica pre-column when using silica-based analytical columns, but include a filter to prevent fragments from the guard column from damaging the injector or subsequent columns.

Detectors

There are two types of detector in common use for radiopharmaceuticals, the ultraviolet/visible (UV/VIS) mass detector and gamma detectors. UV/VIS detectors are relatively unselective mass detectors. They respond to all species absorbing at a particular wavelength. For quantitative measurements they require knowledge of the extinction coefficients of the species detected. The determination of radiochemical purity is normally the most important measurement to be made and is often the only measurement that can be made on no-carrier-added (nca) preparations, because the small masses involved do not provide a large enough signal for common mass detectors. Radioactivity detection is normally made to be specific for a desired nuclide, but multiradionuclide mixtures are uncommon. Radioactivity detectors can directly quantitate proportions of the radioactive species present. Radioactive detectors require knowledge of specific activities if they are to be used to determine mass (the amounts of the radioactive species present).

UV/VIS Detectors

These detectors are flow-through versions of the standard bench spectrophotometer, which normally analyzes samples in a batch mode. They cover wavelengths of 200–750 nm. They differ in two important ways from the bench instruments. The first is that cell volumes are about 10 μl rather than 1–2 ml. The second is that their response times are much faster. Bubble formation in the cell is the biggest problem with the flow-through detectors. Most have ascending solvent paths to try and sweep any bubbles formed out of the light path.

The normally colorless ligands have absorption maxima in the UV region of 200–350 nm, whereas transition metals have their characteristic (d–d) absorption maxima in the visible region from 350–750 nm. When working with ^{99m}Tc, one can easily distinguish between the free ligand and its technetium complex by comparing the radioactive and the absorbance traces. With ^{99}Tc it is not so easy unless the preparation can be spiked with ^{99m}Tc, and this may not be feasible when one is examining material that takes a long time to prepare. An answer is to compare the absorbance at wavelengths characteristic for the metal and for the ligand or complex. The cheapest detectors are fixed wavelength detectors, which are normally set at 254 nm; many can also be set at a limited number of other wavelengths by using appropriate filters. The next step up is a variable wavelength UV detector with a range of about 200–350 nm, and the final step is a UV/VIS detector with a range of 200–750 nm.

It is very easy to change the wavelength with a variable wavelength detector *within* the UV or visible ranges, but it is often quite time consuming to change *between* the visible or UV ranges. This is because different sources are needed for the two ranges; a deuterium lamp for the UV and a tungsten lamp for the visible. The lamps must be aligned correctly in the instrument to obtain the optimum performance, and furthermore the deuterium lamps typically have a warm-up time of about 30 min. One manufacturer's solution is to mount the lamps in prealigned holders and fit these holders into the instrument on an optical rail. Another method is to have both lamps permanently mounted in the instrument with a mechanism to select the appropriate lamp. Each of these solutions removes the alignment problem but does not obviate the need for deuterium lamp warm-up times. These lamps are expensive and have a limited life and so cannot be left on at all times. There is some advantage to being able to switch rapidly from visible to UV and vice versa when analyzing ^{99}Tc complex mixtures.

Photodiode-array detectors have recently appeared on the market. These detectors may eventually replace the current design of UV/VIS detectors. However they still have some limitations, not the least of which is their expense. Their active system is an array of a hundred or so photodiodes, each examining a narrow band of light (normally a few nanometers), instead of a single photomultiplier tube. Photodiode-array detectors use a broad spectrum source with many narrow spectrum receivers (photodiodes), unlike traditional systems that use a narrow spectrum source with a single wide spectrum receiver (photomultiplier). At present, photodiode-array detectors do not cover the whole visible range. One of their main advantages is their ability to make very fast complete scans of the whole energy range. This allows a full energy spectrum of each peak or part of a peak to be obtained without stopping the chromatogram. Similar full-range spectra can be obtained with motorized versions of the old style variable wavelength detector, but at much slower speeds (so that flow must normally be interrupted), and with less sensitivity.

Radioactivity Detectors

For a long time manufacturers have ignored the flow-through gamma detector market inherent in radiopharmaceutical HPLC. This did not impede progress, because most people in the field had enough knowledge to build their own using nuclear spectroscopy/gamma camera equipment. Recently, a number of commercial gamma detectors have become available. Their design reflects their origins in beta detector technology and may be quite different from the homemade versions that have been in use for a decade or so. If the reader is considering purchasing one of these systems, the checks described at the end of this section should be done to ensure that it can perform satisfactorily at the desired activity levels.

Most homemade detectors have as their active part a $2 \times 2''$ or $3 \times 3''$ well-type thallium-activated sodium iodide crystal that produces a signal which is proportional to the energy of the emission and its rate, as shown in Figure 5.1. The signal can be processed in a number of ways. The simplest is to feed the signal to a digital ratemeter. This gives a quantitative measure of the radioactivity present in the form of a step-function chart trace, and also

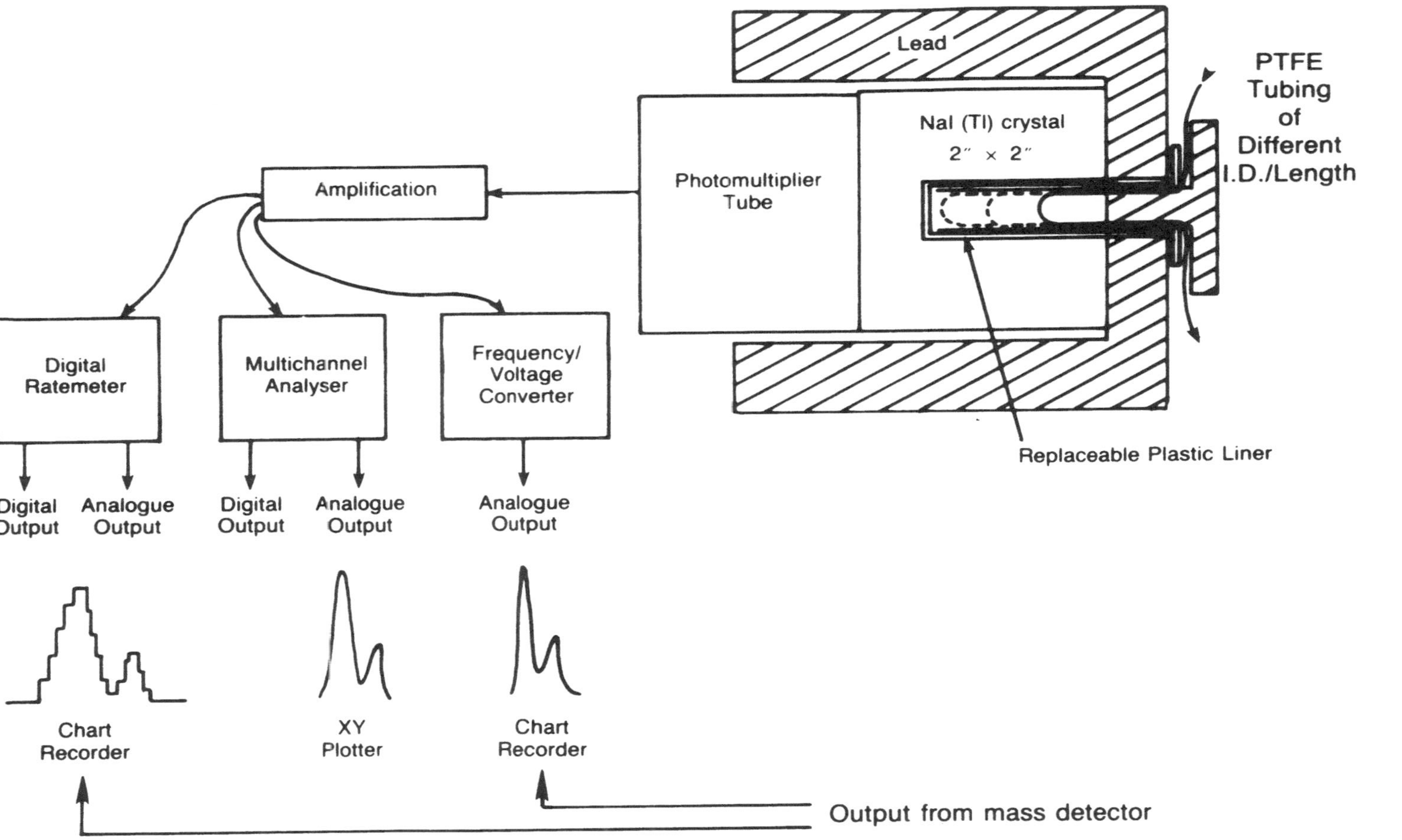

FIG. 5.1. A flow-through gamma detector for HPLC. One design of the active part of the system is shown, together with a variety of means of handling the signal.

gives digital information in the form of counts per unit time or total counts. Alternatively, the signal can be fed into a nuclear spectroscopy grade MCA, which is used in multiscaler mode. In this mode, counts are collected and displayed per unit time rather than in pulse-height analysis mode, when they are collected and displayed depending upon their energy. Count (dwell) times of about 2 sec per channel are most often used. When adding to existing systems, one can pass the eluant from two systems through a single detector by using cheminert® slider valves, which do not contribute much to band spreading. GM survey meters with recorder (ratemeter) outputs can be adapted to make crude HPLC detectors by simple matching of input/output voltage. Although some sensitivity and quantitation accuracy is sacrificed, such detection is an inexpensive means of adding additional detectors as HPLC systems are enlarged.

Hard copy can be obtained in analogue form (off line), by using an *XY* plotter, and in digital form by using a printer. Analogue copy can also be obtained by using a frequency-to-voltage converter between the amplifier and a chart recorder; this can be as simple as a ramp (capacitor/resistive circuit), because it is not used for quantitation if an MCA is present. MCAs are more flexible than digital ratemeters, both in their method of collecting data and in their ability to work on the data after it is collected. There are some MCA boards that have been made to be fitted to microcomputers such as the Apple; these do not as yet have the desired high acquisition rate needed for gamma radiochromatography.

The sensitivity and accuracy of such a radioactivity detector can be varied in a number of ways. The easiest is to change the active volume of the sample cell (at Squibb we routinely use 10 μl). The cells are made from loops of PTFE tubing of different lengths and internal diameters. They are joined to the system by $\frac{1}{4}''$ × 28 tpi (threads per inch) flange fittings. They are held in a fixed geometry inside a 5-ml syringe body. This is then pushed into a 10-ml syringe body, which has had its luer fitting removed and the hole sealed, fixed in the well of the crystal. This prevents contamination of the crystal in the event of a line linkage. It takes about one minute to change the loops.

Apart from changing the sensitivity of the system, loop changes may also be required to eliminate *memory* (sorption of activity to the tubing). This is not normally a problem with ^{99m}Tc, but may be troublesome when one is analyzing mixtures containing the longer-lived iodine radionuclides, especially when iodide is present.

More counts but longer analysis times can be obtained by lowering the flow rate, thus increasing the residence time in the detector. The degree of amplification can also be changed. Specificity is achieved by setting a window around the desired gamma energy, but this, of course, is not a problem for monoradionuclidic solutions.

The accuracy of the detectors is limited at the low end by signal-to-noise ratios and at the upper end by dead time, which is the time taken by the electronics to sort the pulses. Maximum count rates of the order of 20–100

TABLE 5.1. Reproducibility of Detector

Volume (*μl*)	*μCi*	*Counts* (*n*)	% *S.D.*
1	3.9	20,935 ± 745(3)	3.6
2	7.6	41,414 ± 290(3)	0.7
4	14.7	81,165 ± 821(3)	1.0
6	21.4	121,401 ± 452(3)	0.4
8	27.4	160,344 ± 415(3)	0.3
10	32.6	199,391 ± 623(3)	0.3
12	39.2	237,004 ± 405(3)	0.2
14	44.5	272,850 ± 909(3)	0.3
16	49.3	309,128 ± 651(3)	0.2
18	53.4	346,062 ± 212(3)	0.1*

* Maximum counts per 4 second channel = 75,000.

MHz can be handled, depending upon whether a fast analogue-to-digital converter or digital ratemeter is used together with a fast front end, but useful peak maximum count rates are of the order of 100 kHz with a sodium iodide detector.

Whatever the type and source of the detection system, a number of tests need to be performed to determine its capabilities. Linearity, reproducibility, and range are probably the most important, as they are dominated by the detector. They are easily determined by injecting known amounts of an unretained radioactive compound such as pertechnetate into the complete system as shown in Table 5.1. The reproducibility is mainly a function of the volume of the injection and is <0.5% for an average injection.

The linearity is also very good, as shown in Figure 5.2, and the range somewhat surprising until it is realized that the residence time of a point source in a 10-μl detector at 1 ml/min flow rate is less than 1 sec. Thus this system can easily handle 250 μCi under dynamic conditions, but only 2 μCi under static conditions.

The resolution of a system is less a function of the detector than the other equipment. The intrinsic resolution can be measured by multiple, rapid injections of an unretained radioactive species, as shown in Figure 5.3. With this system, the larger loop does not allow the chart recorder to return to base line as fast as the smaller loop. Such calibration curves should be obtained for all radionuclides and detector configuration combinations.

However good the detector is, it should never be forgotten that the separation system, as a whole, ultimately determines whether one can detect and quantitate a particular species.

HPLC columns are filters *par excellence*, and so most particulate materials will be retained by them. This is a problem in analyzing technetium-containing material because of the tendency of technetium to form colloids (reduced hydrolyzed technetium). The retention of radioactivity anywhere

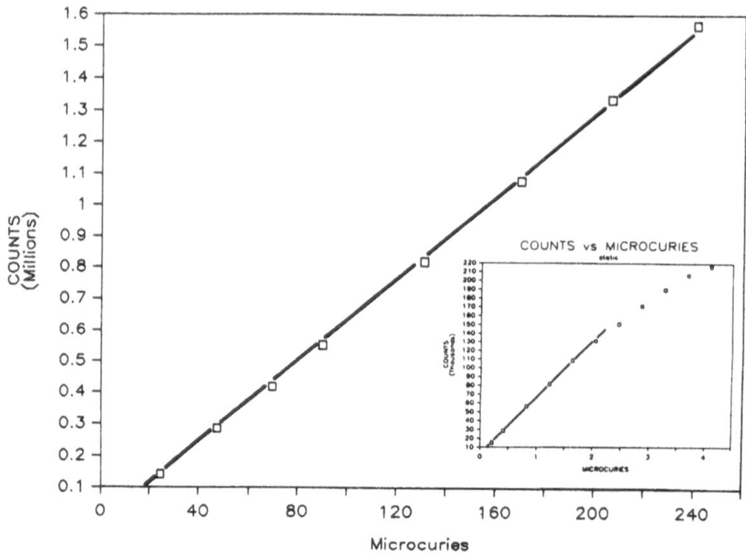

FIG. 5.2. Linearity of the detector.

in the system by any means is a problem, especially if it is selective, because proportional rather than absolute radiochemical purities are normally obtained. It is therefore, necessary to perform a mass balance when developing separations. This can be done in a number of ways:

1. Include a detector loop in the lines between the injector and the first column head. This gives an initial peak on the radiochromatogram

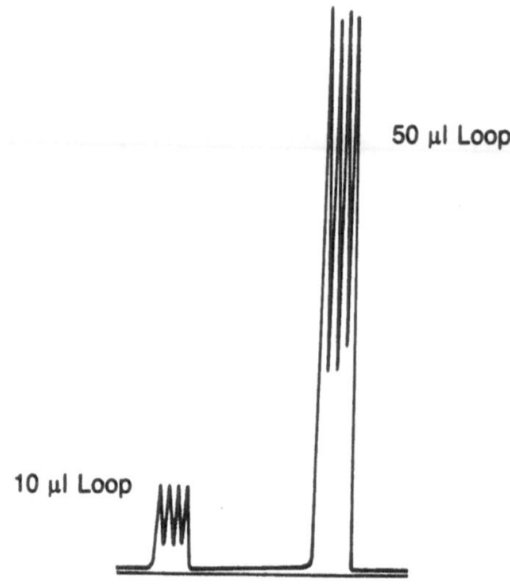

FIG. 5.3. Resolution of the detector.

containing the total counts injected. The total counts eluted can then be compared with the counts injected to determine recovery. There are some drawbacks to this technique. The first is that the increased swept volume in front of the column will contribute to band broadening and give reduced resolution. This can be minimized by having a small sample loop, but the connection will also contribute swept volume. (The loop must be able to withstand the high pressure that can occur in this part of the system.) The second drawback is that the counting equipment may handle the sharp, high activity peak in a different manner than the more diffuse lower activity peak of the chromatogram, i.e., the measured injected counts may be too low because of dead time losses that do not occur with the chromatogram peaks. This could give artificially high recoveries. Finally, the requirement for two loops of similar sizes in the same detector but separated by a column and possibly another detector, or a single high pressure loop with a high pressure switching valve, puts mechanical constraints on the design of the system.

2. Compare the total counts recovered in the sample radiochromatogram to those comprising the radiochromatogram of an unretained pertechnetate solution. To be accurate, the activity per unit volume of the sample and the standard must be the same. The errors will then come from injection volume differences. The method assumes that there is no loss of pertechnetate on the column. If the pertechnetate and the sample are injected close in time to each other, decay can be disregarded.
3. Collect the eluant from the column and compare the counts to a second aliquot of sample (diluted to the same volume) with a well counter. Decay can be disregarded, the errors arise from differences in the initial sample aliquot.

All of these methods have an accuracy of about ±5% when operated under optimum conditions. They should not be used as an assay for reduced hydrolyzed material in technetium kits. Two-dimensional chromatography is probably the most reliable assay for this purpose and should be used in conjunction with HPLC methods.

Other Mass Detectors

Electrochemical detectors are the only detectors among a number of alternative mass detectors that have found some use in radiopharmaceutical chemistry. To be useful, at least one of the species being separated must have an attainable redox region. They are useful as the second of a pair of detectors, working on different principles, that can be used to pinpoint particular chemical species in a complex mixture of similar compounds. They are, of course, mass detectors and cannot be used with nca ^{99m}Tc. Fluorescence or refractive index detectors suffer from the same problems, but are useful for detecting carrier quantities of materials present in nca ^{99m}Tc solutions and when dealing with other radionuclides.

Using two detectors working on different principles is common in radiopharmaceutical chemistry because of the different characteristics of the common species, i.e., radioactive/nca vs. nonradioactive/carrier. The radioactivity detector is normally placed after the mass detector because it is less affected by bubbles, thermal effects, etc. In the case of bubbles, it is beneficial because it provides some back pressure, which prevents their formation in the mass detector.

Controllers/Integrators

Five years ago, microprocessor-controlled HPLC pumps were the top of the line. Now computer-controlled systems are the norm and *smart* systems are on the horizon. One drawback of some computer systems is that they can be slower to program, etc., than the dedicated microprocessor systems. There is still a place for single-pump (uncontrolled) systems, especially for routine applications. But for separations development a two-pump controlled system makes solvent optimization much easier. Current controllers are a combination of pump controller, integrator, and short- to medium-term data storage. The last two operations can be done by an MCA. At present, the controllers cannot fulfill the integrator function of an MCA, unless they are one of the newer computer-based systems. Thus it appears that one is forced into purchasing both systems. What use then, can be made of controlled systems in radiopharmaceutical HPLC?

Unattended operation is the main attribute of controlled systems. To nonradiopharmaceutical chemists, this may mean overnight runs of 50 or more samples with some form of data workup and the results ready in the morning. To the radiopharmaceutical chemist, it should mean stability studies and automatic sequestration of one peak (purification) in a multicomponent mixture.

The first requirement for unattended operation is automatic sample injection. There are a number of fully automated systems on the market, but they are geared to multiple, large volume, air-stable, nonradioactive samples. They do not lend themselves to lead shielding nor to the repetitive injection of small volumes of air-sensitive samples. Homemade systems are relatively easy to make, as shown in Figure 5.4. They need an automatic sample injector such as the Rheodyne 7126. Its operation is controlled by the HPLC controller through relays that regulate the supply of compressed gas needed to actuate the injector. The sample loop of the injector must be partially or fully filled at the appropriate time. This can be done by using a sample syringe actuated by an electrically operated lead screw, which is turned on and off by the position-sensing switch of the injector. The lead screw is pushing on the sample syringe plunger all the time that the injector is in the load position, so it is best to completely fill the loop. In this way, multiple injections of a radioactive sample stored in a syringe can be made. This is

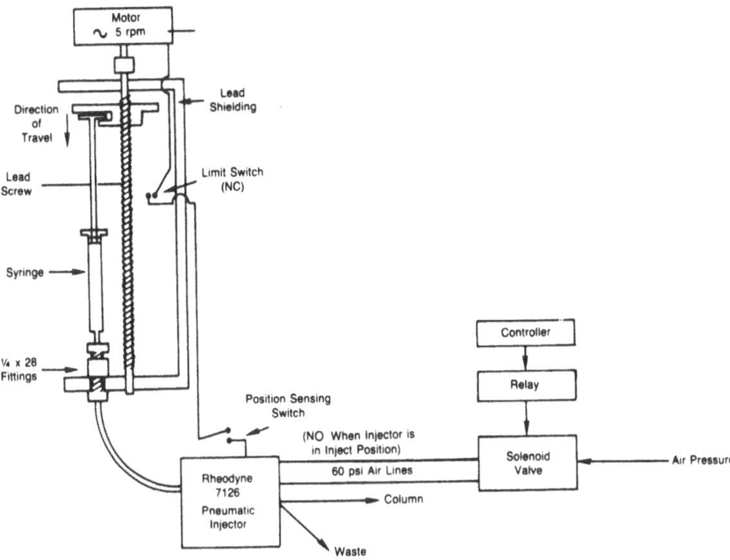

FIG. 5.4. An easy-to-make and operate small volume sample injector for radioactive samples.

ideal for stability studies, and at Squibb we have routinely run such studies overnight, injecting every hour or so. Data acquisition is by chart recorder and MCA. This gives both UV data on the chemical components of the sample and the desired RCP. The MCA is normally set up with multiple regions of interest so that the counts in each peak are integrated and printed out automatically. (Digital ratemeters do not have this option.) The chart recorder and the MCA are controlled by the HPLC controller. When a sample is not on the column, the chart recorder is off, and the pumps are only running at 0.1 ml/min.

There are a number of detectors that have peak stripping options incorporated into them. Most separate peaks into different tubes of a fraction collector. The radiopharmaceutical chemist is often only interested in one peak at a time, and often it is the most abundant peak. Homemade peak strippers are very easy to make and are inexpensive. They require an Altex slider valve or the like, a microswitch, and a straw, in addition to the normal HPLC equipment. The slider valve is air actuated and is placed immediately after the radioactivity detector, shown in Figure 5.5. Its outlet is to waste or into the collecting vial via a needle through the septum of the vial. The air to the valve is controlled by a solenoid valve that can be controlled manually or automatically. Automatic control is achieved by modifying a chart recorder so that as the pen moves over the chart it trips the solenoid valve on or off. The straw is glued to the top of the pen holder and keeps the microswitch closed for the duration of the peak. The collecting time is determined by the

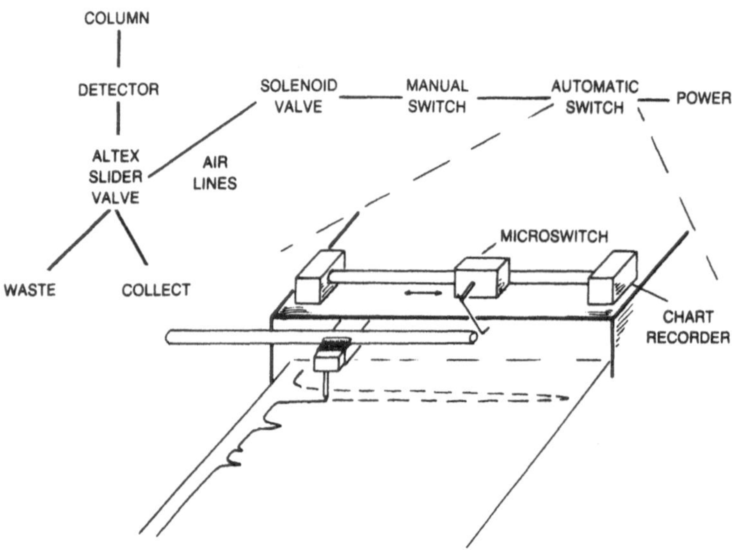

FIG. 5.5. A simple automatic/manual peak separator.

position of the microswitch on the chart paper. Obviously, such a system will only give pure cuts of peaks higher than all the others on the chromatogram.

With such homemade devices and an MCA, a very simple controller is all that is required. Unfortunately, most of the controllers are now computers but without fast count rate capability, so an MCA will be needed until the computers have faster ADCs.

CONCLUSION

After reading this chapter and Chapter 9 on applications, the reader should understand why a reversed-phase column is so popular: It can be operated in reversed phase, normal phase, ion exchange, and ion pairing modes. A basic understanding of bonded phases and their separation mechanisms is thus necessary for efficient method development and troubleshooting.

The novice or experienced chromatographer must realize that with so many options there will be differences of opinion as to which is the best collection of components to use. Our recommendations are inevitably biased toward our personal experiences.

We believe that it is most important for chromatographers to use the system(s) with which they are most comfortable. Familiarity is the key ingredient for reducing problems and for successfully adapting systems to fit the need. This applies equally to the mechanical and the separative mechanisms. In our experience, most people can use HPLC. However, only a small percentage of people have an innate ability with HPLC.

ACKNOWLEDGMENTS

The support of Department of Energy contract DE-AC02-83ER 60140 to A. R. Fritzberg is acknowledged.

REFERENCES

1. Simpson, C.F., ed.: *Techniques in Liquid Chromatography*. John Wiley and Sons, New York, 1982.
2. Snyder, L.R. and Kirkland, J.J.: *Introduction to Modern Liquid Chromatography*. John Wiley and Sons, New York, 1979.
3. Krstulovic, A.M. and Brown, P.R.: *Reversed Phase High Performance Liquid Chromatography*. John Wiley and Sons, New York, 1982.
4. Hamilton, R.J. and Sewell, P.A.: *Introduction to High Performance Liquid Chromatography*. Chapman and Hall (Wiley), London, 1978.
5. Parris, N.A.: *Instrumental Liquid Chromatography: A Practical Manual*. Elsevier, New York, 1976.
6. Horvath, Cs.: *High Performance Liquid Chromatography—Advances and Perspectives*. Vol. 1, Academic Press, New York, 1980.
7. Pryde, A. and Gilbert, M.T.: *Applications of High Performance Liquid Chromatography*. Halsted (Wiley), New York, 1978.
8. Tsuji, J. and Morozowich, W., eds.: *GLC and HPLC Determination of Therapeutic Agents*. Part 1, Marcel Dekker, New York, 1978.
9. Snyder, L.R. and Kirkland, J.J.: *Introduction to Modern Liquid Chromatography*. John Wiley and Sons, New York, 1979, pp. 183–202.
10. Horvath, Cs.: Bonded phase chromatography. In *Techniques in Liquid Chromatography,* Simpson, C.F., ed., John Wiley and Sons, New York, 1982, pp. 229–301.
11. Done, J.N.: Adsorption chromatography. In *Techniques in Liquid Chromatography,* Simpson, C.F., ed., John Wiley and Sons, New York, 1982, pp. 185–198.
12. Horvath, Cs.: Bonded phase chromatography. In *Techniques in Liquid Chromatography,* Simpson, C.F., ed., John Wiley and Sons, New York, 1982, p. 244.
13. Snyder, L.R. and Kirkland, J.J.: *Introduction to Modern Liquid Chromatography*. John Wiley and Sons, New York, 1979, pp. 405–407.
14. Glajch, J.L., Kirkland, J.J., Squire, K.M. and Minor, J.M.: Optimization of solvent strength and selectivity for reversed-phase liquid chromatography using an interactive mixture design statistical technique. *J. Chromatogr.* **199:**57–79, 1980.
15. Krstulovic, A.M. and Brown, P.R.: *Reversed Phase High Performance Liquid Chromatography*. John Wiley and Sons, New York, 1982, pp. 89–105.
16. Snyder, L.R. and Kirkland, J.J.: *Introduction to Modern Liquid Chromatography*. John Wiley and Sons, New York, 1979, pp. 270–281.
17. Melander, W.R. and Horvath, Cs.: Reversed phase chromatography. In *High Performance Liquid Chromatography—Advances and Perspectives,* Vol. 2, Horvath, Cs., ed., Academic Press, New York, 1980, p. 113.

18. Karch K., Sebastian, I. and Halasz, I.: Preparation and properties of reversed phases. *J. Chromatogr.* **122**:3–16, 1976.
19. Horvath, Cs.: Bonded phase chromatography. In *Techniques in Liquid Chromatography*, Simpson, C.F., ed., John Wiley and Sons, New York, 1982, pp. 267–276.
20. Snyder, L.R.: Classification of the solvent properties of common liquids. *J. Chromatogr. Sci.* **16**:223–234, 1978.
21. Snyder, L.R. and Kirkland, J.J.: *Introduction to Modern Liquid Chromatography*. John Wiley and Sons, New York, 1979, pp. 264–265.
22. Bakalyar, S.R., McIlwrick, R. and Roggendorf, E.: Solvent selectivity in reversed-phase high-pressure liquid chromatography. *J. Chromatogr.* **142**:353–365, 1977.
23. Snyder, L.R. and Kirkland, J.J.: *Introduction to Modern Liquid Chromatography*. John Wiley and Sons, New York, 1979, pp. 410–452.

CHAPTER 6

Overall Radio-HPLC Design

CHESTER A. MATHIS, REESE M. JONES, and JOSEPH H. CHASKO

INTRODUCTION

The separation and detection of radioactive tracers by high performance liquid chromatography (radio-HPLC) incorporates many of the same design considerations of conventional (nonradio) HPLC. There are several excellent books that describe the principles of HPLC and contain additional specialized applications and considerations not covered in this short chapter [1–4]. Some design features described in the following pages are not unique to radio-HPLC, but particular emphasis will be placed on considerations that may not be important in conventional HPLC. This chapter is arranged so that some descriptive chromatography topics relevant to later discussion are contained in the Introduction. The essential constituents of a radio-HPLC system are presented in the section entitled Components of the Basic System, and some design variations are discussed in A More Complex Radio-HPLC System. Several design options are examined in the section entitled Additional Considerations, and the decision processes involved in choosing specific options are illustrated in A Specialized HPLC System.

The advantages of conventional and radio-HPLC over low pressure liquid chromatography include: speed of analysis, peak resolution, detection sensitivity, and facile sample recovery. The coupling of radioactivity detectors with conventional mass detectors (UV, refractive index, fluorescence, electrochemical, conductivity, etc.) can add a great deal of versatility to a separation system. Co-eluting nonradioactive peaks do not interfere with

---

Abbreviations: high performance liquid chromatography, HPLC; ultraviolet, UV; radioactive half-life $t_{1/2}$; column efficiency, N; column selectivity, α; column capacity factor, k'; resolution, R; photomultiplier tube, PMT; single channel analyzer, SCA; nuclear instrumentation module, NIM; multichannel analyzer, MCA; analog-to-digital, A/D; voltage-to-frequency, V/F; relative counting efficiency ratio, E; counts per second, cps; transistor-transistor logic, TTL.

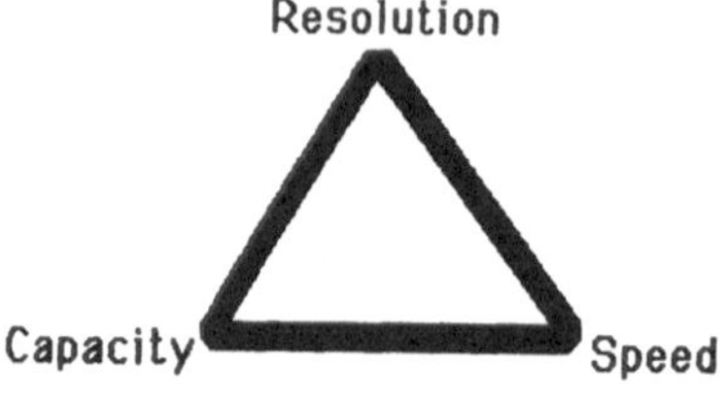

FIG. 6.1. The chromatographer's triangle. The improvement of one aspect in the triangle is at the expense of the other two.

gamma-ray radio-HPLC detectors as they do with other detectors. A radioactivity detection system can increase sensitivity limits to 10^{-20} moles (~1 nCi of ^{13}N) and provide an unusually wide dynamic range for flow detectors (1 nCi/ml to 1 mCi/ml or 10^6 units).

The following discussion is directed primarily to the gamma-emitting isotopes whose decay can be detected with sodium iodide doped with thallium (NaI(Tl)) scintillation crystals. Many of the design considerations that apply to gamma emitters also apply to negatron (β^-) emitters, such as tritium and ^{14}C, except that their decay must necessarily be detected by liquid or plastic scintillation detectors [5]. Positron emitters (β^+) can be detected by NaI(Tl), liquid or plastic scintillation detectors [6].

A successful HPLC separation involves making a compromise between analysis speed, chromatographic resolution, and sample capacity. These three factors form the *chromatographer's triangle* [2] as illustrated in Figure 6.1; an improvement in one of these factors results in the degradation of the other two.

Analysis speed has special significance to radio-HPLC because chromatographic separations of species labeled with short-lived radionuclides such as ^{13}N ($t_{1/2}$, 10 min), ^{11}C ($t_{1/2}$, 20.3 min), and ^{18}F ($t_{1/2}$, 109.8 min) require very rapid analyses. Separations of the in vivo metabolic products of ^{13}N-, ^{11}C-, and ^{18}F-labeled compounds necessitate special sample handling and data acquisition procedures as described later.

Chromatographic resolution is itself a function of three factors: efficiency, selectivity, and capacity factor (not to be confused with sample capacity, described later). Column efficiency (N) depends on various column and hardware-related parameters such as the column packing material, packing particle size, column packing method, solvent flow rates, solvent viscosity, and band spreading owing to the injection system, column fittings, connecting lines, and detector system(s). Once an HPLC system is in place, the efficiency can be changed most easily by changing the length of the column. Column selectivity (α) is a function of specific chemical interactions between the solute molecules and the stationary and mobile phases. Selectivity is affected by changing the column packing material and/or changing the eluent. The column capacity factor (k') is determined by the ratio of the volumes of the stationary and mobile phases in the column—low values of k' mean that the solute is not retained by the column. The capacity factor can be changed by making the mobile phase weaker or stronger. Overall resolution (R) is a function of all three of these parameters and is described by Eq. (1).

$$R = \frac{1}{4}\sqrt{N}\left(\frac{\alpha - 1}{\alpha}\right)\left(\frac{k'}{k' + 1}\right) \tag{1}$$

The best way to increase resolution with isocratic elution methods is to increase efficiency (N), because an increase in N does not lead to longer retention times. Many of the design choices described later in this chapter affect the chromatography efficiency function. Increasing selectivity (α) and capacity factor (k') with an isocratic system leads to longer elution times and wider solute elution peaks, which are harder to detect. Gradient chromatography (changing eluent composition during a separation) can decrease the peak-broadening effect of long retained peaks and can add great flexibility to a separation system. Selectivity and capacity factor choices deal primarily with qualitative column and eluent decisions that are not the primary concern of this chapter. The operation of an efficient radio-HPLC system must take into consideration column and eluent selection, but the reader is best referred to chromatographic journals that deal with specific separations for use with individual radio-HPLC applications.

The column sample capacity (*not* k') is important in design consideration if large quantities of material must be injected so that a radioactive peak can be distinguished from the background noise. Sample size is of concern in analyzing tissue extracts for metabolic product identification due to the large tracer dilution or for preparative or semipreparative chromatography where large chemical amounts are injected and resolved fractions are collected. In general, increasing the column capacity presents little difficulty because longer, bigger columns are readily available.

The design characteristics of a particular radio-HPLC system are determined by the specific application of the system. Applications as diverse as metabolic product identification with short-lived radiotracers, routine radiopharmaceutical quality control with long-lived radionuclides, and radiochemical synthesis using preparative HPLC for product purification require very different design considerations. A nondedicated system capable of facile conversion to one of these specific applications requires a more flexible design. One caveat should be foremost in designing a radio-HPLC: "Keep it simple." A flexible system with switching valves, multiple injectors, columns, detectors, and numerous fittings will lose resolution because of band spreading; a more complex system is also more likely to malfunction. The specific circumstances of the desired separation(s) must be balanced against a loss of resolution; the convenience of a flexible system may well outweigh the adverse effect on resolution of the extra lengths of tubing and valves. If most of the separations performed are simple (such as determining the amount of free $^{123}I^-$ in iodoamphetamine preparations or unbound $^{99m}TcO_4^-$ in ligand preparations), a versatile system with lower resolution is often acceptable. If a complicated metabolic product spectrum requires separation, one should not sacrifice resolution for versatility. Individual design preferences vary from one chromatographer to another, and the choices are often not clear and distinct. Compromises in resolution, speed, and capacity may result from specific design needs for detection sensitivity, rapid analyses, sample size constraints, and a host of other variables. The design factors considered below are by no means complete and are presented only as examples of some factors influencing available options.

COMPONENTS OF THE BASIC SYSTEM

The constituents of the basic radio-HPLC system are illustrated in Figure 6.2. There are very few components of this basic system that can be eliminated and still maintain the function of radio-HPLC. The solvent reservoir, pump, injector, column, mass detector, and chart recorder are all standard HPLC components available from a variety of sources [7]. The price of these items can vary widely with quality. A modular system with an inexpensive injector, pump, fixed wavelength UV detector, replaceable cartridge columns and dual-pen chart recorder could cost as little as \$4000. A well-type NaI(Tl) crystal for gamma-ray detection with associated electronics (e.g., a photomultiplier tube (PMT), preamplifier, amplifier, single channel analyzer (SCA), ratemeter, high voltage power supply, and nuclear instrumentation module (NIM) bin can be purchased for about \$2500. It is possible, therefore, to construct a functional system for as little as \$6500. A commercial flow scintillator unit for β-detection is considerably more expensive (\$8,000–15,000) than the basic gamma detector unit. Higher quality

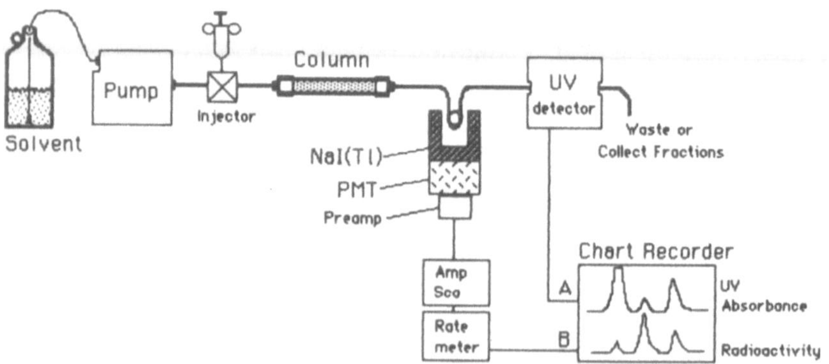

FIG. 6.2. The components of a basic radio-HPLC system. The constituents include: a solvent reservoir, high pressure pump, sample injector, HPLC column, NaI(Tl) well-type detector with associated electronics, UV detector, and dual-pen chart recorder.

pumps, injectors, and detectors are often worth the investment because they provide improved reliability and performance. A good quality system with the components shown in Figure 6.2 could cost $10,000–20,000. Some additional pieces of equipment such as a solvent filtration system, fittings, tubing, syringes, and high quality solvents are also necessary to build and maintain the system.

It is best to protect the pump by filtering the solvent before it enters the pump and to protect the injector and column by filtering the sample prior to injection. Routine sample preparation might include sample filtration with a 0.45-μm syringe-mounted filter, but care should be taken to ensure that the filter and syringe do not selectively retain any of the compounds of interest. In addition, an inexpensive guard column can be inserted in the flow stream between the injector and column to remove any particulates that would ruin the more expensive analytical column. Pump speeds of 0.1–5 ml/min are usually sufficient for analytical applications. Pump head pulse dampening is important for maintaining a constant flow to obtain an accurate detection response, and pump pressure operating limits of ~6000 psi are useful with long columns or small particle packing materials that can increase column back pressures. Preparative systems may require pump flow rates of 5–30 ml/min. If the columns on a particular system are changed often so that different classes of analyses can be performed, it may be convenient to use a column-changing system that does not involve removing the column end fittings. There are column-holding systems available from several manufacturers such as the Waters Radial-Pak® system or the Brownlee column holder. Once a column-holding system is purchased, the cost per column is usually less than a conventional end-fitting column, but a particular manufacturer's holding system may limit the choice of available column packing materials.

The detector should be as close to the injector as is physically possible, and all intervening plumbing and hardware should be minimized. Column-switching valves can be installed to provide facile column changes, but the number of such valves should be limited because of their adverse effect on resolution. With most radio-HPLC applications, a few switching valves and several feet of tubing do not adversely affect peak resolution enough to warrant their exclusion. The relative positioning of the radio-detector and mass detector is determined by which detector output is more important to the chromatographer. One restriction is that refractive index detectors are often fragile and sensitive to back pressure and pulsation, and they should be put last in the flow sequence. The UV detector is the most versatile detector, but specialized detectors such as electrochemical can increase sensitivity in particular applications. Detector signals can be monitored with a dual-pen chart recorder using one channel for mass detection and the other channel for radioactivity. Crude quantitation can be made either by comparing peak heights and areas on the chart paper or by using a chart recorder containing its own integrator (e.g., Linear).

In addition to cost, space may be an important factor in deciding whether to build a modular system by putting together components from different manufacturers or purchasing an integrated system from one source. In general, a modular system is more flexible than an integrated system since design modifications are more easily performed, and a modular system can be adapted to the space limitations of a particular laboratory.

A MORE COMPLEX RADIO-HPLC SYSTEM

An HPLC system containing some available options is shown in Figure 6.3. These options illustrate variations in design, and the individual chromatographer can tailor the specific design characteristics to meet the requirements of his/her analyses.

A gradient programmer adds tremendous flexibility to a separation system. Long-retained, spread-out peaks in an isocratic system can be eluted as sharp peaks in a short time with a gradient system. Some detectors (such as refractive index) may be sensitive to the gradient and complicate the analysis, but the background base-line change attributable to the gradient can be subtracted when using a computer-based data acquisition system. The radio-HPLC gamma detector is insensitive to interfering solvent effects such as these.

A guard column between the injector and column increases the column life, but may reduce system efficiency. However, a guard column of material identical to that of the analytical column can sometimes increase overall column efficiency. The selective adsorption and retention of compounds labeled with high energy gamma emitters by the guard column or analytical column can be detected with a radiation survey meter, but this method gives only a qualitative indication of activity retention. More quantitative measurements of activity loss on the columns can be made by external assay or by using an additional detection loop placed prior to the column (see below).

A thermostatic jacket surrounding the column can be used to improve the resolution in some separations, but it is generally more convenient to change the selectivity or capacity factor than to operate most columns above ambient temperature.

The NaI(Tl) scintillation crystal is often used as a gamma-ray detector because it is relatively inexpensive, commercially available from a number of sources, and readily adaptable to most radio-HPLC applications [8]. The sensitivity of the NaI(Tl) crystal in radio-HPLC applications (~1 nCi) results in good part from the large volume crystals that are used (~200 ml). Large crystal volumes often lead to increases in background noise that can be partially compensated by using narrow energy windows on the SCA and by thick lead shielding surrounding the crystal. A large NaI(Tl) detector, a large photomultiplier tube, and 5 cm of lead shielding can create problems if laboratory space limitations are severe. A semiconductor crystal such as CdTe does not require extensive shielding or a PMT and may thus help

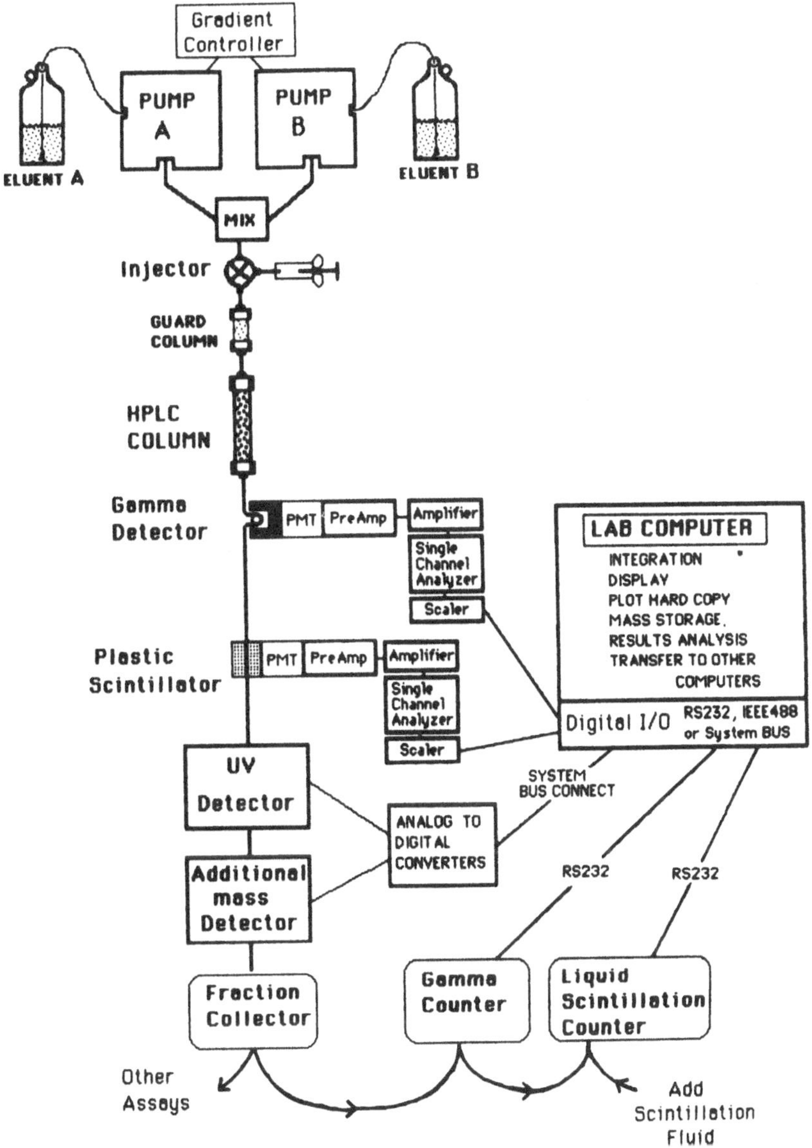

FIG. 6.3. A more complex radio-HPLC system with some options. In addition to the basic components shown in Figure 6.2, some options shown here include: a gradient elution controller and another pump, a guard column, a plastic scintillator for β^- detection, a fraction collector, and a computer-based data acquisition system.

alleviate space limitations. The use of a CdTe crystal in radio-HPLC applications has recently been described [9]. The detection sensitivity of CdTe per unit volume is greater than NaI(Tl). The background count rate of a small (5–10 mm^3) CdTe detector is more than a factor of ten lower than NaI(Tl), but the small crystal volume of CdTe limits the absolute counting sensitivity of this detector, especially at gamma energies greater than 150 keV. The small detector flow cell volume (15 μl) permits greater peak-to-peak resolution, but also decreases the residence time of the radioactivity in the detector. The CdTe detector may be useful for radio-HPLC applications where low energy gamma rays are detected, detector space is limited, and detection sensitivity is not a primary design consideration.

Three types of detectors coupled in series are shown in the more elaborate HPLC system of Figure 6.3. The resolution decreases the further downstream the detector is from the column, and two detectors usually are the most that should be placed in series. Manual or electrically controlled stream splitters are sometimes used in amino acid/protein analyses to divert a fraction of the eluent for derivatization prior to detection. It is possible to split the eluent stream and send a fraction of the stream to the UV detector and a fraction to the radioactivity detector, creating a parallel analysis system with little loss of resolution resulting from band spreading. Stream splitting results in some loss of signal to each detector; if resolution is a more important criterion than detection sensitivity, a stream splitter may be worth considering.

A fraction collector is an inexpensive alternative to a beta flow detector. Peaks can be collected, and a liquid scintillation counter (if available) can be used to quantitate the collected fractions of long-lived species. An off-line liquid scintillation counter is much more sensitive than an on-line counter because of its longer sample counting times. The use of a fraction collector to increase detection sensitivity must be balanced against the inconvenience of individual sample collection and counting requirements.

A radio-HPLC system may contain numerous radiation and mass detectors on each flow system and may consist of several parallel flow systems to achieve flexibility and sample analysis throughput. The data generated by such a system must be accumulated, quantitated, and displayed for analysis. It is often desirable to store the data for later retrieval and manipulation. Several methods may be used to acquire and display the data, ranging from simple chart recording to expensive and sophisticated computerized acquisition systems. A computer-based system can add complexity and expense, but may be very useful and well worth the cost. If the chromatographer performs numerous analyses or several chromatography systems are used simultaneously, a computer-based system can be essential. Systems can be built around some common inexpensive microcomputers such as the Apple IIe or IBM PC [10–12]. Computerized multiport data acquisition systems are available from several manufacturers for collecting, storing, and analyzing chromatograms (Nelson Analytical and IBM). The large dynamic detection range often encountered with radio-HPLC measurements places constraints upon systems designed for recording signals from conventional detectors. These considerations are briefly described below and should be kept in mind when adapting a system designed for conventional detectors for use with radio-HPLC applications.

Radioactive counting data can be collected in two ways subsequent to gamma energy discrimination: (1) direct pulse counting by some type of digital counter, or (2) pulse rate detection by a count ratemeter. A scaler is a device that counts pulses for a timed interval and then transmits the sum to a storage device; scalers are available as single or multichannel stand-alone modules or computer cards. Pulse rate to voltage conversion is usually done with single channel ratemeters.

A count ratemeter is a device that displays a counting rate rather than the total number of counts in a fixed time period. The advantage of a ratemeter is that it continuously records the count rate, and its analog output can be plotted conveniently on a chart recorder. Ratemeters are designed so that each incoming pulse adds a charge to a capacitor. The capacitor leaks its accumulated charge across a resistor, producing a voltage output that is proportional to the incoming pulse rate. True voltage output proportionality exists only when the incoming charge rate and the capacitor leak rate reach an equilibrium. Ratemeters contain selectable time constant settings that control the time interval over which the charge is allowed to accumulate on the capacitor. If the selected time constant is too short, the output signal will be erratic. If the time constant is too long, the ratemeter will respond slowly to changes in the count rate and true peak shape of radiochromatograms will not be properly represented. Even with short time constants (~1 sec), the true peak shape may be smeared or clipped owing to the rapidly changing count rate over wide dynamic counting ranges. The output of the ratemeter may be plotted on a chart recorder or sent to a computer for storage after redigitizing the signal. A chart recorder is a simple inexpensive device that has the advantage of low cost and prompt display for perusal and rough quantitation. The dynamic range of the output of a ratemeter is limited to about 1 : 100 unless range switching or a logarithmic scale ratemeter is used. Manual range switching requires close supervision and a log display complicates integration.

The analog (voltage) output from a ratemeter or a mass detector may be digitized and fed to a computer for storage. An integrator or computer using a ratemeter as a "front-end" device for pulse counting suffers from the limitations imposed by the digital-to-analog and analog-to-digital (A/D) conversion processes at the ratemeter and the A/D converter interfaces. The limitations are those of dynamic range and resolution time. The dynamic range should usually be 1 : 1,000 or 1 : 10,000 for use with radio-HPLC. The A/D converter typically digitizes to 8, 12, or 16 bits, which yield dynamic, signed ranges of 1 to 128, 2048, and 32768 respectively (and twice as much without a sign bit). Eight bits provides far too little range, 12 bits marginally enough, and 16 bits is more than adequate for good quantitation. The effective range can be extended by changing the gain on either the ratemeter or the A/D converter, but this can cause a base-line shift. The A/D converter can also average the ratemeter signal over a programmable interval. A short time interval is the most accurate and is best followed by averaging the signal with software after collection. A more accurate conversion of the analog signal to a digital signal can be made with a voltage-to-frequency (V/F) converter (Nelson Analytical). The analog signal is converted to a frequency that is proportional to the input voltage over a wide dynamic range (e.g., 10^6). The frequency output can be integrated over a variable interval (0.01–5 sec) using a scaler, and the data can be transferred to the computer for storage. The V/F converter has the advantage over an A/D converter of true signal integration without the requirement of constant signal sampling by the computer.

The least expensive way to "computerize" analog output is with a stand-alone digital integrator that performs A/D or V/F conversion (Hewlett-Packard or Spectra-Physics). Integrators capable of peak processing, reporting, and networking are available for ~$2000. Such integrators have the advantages of ease of installation and use; however, they usually lack the capability of performing decay corrections and are often limited to two channels of input information. Some integrators filter the input data signal to achieve a noise-free base line and small peaks (<1%) can be filtered out and lost. For some applications, it may be necessary to collect the data and analyze it before filtering so that small peaks can be quantitated. More complex integrators are available as part of HPLC controller systems costing >$10,000, and these controllers may permit control of data acquisition and allow for further data analysis by BASIC programs.

Conversion of an analog output signal from a ratemeter with a V/F converter still suffers from the signal conversion limitations imposed by the ratemeter at the front end. The highest data collection accuracy can be obtained with a scaler-timer combination sending digital information directly to the computer. Pulse counting with external scalers involves summing the pulses over the scaling interval and transferring the data to a computer for storage. The accumulation of a sequence of these sums over time in separate computer storage registers is referred to as multiscaling. The collection interval may range from 1 msec to 10 sec, depending upon hardware and software constraints. The maximum number of counts that can be stored in the scaler is the storage capacity divided by the scaling interval; a system should be able to store the maximum rate (10–30 kHz) of a typical counting system. A scaler with 8-bit storage registers (256 counts) must transmit to the computer at 13-msec intervals to manage 20 kHz, but a 16-bit scaler can accumulate 65,536 counts before overflow and would be required to transmit every 3 sec at 20 kHz. If desired, the data can be accumulated and averaged over a longer interval with software before storage. A collection time interval of ~1 sec provides adequate resolution for most radio-HPLC systems. Longer collection intervals are appropriate for slow HPLC (>20 min), and 0.1–0.5 sec intervals for ultra-high speed HPLC may be required. The maximum acceptable collection interval is determined primarily by peak resolution and should be less than half the minimum base-line separation between peaks to retain resolution. Peak-to-peak separation is determined by the overall resolution of the system and by the detection loop size (see below).

Since a timer-scaler integrates the true counts over the collection interval, it is much more accurate in quantitating peak areas than a ratemeter. While the true shape of the peak may be lost with larger collection intervals, the integrated area under the peak remains accurate as long as the counting interval permits no peak-to-peak overlap. Most chromatographers require that the peak shape be adequately displayed, and at least 10–20 data points per peak are needed to properly represent peak shape [13].

Data collection with scalers can be accomplished three ways: (1) via a multichannel analyzer (MCA) configured for multiscaling, (2) with an external single or multichannel scaler communicating with a computer via serial or parallel character transmission, or (3) with a multichannel scaler within the computer communicating directly with computer memory.

A multichannel analyzer can conveniently be configured to sort pulses by time rather than energy difference and store the data points in its internal registers (generally 1024–4096 capacity). The MCA acts both as a timer-scaler and a storage device, and this process is referred to as multichannel scaling. This is a good alternative when an MCA is available, and only a single channel of counting information requires collection. An MCA can be configured to count pulses from an amplifier/SCA at regular intervals for a selected period, and then either process the data (if computer based) or simply display the information. The processing capabilities of MCAs are usually limited, and may not include the ability to handle multiple signal ports or to perform decay corrections.

Suzuki has recently described a method of using a MCA as a sorting device to simultaneously monitor the eluent of an HPLC for the gamma energy spectra of several different radionuclides and perform multiscaling operations in the different energy regions [14]. This technique uses a Ge(Li) detector because of the requirement of good gamma energy discrimination and requires a separate computer to assist with the data storage, decay corrections, and background analyses. It illustrates the versatility of a computer-based system and may be useful for analyzing radiochromatography data involving several radionuclides.

An external stand-alone scaler can handle many channels of pulse input and transmit the information to a computer. The scaling interval is often switch set and may be the same for all channels. The counts are collected for the scaling interval and transmitted to a computer or storage device for later processing. Timer-scalers can output digital information and collect data in the next time bin simultaneously (termed a "buffered" time-scaler). The digital output of commercially available timer-scalers may have to be made compatible with the input ports of microcomputers, and devices are available to provide this conversion (Ortec, Canberra).

A scaler card can also be attached directly to the computer's data bus and function much like an external scaler with scaling intervals programmable from software. The card should handle most of the collection without intervention by the computer except to interrogate or control. This is particularly important on a multitasking computer, which might be unable to directly control the counting process because of other user requirements. The data-bus-connected scaler-timer card offers the greatest flexibility of use and software control, but usually requires a custom-fabricated scaler card.

Several factors might be considered when choosing a computer-based data acquisition system for incorporation into a radio-HPLC system. The system should be able to accept at least two signals from each HPLC detection system—one channel from the mass detector and one channel from the radioactivity detector. Additional communication ports (RS232 or IEEE 488) for connecting to other counters and computers may be useful. If multiple HPLC systems are to be used, the system should allow each channel to be started and stopped independently. Additional hardware options include a medium/high resolution display and a plotter-printer for producing hard copies of chromatograms. Some form of fast mass storage is also recommended so that the raw data can be stored for later analysis and reference. For radio-HPLC of short-lived radionuclides involving difficult and expensive experiments, it is often desirable to collect and store chromatography data in as raw a form as possible and perform scaling and filtering with software rather than hardware. The various software filtering techniques can then be compared without the need for performing multiple chromatograms of the same sample. Filtering involves a trade-off between time resolution and signal-to-noise ratio. Shorter averaging times lead to greater noise and degraded peak shape (particularly for low activity peaks) but better peak resolution. The amount of data storage required increases sharply with shorter averaging times. If advanced statistical peak-fitting algorithms are used, it is more rigorous to perform the fitting and integration operation on raw data. The data may then be displayed either filtered or reconstructed from the analytical fitting functions.

For the greatest flexibility, the computer-operating system should be able to support high level programming languages such as BASIC or FORTRAN (with access to the I/O ports, internal clocks, and screen graphics), so that programs can be written for controlling runs, taking data, and analyzing chromatograms. Customized software designed for the individual user's needs may require a good deal of programming time to institute. While customized programming provides the greatest capabilities, it may also demand considerable software expertise. Several companies offer computer based systems with packaged software for chromatography data acquisition and chromatogram analysis. Since most commercially available computer-based chromatography data acquisition systems are designed for conventional applications, care should be taken in selecting a system with enough flexibility for use with radio-HPLC applications.

Several commercial data acquisition systems are available that have enough flexibility to meet the needs of a radiochromatography system. An example of such a system is built around the IBM CS9000 supermicro computer [15]. The system can be purchased as a preconfigured chromatography data acquisition system with a four-channel analog/digital input card, printer-plotter, mass-storage, and chromatography control software. The computer is easily interfaced with other laboratory instruments, and several high level programming languages are available for controlling input/output and analyzing chromatography data. The 16-bit data bus allows comfortable data acquisition over the wide dynamic range required of a radiochromatograph.

ADDITIONAL CONSIDERATIONS

A useful technique for removing an interfering peak(s) without the need for gradient programming is the use of a *stripper column* between the injector and separation column in the position usually occupied by a guard column. A stripper column is physically similar to a guard column, but it has a different purpose. A guard column is used to protect the analytical column from large particles and other materials that might degrade column performance; a stripper column selectively retains compounds that might interfere with an analysis. A stripper column can retain material for subsequent recovery [16] or irreversibly bind problem peaks and impurities [17]. Figure 6.4 shows the usefulness of a stripper column that retains anionic material and allows elution of neutral and cationic species.

Preseparation detection of radioactivity by a *pre-loop* is unique to radio-HPLC. The pre-loop is located between the injector and separation column as shown in Figure 6.5. The pre-loop provides a measure of radioactive mass balance between the amount of activity injected onto the column and the amount eluting in peaks from the column; an activity imbalance quantitatively indicates activity bound to the column. This device is important when measuring radiochemical purity for quality control because some impurities might bind to the column and would otherwise not be detected. This feature is also extremely useful when working out a new separation or when using a stripper column to retain activity (Fig. 6.4). The amount of activity bound to the stripper column can be quantitated by placing the pre-loop on the upstream side of the two columns; activity retained by the stripper column can be determined by the activity imbalance between the amount of activity that passes through the pre-column loop and the amount that elutes from the analytical column.

The pre-loop can also be used to indicate detector overload resulting from deadtime loss in the radioactivity detector. A flattened or concave pre-loop peak shape is indicative of detector overload. A pre-loop volume of 20–50% of the post-column sample loop results in maximum count rates about 50% higher than in the peaks themselves (depending upon sample size and peak sharpness) and provides a warning of detector overload. This volume may need to be decreased for high efficiency, low capacity, chromatography where band spreading would be objectionable.

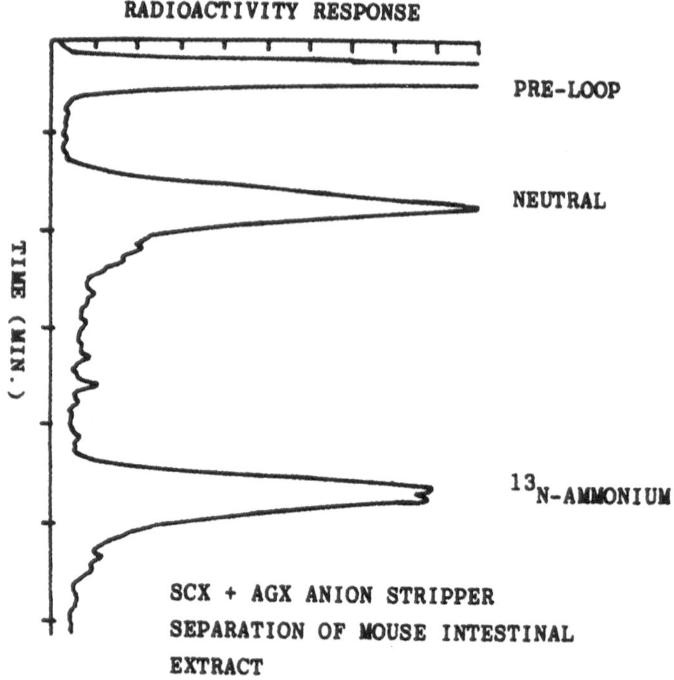

FIG. 6.4. An ^{13}N radiochromatogram showing the use of a stripper column to remc interfering compounds [17]. The labeled intestinal contents of a mouse 20 min af intratracheal instillation of $^{13}NO_2^-$ are shown. The 511 keV positron annihilati photons from ^{13}N were detected with a well-type NaI(Tl) detector and a 0.3 sample loop. The radio-HPLC separation was achieved with a 0.46 × 25 cm Wh man Partisil-10 SCX® column eluted with a 30 *mM* sodium phosphate buffer (pH 5 and a flow rate of 3 ml/min. Anions eluting at the void volume and masking t neutral species were removed prior to the analytical column with a 3-cm strip column containing AG1-X8® anion-exchange resin (Bio-Rad). The neutral spec were subsequently determined to be primarily ^{13}N-urea and ^{13}N-glutamine. The p loop (Figure 6.5) peak was used to determine radiochemical mass balance, and t maximum is off scale on this chromatogram.

The relative counting efficiency ratio (E) of activity flowing through t pre-loop to that of the same activity flowing through the sample detecti loop must be determined to calibrate the system. This can be done with injection of a sample known to be completely eluted from the column or w a delay loop in place of the column. The value of E is highly dependent up the position of the loops within the crystal well; the pre-loop should be firn fixed to the sample loop and both loops should fit snugly within the crys well. Once the system has been calibrated for a particular pre-loop a sample loop size, the value of E should not vary if the positions of the t loops within the crystal well remain the same and if the same energy gam rays are detected. If different radionuclides are used with widely differ

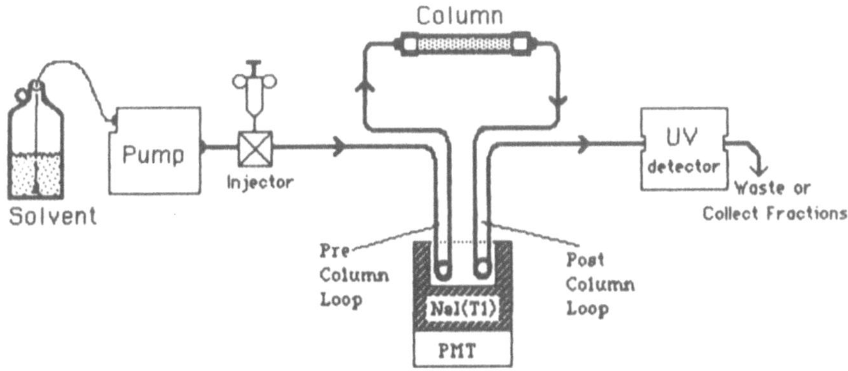

FIG. 6.5. An activity pre-loop installed between the injector and column. The pre-loop provides radioactivity mass balance for stripped or retained peaks (Figure 6.4) and also indicates detector overload.

gamma energies, it is best to construct the pre-loop and sample loop out of the same material so that the gamma ray attenuation will be the same in both loops. The pre-loop must necessarily be made from stainless steel tubing because of the high pressure upstream from the column. For high energy (>100 keV) gamma emitters, the detection loop should also be constructed from stainless steel tubing. With applications using low energy gamma emitters such as ^{125}I, a teflon tubing detector loop will increase the counting sensitivity; a new value of E must be determined if a teflon detection loop is inserted in place of a stainless steel loop. In practice, the activity peak associated with the pre-loop multiplied by $(1/E)$ should be equal to the sum of the decay-corrected activity from the eluted peaks. Activity losses of as little as 5% can be routinely quantitated with a pre-column detection loop (unpublished data). Summed peak recovery less than the scaled pre-loop peak indicates a noneluted component(s); summed peak recovery greater than the scaled pre-loop peak and a flattened or concave pre-loop peak are indicative of detector overload. Typical maximum count rates are 20,000–30,000 cps for standard NIM bin electronics, and it is best to keep count rates below 10,000 cps to minimize deadtime losses.

External sampling may be employed as an alternative to pre-loop detection. The activity from all the eluted peaks should be collected and an aliquot of the collected pool counted using a gamma counter. The total activity in the pool can then be determined and compared with the activity injected onto the column. This procedure is applicable with long-lived radiotracers whose low gamma energies would preclude pre-loop detection. Routine external counting to determine activity imbalance is more inconvenient than using a pre-loop detection system, but it may be necessary for use with low energy radionuclides or as a verification procedure for quantitating the pre-loop detector efficiency.

Several factors should be considered when attempting to optimize the sensitivity of a NaI(T1) detector. The intrinsic photopeak efficiency of a cylindrical 7.6 × 7.6 cm (1 × d) NaI(T1) crystal is optimum for gamma rays in the range of 90–160 keV, and positron emitters with 511 keV gammas are detected with ~40% the efficiency of 120 keV gamma rays. A NaI(T1) detector containing a "well" into which the coiled sample loop can be inserted results in good counting geometry. A cylindrical 7.6 × 7.6 cm NaI(T1) well crystal can give 10–30% overall counting efficiency for 100–600 keV gammas. Background noise and dark current can be minimized with a 5-cm lead shield surrounding the crystal and a threshold discriminator or single channel analyzer on the amplifier output. The optimum detector loop size is a function of several factors including flow rate, column size, desired sensitivity, resolution between peaks, and activity in the sample. A larger detection loop provides longer residence time in the crystal well and increases the total counts in a peak, but also decreases peak resolution as a result of time averaging, band spreading, and peak overlap. The volume must be large enough for the minimum sensitivity required, but not so large that high activity samples would overload the detector. Experimentation is necessary to find the best balance of sensitivity and resolution, but a volume of ~10% of the peak elution volume is a good compromise for low activity (<10 μCi/ml) samples; this volume should be decreased to 1–5% for high activity (>50 μCi/ml) samples. We found detector volumes of 0.2–0.4 ml best for low activity samples on a 0.46 × 25 cm 10-μm analytical column with 3–20 min chromatograms at 2–3 ml/min flow rates. It was necessary to inject <1 μl of samples containing 1 mCi/ml to avoid detector overload. A detector volume of 0.05–0.2 ml would be more appropriate for flow rates of 0.1–1.0 ml/min on small-bore columns (≤2 mm) or for use with samples containing >1 mCi/ml of activity.

A SPECIALIZED HPLC SYSTEM

An HPLC system was constructed to measure the metabolic fate of short-lived tracers after administration of labeled precursors [17,18]. The chromatographic separations of the expected metabolites had not yet been developed, so a variety of different column types were purchased to offer several choices for methods development. Most of the metabolites were either anions, cations, amino acids, or related compounds with a net charge under the appropriate elution pH conditions. Several different weak and strong ion exchange columns (NH_2, CN, SAX, and SCX) were selected along with a general purpose C-18 reversed-phase column (e.g., Waters, Whatman and Altex). A programmable gradient elution system (Waters) was determined to be essential owing to the analysis speed and high resolution necessary to separate the metabolites labeled with short-lived radionuclides.

For the methods development, chemical standards could be detected by several types of mass detectors; a multiwavelength UV absorbance detector (Waters) was selected because of its flexibility for detecting a variety of compounds in this application. The retention volumes of expected metabolic products were determined on several columns using the multiwavelength UV detector. During the developmental stages of one experiment, a ^{14}C-labeled amino acid was used in animals to determine which potential metabolites would be labeled and in what proportions [17]. An on-line flow-through β-detector was considered, but was not selected; the convenience of the flow detector did not outweigh its high cost and limited sensitivity. An automated fraction collector (ISCO) combined with an available liquid scintillation counter in which fractions could be counted for long periods of time to maximize ^{14}C detection sensitivity was chosen.

One experimental protocol consisted of injecting a mixture of a ^{14}C-labeled and ^{11}C-labeled amino acid into an animal (Jones and Sargent, unpublished results). The large tracer dilution in the animal required that mCi amounts of ^{11}C-labeled compound be injected so that metabolites could be detected 60 min later. Serial blood samples were collected for assay at the same time the distribution of the ^{11}C tracer activity was imaged with a positron emission tomograph. The 20-min half-life of ^{11}C limited the ability of radio-HPLC to follow the slow transfer of the label from one metabolic pool to the next to about an hour post-injection. With the co-injected ^{14}C tracer, the same metabolism could be followed independently for many hours, as shown in Figure 6.6.

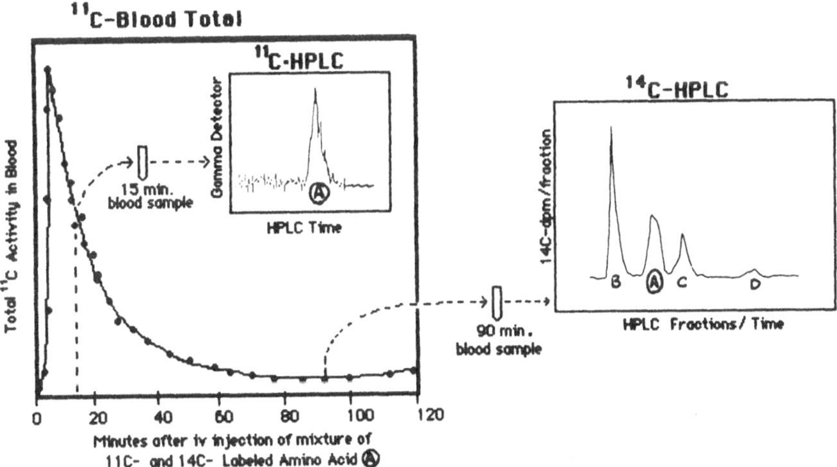

FIG. 6.6. Labeled metabolites in plasma following intravenous injection of a ^{11}C- and ^{14}C-labeled amino acid (A). Serial blood samples were drawn to determine the total activity blood clearance curve. Selected early blood samples were prepared for HPLC injection and ^{11}C-labeled metabolites were chromatographed and detected with a flow-through NaI(Tl) detector. Fractions were collected and after the ^{11}C activity had decayed, scintillation fluid was added for counting in a liquid scintillation counter. Several labeled metabolites (B, C, D) are clearly visible in the ^{14}C chromatogram of the 90-min plasma sample, but had not formed in detectable quantities at the time of the first ^{11}C chromatogram at 15 min (Jones and Sargent, unpublished data).

In another group of experiments, the metabolic fates of $^{13}NO_2^-$ and $^{13}NO_3^-$ were determined in animals, bacteria, and plants [19]. In the animal experiments [17,18], blood samples and gut contents were analyzed by radio-HPLC to determine the fate of the label, as shown in Figures 6.7 and 6.8; plant extracts (J. Thayer and R. Huffaker, unpublished results) were analyzed in a similar manner, illustrated in Figure 6.9. The short half-life of ^{13}N (10 min) required a rapid sample preparation procedure and analysis.

Pre-HPLC sample processing with short-lived radionuclides was minimized in order to save time. The short half-lives of the ^{13}N and ^{11}C isotopes permitted very little time for plasma extraction or sample concentration

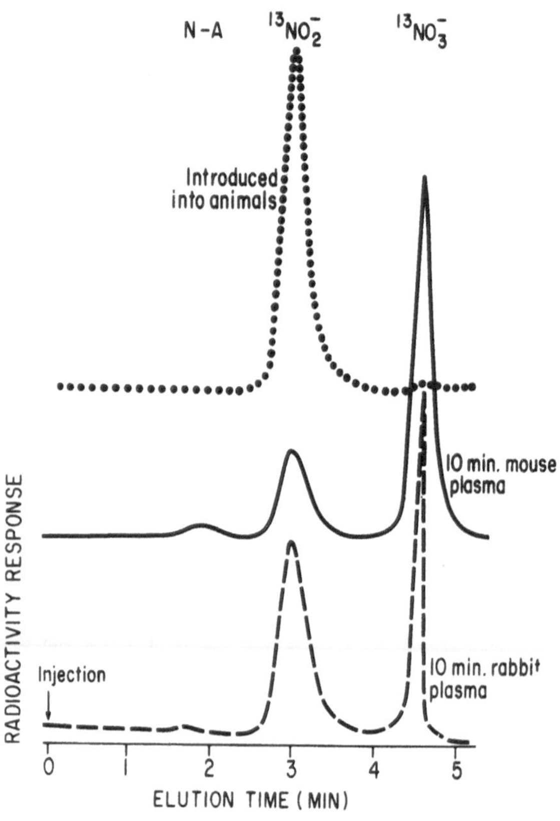

FIG. 6.7. Separation of the oxidation products of administered $^{13}NO_2^-$ in mouse and rabbit plasma with radio-HPLC [18]. The separation conditions are: Whatman 0.46 × 25 cm Partisil-10 SAX column eluted with a 30 *mM* sodium phosphate buffer (pH 3.1) and a flow rate of 3 ml/min. The figure legend is:

...... $^{13}NO_2^-$ administered to the animals
—— mouse plasma from blood drawn 10 min after lung instillation
--- rabbit plasma from blood drawn 10 min after IV injection

(From [18]. Copyright 1981 by the AAAS.)

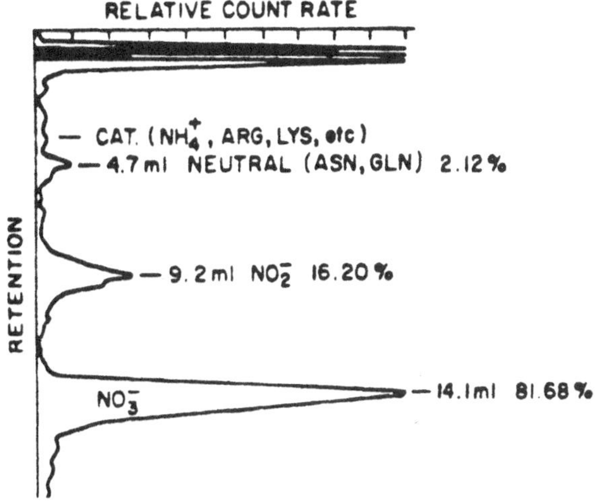

FIG. 6.8. Radio-HPLC of mouse small intestine contents after intratracheal instillation of $^{13}NO_2^-$ [17]. Column conditions are the same as in Figure 6.7.

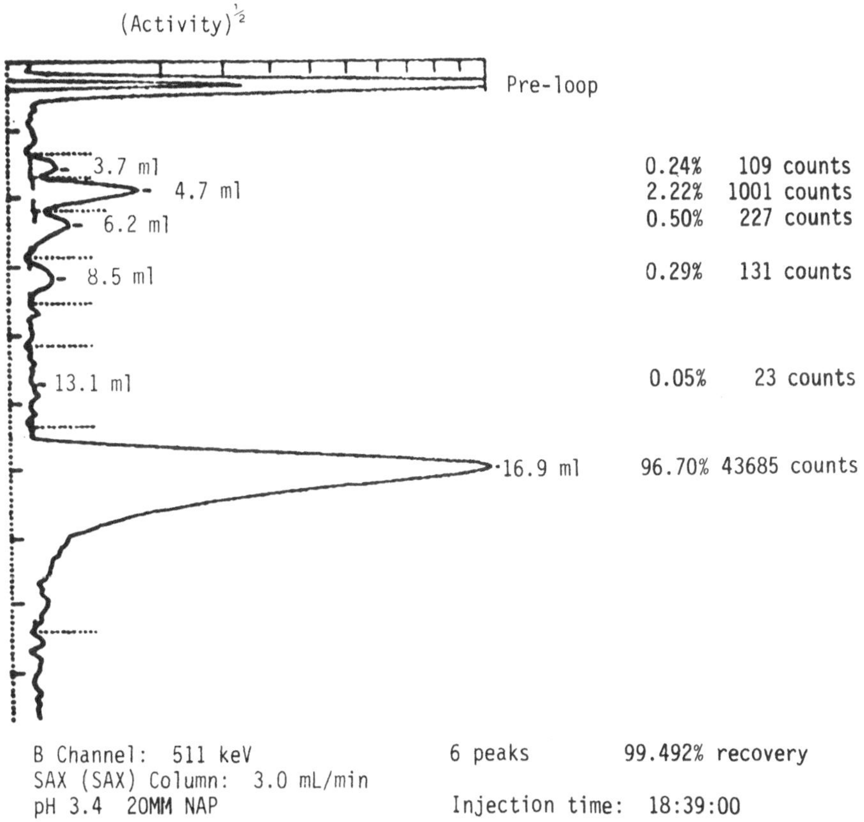

(Activity)$^{1/2}$

Pre-loop

3.7 ml	0.24%	109 counts
4.7 ml	2.22%	1001 counts
6.2 ml	0.50%	227 counts
8.5 ml	0.29%	131 counts
13.1 ml	0.05%	23 counts
16.9 ml	96.70%	43685 counts

B Channel: 511 keV 6 peaks 99.492% recovery
SAX (SAX) Column: 3.0 mL/min
pH 3.4 20MM NAP Injection time: 18:39:00

FIG. 6.9. Separation of ^{13}N-labeled ammonium, glutamine, glutamate, aspartate, nitrite, and nitrate from an extract of barley leaves incubated with radioactive nitrate for 20 min in the dark (Thayer and Huffaker, unpublished data). Note the square-root scale on the ordinate, which permits the display and analysis of small radiolabeled peaks.

procedures to improve sensitivity and extend the life of the HPLC column. A blood sample for HPLC injection was prepared in less than 5 min by centrifuging the whole blood to remove cells, precipitating large proteins by denaturation, and performing another centrifugation [18]. In addition to the denaturation and spin-down procedures, a guard column was used to extend the useful life of the more expensive analytical column. The minimal loss of sample resolution caused by the guard column was compensated by the extension of the lifetime of the analytical column.

It was necessary to inject as much as 0.5 ml of the denatured plasma supernatant sample onto the radio-HPLC so that metabolites could be detected an hour after the start of the experiment. This large injection volume required the use of relatively high capacity 0.46 × 25 cm columns to fully resolve peaks.

A radioactivity pre-loop was used to ensure that all radioactive peaks were eluted from the column or to quantitate the effects of a stripper column (Figure 6.4). The pre-loop was also used to check each injection for detector overload and deadtime loss. The minor loss in resolution attributable to the pre-loop was deemed to be much less important than radioactivity balance and overload checks.

The radioactivity detector (Harshaw crystals/bases and Ortec and Canberra electronics) was placed in-line before the UV detector since resolution of the radiolabeled compounds was considered to be more important than peak resolution detected by UV absorbance in these experiments. A 0.3-ml sample loop in the NaI(Tl) radioactivity detector was determined to be optimum for the 3 ml/min eluent flow rates, low sample activity, and metabolite product resolution for these analyses. The detection limit was approximately 1 nCi.

A complex metabolic product spectrum sometimes dictates that two different columns be used to completely separate all the labeled metabolites. Some compounds can be isolated by both assays and thus afford independent chromatographic verification. For short-lived metabolite studies, three parallel HPLC systems were constructed so that the different assays could be performed simultaneously, as shown in Figure 6.10. This obviated the requirement to switch and equilibrate the different columns with new eluent as would have been required if only one system were used. Three HPLC systems operating simultaneously allowed many samples to be analyzed in a short time.

For an application with two or more parallel HPLC systems operating simultaneously, the data acquisition system is an important consideration. A computer-based acquisition system was selected for this application with several factors in mind. The system had to be flexible and contain the following characteristics: at least four input/output interfaces, a multitasking realtime operating system, sufficient memory and data storage facilities, graphics display, hard copy plot printout, and software support for data acquisition and analysis.

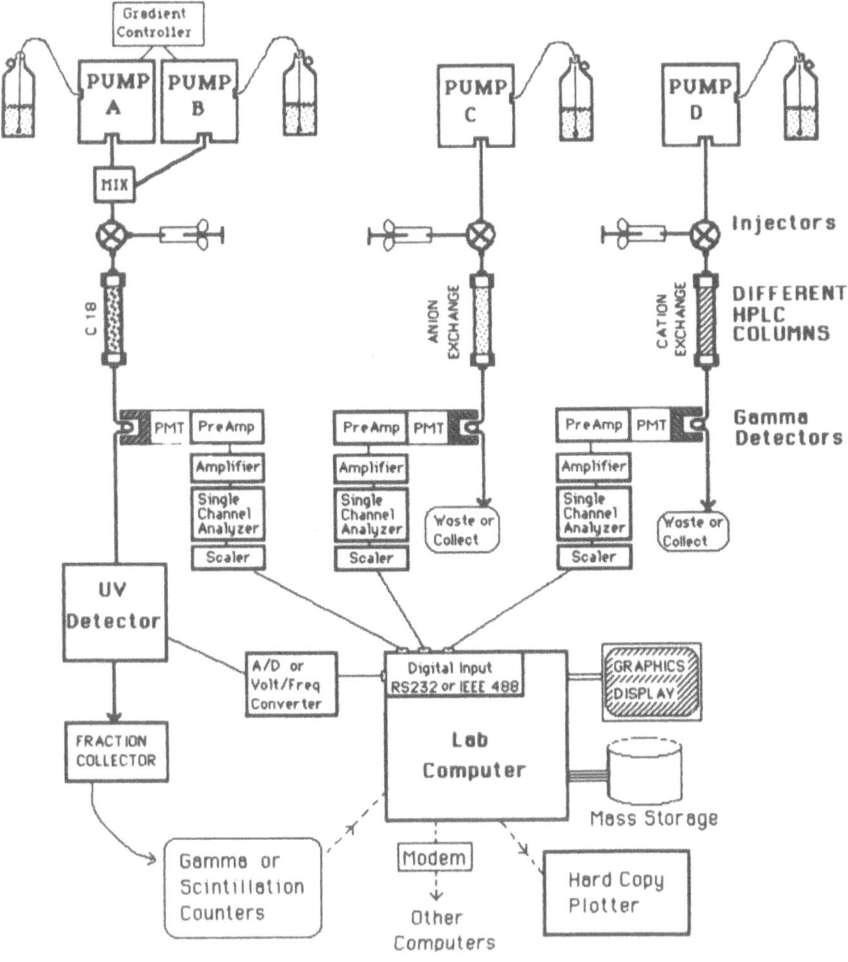

FIG. 6.10. Schematic of parallel radio-HPLC systems. If multiple systems are required for use with short-lived isotopes or large numbers of analyses, two or more chromatography systems can be run simultaneously. A computer-based data acquisition system may be essential for three or more parallel radio-HPLCs.

One system was designed which controlled the computer input from three HPLCs, three NaI(T1) detectors, a variable wavelength UV detector, and a radio-TLC scanner. A PDP 11/44 computer controlled all data collection and analysis under the RSX-11M multitasking, multiuser operating system. Data from gamma detectors were fed to the computer via external and internal scalers, and data from mass detectors (Waters UV/Vis and Wescan conductivity) were sent via a 16-channel internal A/D converter. Total capacity was 12 channels of TTL pulse-counting information and 8 channels of programmable analog inputs. Operation of data collection programs was "user-friendly." After log-in, the user selected the data acquisition type by activating an indirect command file which started the central task and the data processing task. The central data collection task monitored data collection devices and stored data in a multichannel rolling stack. Communication with sister (data handling) and daughter (processing and storage) tasks was accomplished via common storage in areas shared by various tasks, and synchronization was done with global flag events. A parameter file for each user or activity specified acquisition parameters: number and identification of channels stored, collection intervals, HPLC columns, flows, solvents, detectors, etc. The user specified the parameter file and data file names. Control and display of data acquisition was then controlled through keyboard commands. Up to nine separate users could simultaneously run different (or the same) acquisition tasks, and additional users could operate analysis tasks.

This complex system was designed at University of California, Davis to

FIG. 6.11. Photograph of parallel radio-HPLC systems. Modular HPLC design can make efficient use of space; the bench-top space shown in the photograph contains four injectors, four HPLC pumps, a gradient programmer, two UV detectors, and one fluorescence detector. The area behind the pumps contains three shielded NaI(T1) gamma detectors, and the area to the right of the pumps contains the associated NIM bin electronics, chart recorders, and computer terminal; the computer is located in an adjacent room. All the equipment shown in the photograph can be reached and serviced by one operator.

rapidly analyze large numbers of expensive ^{13}N-labeled metabolic products [17,18] and required years of programming development. It has the advantage of great power and flexibility, but required extensive custom programming and an expensive computer. Based on our experience, we advise that one of the commercially available "canned" systems be purchased (e.g., an IBM CS9000). The initial cost of a commercial system may be high and flexibility may be lost owing to the difficulty of modifying a canned system, but this may be compensated by the savings in time and overall cost realized by avoiding the development of a custom system.

SUMMARY

Chromatographic separations involve a compromise between the three factors of the chromatographer's triangle: analysis speed, sample capacity, and resolution. These considerations also affect the design and operation of a radio-HPLC system. In addition, compromises such as resolution vs. sensitivity and specialization vs. flexibility should be considered in design planning. The specific application of a system should determine its design characteristics. Some factors to consider in choosing a design include the use of: a gradient system, a pre-loop, a guard column, a stripper column, an appropriate radioactivity detector sample loop size, a mass detector, a fraction collector, parallel chromatographic systems, a computer acquisition system, and a modular design to fit space requirements. In radio-HPLC design "form follows function."

ACKNOWLEDGMENTS

The authors wish to thank J.R. Thayer and N.J. Parks of Crocker Nuclear Laboratory, University of California, Davis for permission to use Figures 6.4, 6.7, 6.8, and 6.9 and for helpful discussions. In addition, we would like to acknowledge the support and help of K.A. Krohn, T. Sargent III, and Y. Yano. This research was supported by NSF grant DEB 77-01199, California Air Resources Board grants A7-190-30 and A0-031-31, and by the Biology and Medicine Division of the Department of Energy under contract #DE-AC03-76SF00098.

REFERENCES

1. Snyder, L.R. and Kirkland, J.J.: *Introduction to Modern Liquid Chromatography*. Wiley-Interscience, New York, 1979.
2. Johnson, E.L. and Stevenson, R.: *Basic Liquid Chromatography*. Varian Associates, Palo Alto, 1978.
3. Kabra, P.M. and Marton, L.J.: *Liquid Chromatography in Clinical Analysis*. Humana Press, Clifton, New Jersey, 1981.

4. Yost, R.W., Ettre, L.S. and Conlon, R.D.: *Practical Liquid Chromatography*. Perkin-Elmer, Norwalk, Connecticut, 1980.
5. Reeve, D.R. and Crozier, A.: HPLC radioactivity monitors—fact and fiction. *Lab. Pract.* **32**(3):59–60, 1983.
6. Langstrom, B. and Lundqvist, H.: A β-particle flow detector for liquid chromatography. *Radiochem. Radioanal. Letters* **41**:375–382, 1979.
7. See for example: (a) *Guide to Scientific Instruments*, American Association for the Advancement of Science, or (b) *Analytical Chemistry Lab Guide*, American Chemical Society, for more complete lists of manufacturers of HPLC equipment and related hardware described in this chapter.
8. Zodda, J.P., Heineman, W.R., Gilbert, T.W. and Deutsch, E.: Quantitative determination of pertechnetate by high-performance liquid chromatography with UV detection. *J. Chromatogr.* **227**:249–255, 1982.
9. Needham, R.E. and Delaney, M.F.: Cadmium telluride γ-ray liquid chromatography detector for radiopharmaceuticals. *Anal. Chem.* **55**:148–150, 1983.
10. Gates, S.C.: Laboratory data collection with an IBM PC. *Byte* **9**(5):366–378, 1984.
11. Hallgren, R.C.: Putting the Apple II to work. Part 2: The software. *Byte* **9**(5):382–396, 1984.
12. Wyss, C.R.: Planning a computerized measurement system. *Byte* **9**(4):114–123, 1984.
13. Hettinger, J.D., Hubbard, J.R., Gill, G.M. and Miller, L.A.: A new computing integrator for chromatography. *J. Chromatogr. Sci.* **9**:710–717, 1971.
14. Suzuki, K.: Simultaneous monitoring of several radionuclides in a radio-liquid chromatography system. *Int. J. Appl. Radiat. Isot.* **35**:801–804, 1984.
15. Clune, T.R.: The IBM CS9000 lab computer. *Byte* **9**(2):278–290, 1984.
16. Chasko, J.H. and Thayer, J.R.: Rapid concentration and purification of ^{13}N-labelled anions on a high performance anion exchanger. *Int. J. Appl. Radiat. Isot.* **32**:645–649, 1981.
17. Thayer, J.R., Chasko, J.H., Swartz, L.A. and Parks, N.J.: Gut reactions of radioactive nitrite after intratracheal administration in mice. *Science* **217**:151–153, 1982.
18. Parks, N.J., Krohn, K.A., Mathis, C.A., Chasko, J.H., Geiger, K.R., Gregor, M.E. and Peek, N.F.: Nitrogen-13-labeled nitrite and nitrate: distribution and metabolism after intratracheal administration. *Science* **212**:58–61, 1981.
19. Krohn, K.A. and Mathis, C.A.: The use of isotopic nitrogen as a biochemical tracer. In *Short-Lived Radionuclides in Chemistry and Biology*, Root, J.W. and Krohn, K.A., eds., American Chemical Society, Washington, D.C., 1981, pp. 233–249.

CHAPTER 7

Quantitation of Radiolabeled Molecules Separated by High Pressure Liquid Chromatography

MICHAEL J. KESSLER

INTRODUCTION

The technique of high performance liquid chromatography (HPLC) has developed the last 15 years into a rapid and accurate method of separating and quantitating chemical and biological molecules. This quantitation has been easily accomplished for compounds that absorb in the ultraviolet [1] or visible spectrum [2], have different electrochemical properties [3], have specific fluorescent properties [4], are quantitated by a mass sensitivity detector [5], or are quantitated in a mass spectrometer. However, the accurate method of quantitating the radioactive molecules as they elute from the HPLC has been very difficult, costly, and time consuming.

The original method of quantitating these radioactive molecules was the use of the fraction collector/scintillation counter technique [6]. This technique is still used routinely in many laboratories. This involved the setting up of a fraction collector on the exit line of the HPLC system after the last detector. Fractions are collected (0.05–2.0 ml) at preset time intervals (0.1–2.0 min). The tubes containing the collected fractions are removed from the fraction collector after the end of the HPLC run. A specific aliquot (0.01–2.0 ml) is accurately removed with a pipette from each tube and placed into individual scintillation vials. A scintillation solution is added to each vial, the

---

Abbreviations: high performance liquid chromatography, HPLC; counts per minute, cpm; disintegrations per minute, dpm; gas chromatography/mass spectrometry, GC/MS; radioimmunoassay, RIA; photomultiplier tube, PMT; time, T; count rate, R; percent of standard deviation, %SD;

vials are capped and shaken, then the vials are labeled and transported to a scintillation counter where each fraction is quantitated. The data are removed from the counter, and the data obtained from each fraction are graphed. The total radioactivity of each peak is then determined by integrating the area of each. This procedure is a very long, tedious, and costly method for quantitating each peak for every sample analyzed on the HPLC. This method severely restricts the use of radioactivity to identify and quantitate molecules as they elute from the HPLC.

The technique of flow-through counting for on-line quantitation of radioactive peaks eluting from the HPLC was introduced in the early 1960s [7,8]. The original technique made use of a solid anthracene flow cell for the quantitation of the radioactive molecules. The major problems with this technique were that it was very difficult to quantitate low energy beta emitters, the solid flow cell easily became contaminated with radioactivity, and the background was relatively high. These factors contributed to both a rising base line and lower sensitivity.

The next major improvement in the on-line radioactivity detection technology was the use of the homogeneous flow cell [9,10]. This cell mixed the HPLC eluent with a nongelling scintillation solution and quantitated the radioactive peaks using the liquid scintillation counting technique. This allowed the user to obtain greatly improved sensitivity for low energy beta emitters (^{3}H), similar to that obtained by conventional liquid scintillation counters. It also eliminated the problem of solid cell contamination by not using a solid support. The major problem accompanying this technique was that the complete HPLC effluent was mixed with scintillation solution, thus no further chemical or biological techniques could be performed. At about the same time, an electronic stream splitter was introduced to circumvent this problem. This allowed the user to split off a portion (1–99%) of the HPLC eluent to be saved in a fraction collector with the remainder quantitated in the flow-through radioactivity detector. This electronic stream splitter enables the user to obtain the high sensitivity of the homogeneous liquid cell and to recover a certain percentage of the sample for further chemical or biological analysis.

DETECTION OF HPLC-SEPARATED RADIOACTIVE COMPOUNDS

The basic theory of operation used by the flow-through detector and the static counting of collected fractions for quantitating radioactive molecules eluting from the HPLC is scintillation counting. This technique has been used for over 30 years for the quantitation of radionuclides in a static counting system [11]. Scintillation counting involves the transfer of the energy from the beta particle by one of two methods [12]. The first is heterogeneous or solid scintillation counting. This involves the direct transfer of the energy from a beta or gamma particle, emitted by the radionuclide, to a scintillator.

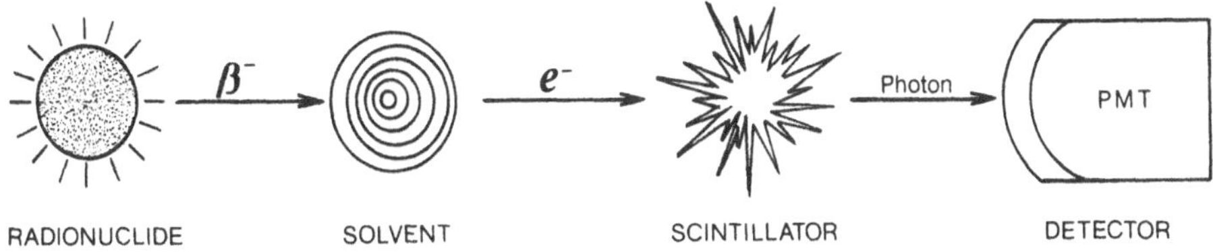

FIG. 7.1. The basic theory of the liquid scintillation counting process.

The scintillator molecule is activated to a higher energy level and emits a blue or violet light flash when it returns to the ground state. The second method is that of liquid scintillation counting, shown in Figure 7.1 [13]. This involves the transfer of energy from a beta particle to activate a solvent molecule. The activated solvent then transfers its energy to a scintillator molecule, which emits a light flash as with the solid scintillator technique. The intensity of the light flashes is directly proportional to the energy of the beta particle emitted by the radionuclide.

Now that the general theory of detection of the flow-through radioactivity detector has been investigated, it is necessary to understand how flow counting differs from standard static scintillation counting.

An excellent model for examining how flow counting is accomplished is shown in Figure 7.2. This model assumes that the light flashes (radioactivity) exist in discrete segments (radioactive peaks) on a train going down the train track at a specific speed (flow rate of HPLC and scintillation fluid). A specific observer (photomultiplier tube) is standing in a train station and counting the light flashes. Two parameters must be determined for the flowing system: background counts and sample counts. In the flowing system, these two are determined in different manners. The background count is composed of electronic noise, cosmic radiation, or radioactivity from the HPLC. These counts are independent of the flow rate of the liquid through the counting cell. This is like a flashing light standing just outside the train station on a stationary pole (not dependent on the speed of the train). The counts seen by the observer as the train passes the station are not static but dependent on two parameters: flow rate (speed of the train) and cell size (size of the train station). These parameters determine the residence time of the sample in the cell, and are used in calculating the actual number of cpm seen for each peak in the sample. The residence time is defined as the cell

FIG. 7.2. The basic theory of on-line radioactivity flow detectors illustrated with the use of a train model.

volume (ml) divided by the flow rate (ml/min) of the sample through the flow cell:

$$\text{Residence Time} = \frac{\text{Cell Volume}}{\text{Flow Rate}} \tag{1}$$

To determine the number of cpm seen by the observer at the station, the following equation is used:

$$\text{cpm} = \frac{\text{Counts} - (\text{Background} \times \text{Residence Time})}{\text{Residence Time}} \tag{2}$$

Therefore, if the observer recorded 30 flashes from the light on the stationary pole at the station in one minute (background), 1015 flashes from the train (flowing sample) with a speed of 5.0 ml/min and a 2.5-ml flow cell size, the number of cpm in the sample would be:

$$\frac{1015 - (30 \times 0.5)\ \text{counts}}{0.5\ \text{min}} = 2000\ \text{cpm} \tag{3}$$

It is important that the background counts are subtracted from the counts before being divided by the residence time, otherwise the background counts are amplified by the residence time of the sample in the flow cell. In general, for subtraction of background in the flowing system, 1.5 to 2.0 times the static background counts are used to eliminate statistical variation of background, which could be seen as low level peaks of radioactivity. With the reprocessing capabilities of the computer system, various backgrounds can be subtracted in order to obtain the best results for each sample quantitated in the flow detector. In order to eliminate any statistical randomness of obtaining counts at short time intervals (2 or 6 sec), special smoothing routines are commonly used in order to smooth the data. These eliminate spurious highs and lows in the background during the HPLC run and during graphing of the resultant data. The effect of using the smoothing routine is shown in Figure 7.3. As can be seen, the nonsmoothed graph shows a ragged base line, whereas the base line on the smoothed graph is very stable. The smoothing routine does tend to cause an apparent loss of resolution of the peaks as they elute from the HPLC. The results using the flow-through radioactivity detector can be expressed as either smoothed or nonsmoothed results since the raw data is stored on disk, therefore, the operator has the option of choosing if the results are to be smoothed using a 3-, 5-, or 11-point weighted, moving average smoothing routine.

The final step in the data reduction process is conversion of the cpm, which represents the number of counts seen by the detector in one minute, to the actual number of disintegrations per minute (dpm) present in the sample. The proportion of number of cpm in the sample to the number of dpm actually present in the sample represents counting efficiency and can be

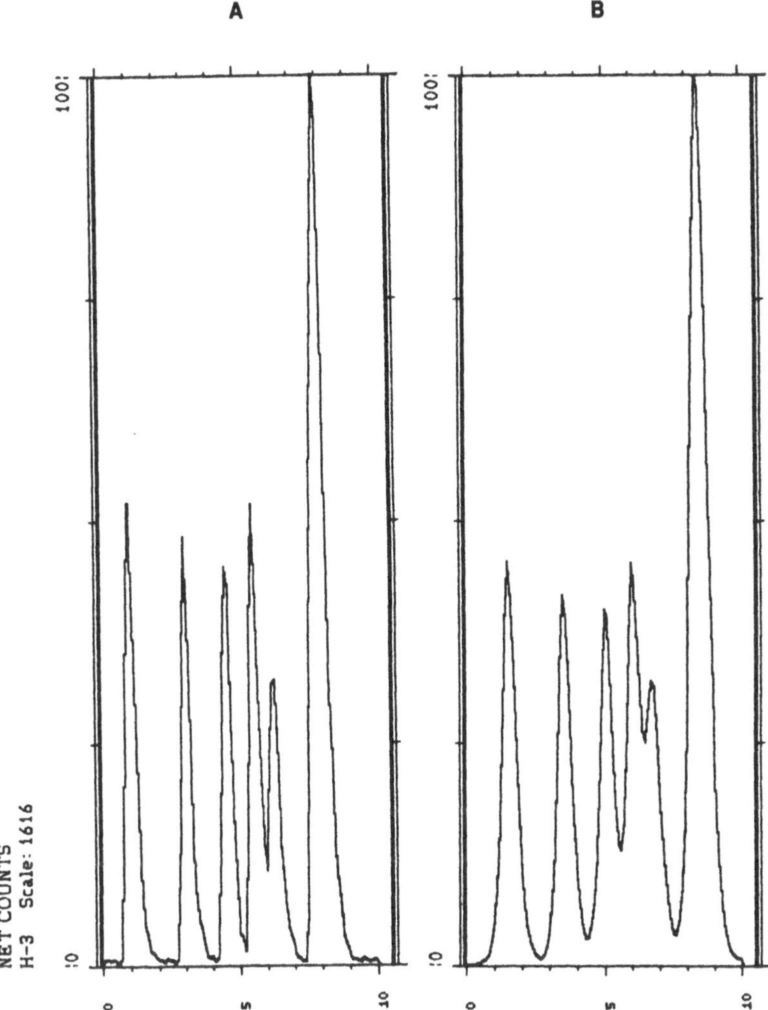

FIG. 7.3. The graphic results of an ^{3}H-progesterone incubation with a human testicular biopsy chromatographed on a reversed-phase ^{18}C column-OSD II (0.45 × 25.0 cm). The chromatogram was developed with a 10-min gradient of 50% methanol/50% H_2O to 100% methanol in 15 min at a flow rate of 1.0 ml/min. The column effluent was mixed with liquid scintillation (FLO-SCINT II) at a ratio of 3 : 1 (FLO-SCINT II: HPLC effluent). Chromatogram A represents the raw nonprocessed data plotted directly. Chromatogram B represents the same data with a special 11-point smoothing algorithm used to eliminate the variability (randomness) of radioactivity counting.

calculated by this equation:

$$\frac{\text{cpm}}{\text{dpm}} \times 100 = \text{Counting Efficiency} \tag{4}$$

This counting efficiency can be easily determined by obtaining a sample containing a known number of actual dpm and determining the cpm in the flow cell of the radioactivity detector using the same conditions (HPLC solvent and scintillation solution) employed during the analysis of the samples.

Two different types of separation methods exist for HPLC: isocratic and gradient elution. For calculating dpm in the isocratic elution mode, the counting efficiency is the same throughout the entire HPLC run and the above calculation can be used. During gradient elution on the HPLC, the counting efficiency can change from the beginning to the end of the run. Thus, for the accurate determination of the dpm in the sample, the efficiency must be calculated at various times during the gradient elution. These efficiencies are stored, as a gradient quench curve, and used to calculate the number of dpm in any peak at any time during the HPLC run. This curve is stored on disk and can be used at any time dpm are required when a particular gradient for an HPLC run is used.

DETECTOR DESIGN

A block diagram of the basic components of a flow-through radioactivity detector is shown in Figure 7.4. The detector is composed of several basic components: (1) counting circuitry, (2) keyboard controlled devices, (3) peripheral components (printer and terminal), and (4) computer for data reduction. Each of these components is examined in detail.

First, the counting circuitry is the very heart of a flow-through scintillation counter; it is where the light flashes from the scintillator are quantitated and sorted based on their intensity. This counting circuitry has the following components: (1) flow cell, (2) detection device, (3) amplification system, and (4) pulse-height analyzer. Each of these components, along with its function, is discussed.

The flow cell (#4, Figure 7.4) is the location where the radioactivity from the HPLC is quantitated by either solid or liquid scintillation counting technique. Three basic flow cell types, shown in Table 7.1, are used in flow-through radioactivity detectors: (1) solid scintillator, (2) liquid scintillator, and (3) high energy isotope.

Three solid scintillator (heterogeneous) cell types are available. Each has a specific use depending on the HPLC solvent used, the isotope to be quantitated, and the sensitivity required. The major advantage of the solid scintillator is that the radioactivity can be quantitated directly (without the addition of scintillation solution). Three major disadvantages exist when using the

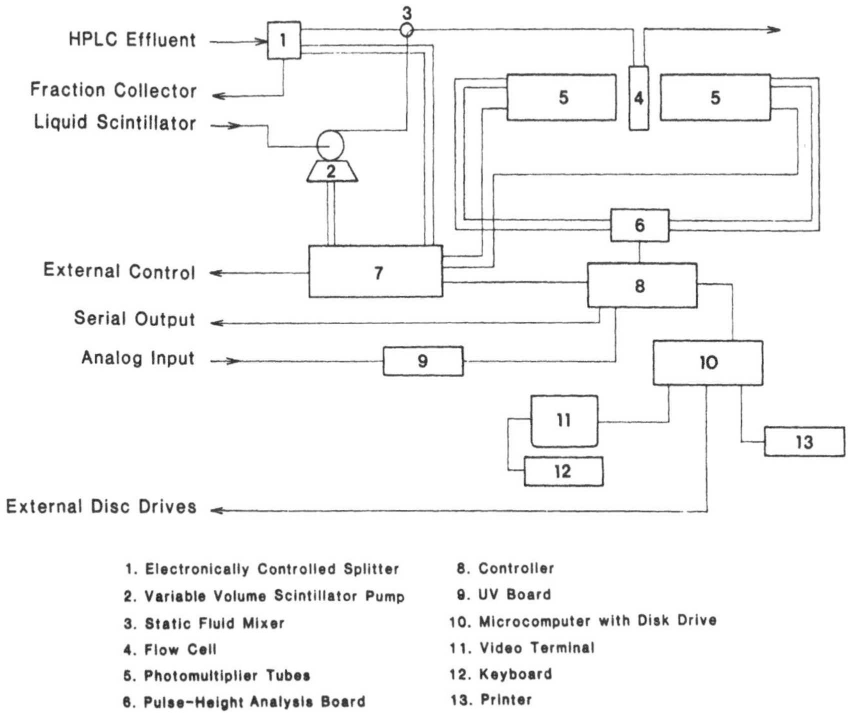

FIG. 7.4. A block diagram of the FLO-ONE®\Beta flow-through radioactivity detector. (Radiomatic Instruments.)

heterogeneous cell. First, the overall sensitivity (counting efficiency) of the solid scintillator for low energy beta emitters is very low (^3H, <1%, ^{14}C, 10–65%). Second, the radioactive molecules bind reversibly to the solid support. This is partially due to the fact that the solid support (glass, etc.) is similar in composition (silica) and particle size (70–200 μm) to the solid support used in many HPLC columns. This causes problems with radioactivity from a previous HPLC run eluting from the solid scintillator at any time during subsequent HPLC analyses. Third, the radioactive molecules bind irreversibly to the solid scintillator. This results in a radioactive contamination of the solid support, which causes an increase in the background and eventually requires the complete replacement of the solid scintillator. Peptides, steroids, lipids, phospholipids, and proteins are common examples of radioactive molecules that bind either reversibly or irreversibly to solid scintillators. Therefore, the solid scintillator cell is most applicable for use by researchers with higher energy beta emitters (^{32}P, ^{14}C, ^{35}S, etc.), large numbers of counts, and containing molecules that do not bind or adsorb to the solid scintillator. This cell is best used for a quality control application where the type of molecules, retention times on HPLC, and number of peaks are consistently the same and reproducible from analysis to analysis.

TABLE 7.1. Types of Flow Cells for the Flow-Through Radioactivity Detector

A. Type 1: Solid Scintillator (Heterogeneous) Cell
   1. Basic types
      Plastic scintillator cell (20–25% efficiency for ^{14}C)—aqueous solvent system only.
      Glass scintillator cell (20–25% efficiency for ^{14}C, 1% for ^{3}H).
      Europium-activated calcium fluoride (75–80% efficiency for ^{14}C, 5% for ^{3}H).
   2. Advantages for solid scintillator cell:
      High salt or buffer concentrate doesn't affect counting efficiency.
      Nondestructive to sample.
   3. Disadvantages of solid scintillator cell:
      Samples may bind reversibly or irreversibly to solid support.
      Low counting efficiency of low energy beta emitter (^{3}H).

B. Type 2: Liquid Scintillator (Homogeneous) Cell
   1. Basic type—teflon tubing.
   2. Advantages of liquid scintillator cell:
      High counting efficiency (75–95%) for high energy beta emitters (^{14}C, ^{35}S, ^{32}P, etc.).
      High counting efficiency (20–60%) for low energy beta emitters (^{3}H).
      No adsorption or contamination of teflon lines.
   3. Disadvantages of liquid scintillator cell:
      The samples must be mixed with a nongelling scintillation fluid, thus no further chemical or biological studies of the material are possible except when an electronic stream splitter is used.

C. Type 3: High Energy Isotope Cell
   1. Basic type—teflon tubing with specially designed cell.
   2. Advantages of cell type:
      No scintillation cocktail required.
      No solid support required.
      High counting efficiency for high energy beta and gamma emitters ^{32}P (40–60%), ^{125}I (10–20%), ^{99m}Tc (20–50%).
   3. Disadvantages of cell type:
      Medium energy beta emitters counted with low efficiency (0.05%) ^{14}C.

The second type of flow-through cell, shown in Table 7.1, is the liquid scintillator (homogeneous) cell. This cell type requires the mixing of the HPLC eluent at a special mixing tee (#3, Figure 7.4) with a special nongelling scintillation solution supplied by a digitally controlled pump (#2, Figure 7.4). The mixed (homogeneous solution) is then counted, on-line, as it passes through the flow cell. This flow cell type has the advantage of exposing the mixture in the flow cell to only teflon tubing, and thus preventing contamination as seen for the solid scintillator. High counting efficiencies are obtained for low energy (^{3}H, 20–60%), high-energy (^{14}C, ^{35}S, 50–95%, ^{32}P, 80–95%) beta emitters, and low energy gamma emitters (^{125}I, 50–90%). The only disadvantage of the liquid scintillator mixing cell is that the radioactive molecules, once they elute from the HPLC, are mixed with scintillation fluid and cannot be recovered for further chemical or biological studies (GC/MS, RIA, enzyme assay, biological assay, or recrystallization). This problem can be circumvented in the flow detector by the use of the built-in electronic stream splitter (#1, Figure 7.4). The eluent from the HPLC is proportioned (1–99%) by the binary electronic stream splitter to the flow detector. This binary splitting technique proportions the sample accurately between two ports, fraction collector and radioactivity detector, regardless of flow rate, viscosity, or resistance of the HPLC flow. Thus, a portion of the radioactive molecules eluting from the HPLC can be collected in a fraction collector, with the remainder quantitated in the flow-through radioactivity detector.

The third type of flow cell is the high energy isotope cell. This cell incorporates the advantages of both the first two cell types in that no solid scintillator or liquid scintillation solution is required for quantitation. This is accomplished by using a standard homogeneous flow cell with a scintillator plastic situated between the teflon coil and the PMTs. The isotopes that can be counted have enough energy to pass through the teflon cell and interact with the scintillator. High counting efficiencies for ^{32}P (40–60%), ^{125}I (10–20%), and ^{99m}Tc (20–50%) are obtained with this flow cell type. Millicurie levels of ^{14}C can also be quantitated using this flow cell type even though the counting efficiency is only 0.05%; thus, it is ideal for preparative separation of ^{14}C-radiolabeled molecules by HPLC. The only disadvantage is that low energy beta emitters cannot be quantitated using this type of flow cell.

The next component of the counting system is the detection device. This is composed of a high voltage power supply (#7, Figure 7.4) and two photomultiplier tubes (#5, Figure 7.4). The primary role of the photomultiplier tubes is to detect light flashes produced by the scintillator. The number of light pulses (3–4 for ^{3}H) is amplified to as high as 150 million in a typical 13-dynode photomultiplier tube. Because of this large amplification factor, the high voltage used to drive the photomultiplier tube must be extremely stable. A 1% change in high voltage can cause as much as a 10% change in response of the photomultiplier. The electronic pulse produced by the photomultiplier tube is then altered (widened), and the pulse distortion removed by a preamplifier. The pulses are then further amplified and shaped by a linear amplifier. The signals from both photomultiplier tubes are then summed, thus reducing the statistical fluctuation in the number of light flashes reaching each tube and producing a less distorted pulse-height spectrum. This spectrum is directly proportional to the energy of the beta particle and the intensity of the light flash seen by the counting circuit. The pulses are then sent from the pulse summation circuitry to the pulse-height analyzer. The pulses are sorted on the basis of their pulse heights/energy of beta particle emitted. This is done by the differential pulse-height analyzer (#6, Figure 7.4). This is composed of two discriminator circuits connected to a coincidence circuit. The two discriminators are used to set windows for the size of the pulses to be counted. The pulse must be above the lower discriminator setting, but below the upper discriminator. Once the pulse meets these two criteria it is sent to the coincidence circuit where the electronics indicate whether both photomultipliers see the light flashes at the same time. If the pulse meets the third criterion, it is counted by the counting circuitry and sent to the controller and microcomputer (#8, Figure 7.4), or a digital ratemeter.

In addition to the counting circuitry, several special optional features can be found in the most sophisticated flow-through radioactivity detectors. These include a keyboard controllable (#11, 12, Figure 7.4) scintillation solution pump (flow rate control), electronic stream splitter, and electronic discriminator settings for isotope selection. These features eliminate the previous problems of manually adjusting all three parameters and allow for a more accurate and complete control of the flow-through liquid scintillation system.

Finally, a sophisticated computer system (#10, Figure 7.4) is incorporated into the system to allow the user to obtain and store data permanently on disk, to plot a graph (#13, Figure 7.4) of the on-line results, to reprocess the stored data with the ability to change various graphic or integration parameters. A complete data reduction package includes permanent storage of the data on disk, camera-ready graphs; processing the data in real time or reprocessing at some later time; count and time scale expansion; peak integration with retention time and percentage of run calculations; variable background value subtraction; dpm correction for results using quench curve correction efficiencies or static efficiency calculations; one-, two-, or three-channel plot of two radioisotope channels and one UV channel; and peak area rejection for elimination of minor peaks from the integrated total. All of these features have been added in order to facilitate the most complete data reduction package available while using an internal computer system. An example of the graphic and data reduction presentation is shown in Figure 7.5.

CONSIDERATIONS IN USE OF DETECTOR

The basic theory of radioactivity detectors and flow-through counts has just been discussed. Now, the use of the flow-through radioactivity counter for the quantitation and separation of radioactive molecules as they elute from the HPLC is examined. The following topics are discussed in detail: (1) resolution, (2) sensitivity, and (3) reproducibility.

First, the resolution that can be obtained with the flow-through radioactivity detector was investigated. The resolution is dependent upon three basic factors: (1) total flow rate, (2) flow cell size, and (3) update time—the detector pulses are summed over a particular time interval. Figures 7.6 and 7.7 demonstrate how changing the cell size and update time can considerably change the resolution of peaks by the flow-through radioactivity detector. Figure 7.6 shows the injection of seven radioactive peaks (^{3}H) separated by the following time intervals: 120, 90, 60, 45, and 30 sec. Figure 7.6A illus-

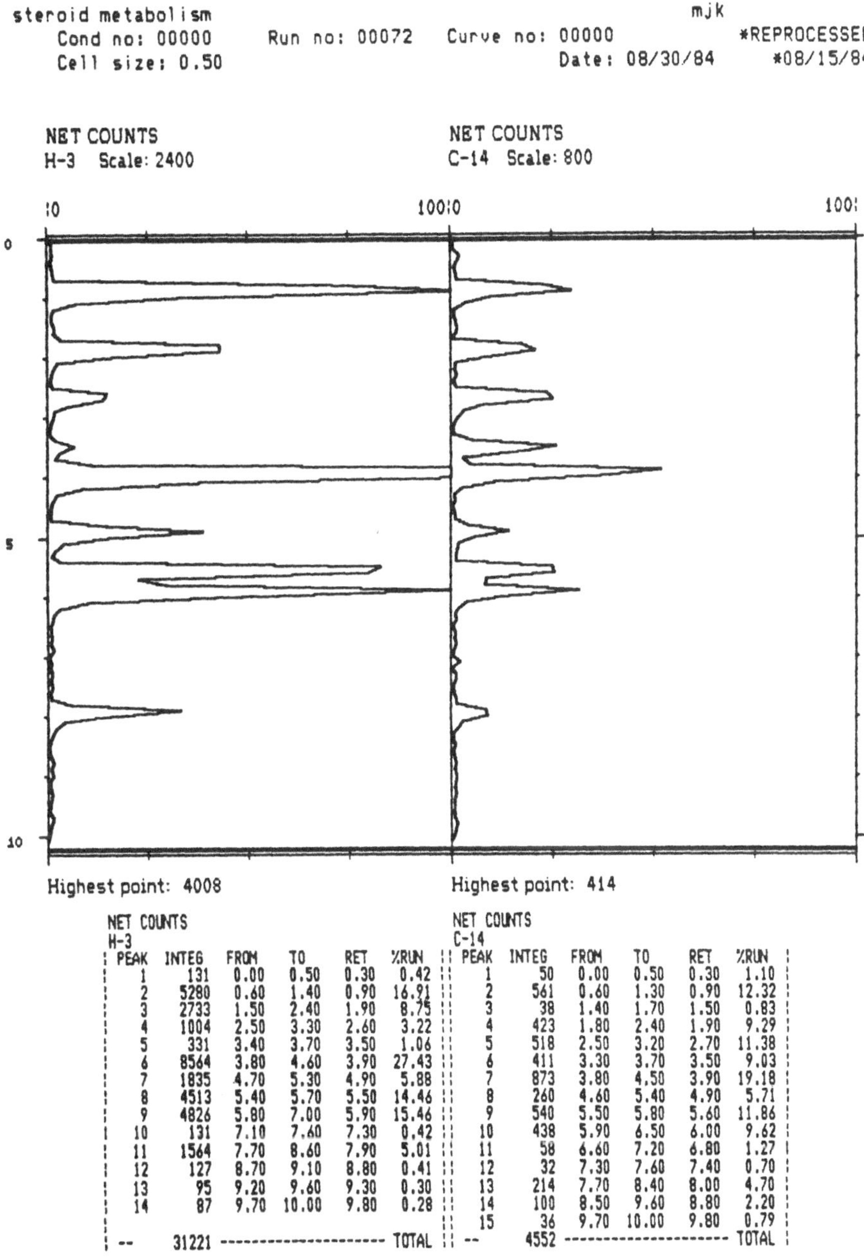

NET COUNTS
H-3

PEAK	INTEG	FROM	TO	RET	%RUN
1	131	0.00	0.50	0.30	0.42
2	5280	0.60	1.40	0.90	16.91
3	2733	1.50	2.40	1.90	8.75
4	1004	2.50	3.30	2.60	3.22
5	331	3.40	3.70	3.50	1.06
6	8564	3.80	4.60	3.90	27.43
7	1835	4.70	5.30	4.90	5.88
8	4513	5.40	5.70	5.50	14.46
9	4826	5.80	7.00	5.90	15.46
10	131	7.10	7.60	7.30	0.42
11	1564	7.70	8.60	7.90	5.01
12	127	8.70	9.10	8.80	0.41
13	95	9.20	9.60	9.30	0.30
14	87	9.70	10.00	9.80	0.28
--	31221	----------	----------	----------	TOTAL

NET COUNTS
C-14

PEAK	INTEG	FROM	TO	RET	%RUN
1	50	0.00	0.50	0.30	1.10
2	561	0.60	1.30	0.90	12.32
3	38	1.40	1.70	1.50	0.83
4	423	1.80	2.40	1.90	9.29
5	518	2.50	3.20	2.70	11.38
6	411	3.30	3.70	3.50	9.03
7	873	3.80	4.50	3.90	19.18
8	260	4.60	5.40	4.90	5.71
9	540	5.50	5.80	5.60	11.86
10	438	5.90	6.50	6.00	9.62
11	58	6.60	7.20	6.80	1.27
12	32	7.30	7.60	7.40	0.70
13	214	7.70	8.40	8.00	4.70
14	100	8.50	9.60	8.80	2.20
15	36	9.70	10.00	9.80	0.79
--	4552	----------	----------	----------	TOTAL

FIG. 7.5. The graph and data obtained from the incubation of ^{3}H-progesterone with Sertoli cell culture for 24 hr with the addition of a mixture of ^{14}C recovery steroids. The same chromatographic conditions as in Figure 7.3 were used. The plot and data reduction of the two isotopes are shown as displayed on the printer.

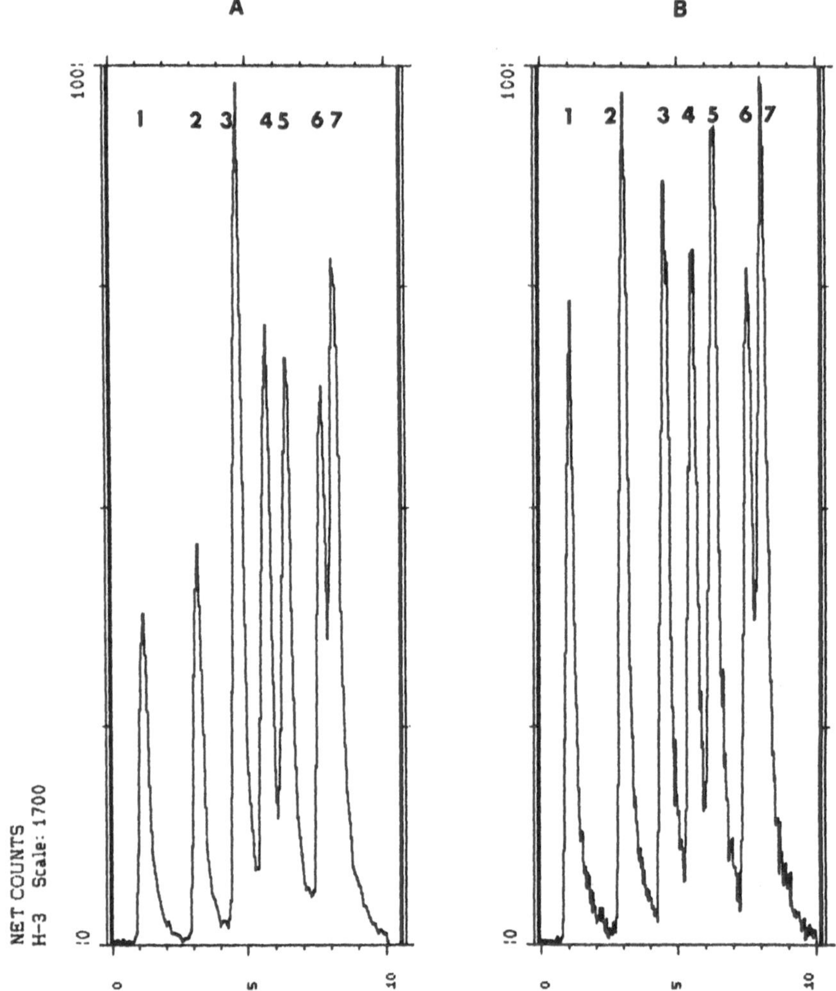

FIG. 7.6. A peak resolution study using the flow-through radioactivity detector with a 2500-μl homogeneous flow cell. An ^{3}H-testosterone sample was injected at intervals of 120, 90, 60, 45, and 30 sec into a 0.45 × 25 cm ODS II column having a flow rate of 1.0 ml/min of methanol. The HPLC effluent was mixed with liquid scintillation solution (FLO-SCINT II) at a ratio of 4:1 (FLO-SCINT II: HPLC effluent). Chromatogram A presents a graphic plot of the results when a 6-sec update time was used; Chromatogram B represents a 2-sec update time. Peaks 1–2 are 120 sec apart, 2–3 (90 sec), 3–4 (60 sec), 4–5 (45 sec), and 6–7 (30 sec).

trates the resolution obtainable using a 2500-μl liquid cell and a 5.0 ml/min total flow rate through the cell, with a 6-sec update time. As can be seen from this figure, a peak separation of 1.5 min is required for base-line resolution of both peaks when equal amounts are injected. Peaks 45 sec apart have a resolution of about 80% base line, and 30 sec shows one-third base-line resolution using these conditions. This resolution comparison is done using

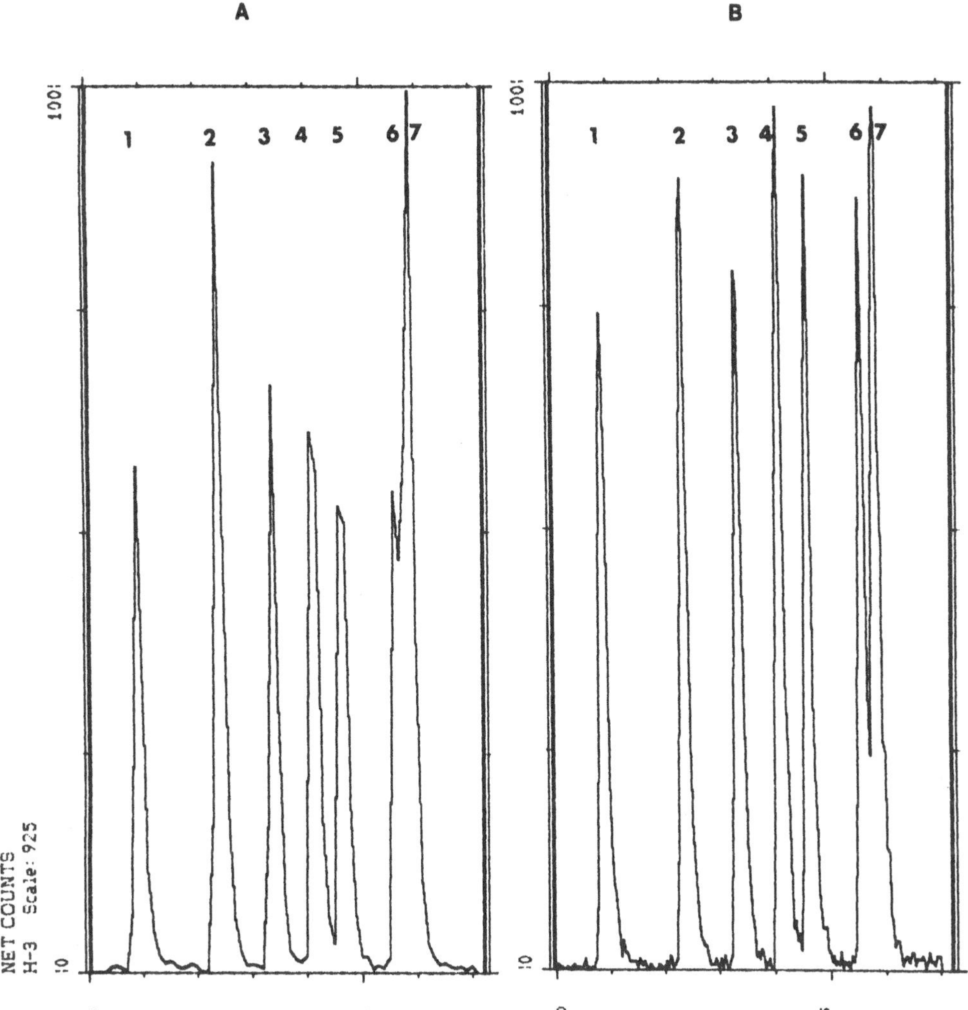

FIG. 7.7. A peak resolution study using the flow-through radioactivity detector with a 500-μl homogeneous flow cell. An ^{3}H-testosterone sample was injected at intervals of 90, 60, 45, 30, and 15 sec using the same conditions as in Figure 7.6. Peaks 1–2 are 90 sec apart, 2–3 (60 sec), 3–4 (45 sec), 4–5 (30 sec), and 6–7 (15 sec).

raw counts, and nonsmoothed data, the smoothing routine (if used) (Figure 7.3) would cause further loss of resolution. Figure 7.6B illustrates the use of a 2-sec update with the same flow rate and cell size. Better resolution is obtained with only 1.0-min separation required for base-line resolution and 45-sec separation showing a 90% of base-line resolution, with about 50% base-line resolution with 30-sec peak separation.

The next parameter to be investigated was changing the flow cell size while keeping the total flow rate at 5.0 ml/min. This can be seen in Figure 7.7 using a 500-μl cell, 5.0 ml/min flow rate, and either a 6- or 2-sec update time.

Figure 7.7 shows seven ^3H peaks injected 90, 60, 45, 30, and 15 sec apart. Note that the time scale has been expanded by a factor of two in order to see better resolution for the peaks 15 sec apart. As can be seen from Figure 7.7, peaks at both 1 min and 30 sec can easily be base-line resolved using a 2- or 6-sec update time, but the 15-sec peaks can be only partially resolved using a 6-sec update time. The 15-sec separated peaks are over 75% base-line resolved using the 2-sec update time. As can be seen, changing the cell size and update time can easily change the peak resolution obtained with the flow-through radioactivity detector. Therefore, it is important to choose the proper flow rate and cell size to obtain the best resolutions for the radioactive peaks eluting from the HPLC without sacrificing the sensitivity of the flow detector. The sensitivity is immensely dependent upon the residence time, thus better resolution in the flow detector requires some loss of sensitivity.

Third, the sensitivity and reproducibility of the flow-through radioactivity detector was examined. Radioactivity is a random event, not a continuous output of radioactivity decays. Thus, statistics are important in determining the sensitivity of the flow-through radioactivity detector and in understanding radioactive counting. Without going into much detail, the following equation:

$$T = \frac{1}{R}\left(\frac{100}{\%\mathrm{SD}}\right)^2 \tag{5}$$

with R equaling the count rate, %SD is the desired level of precision, and T (minutes) is the time required to reach that level of precision. Therefore, for the flowing system, a calculation can be done to determine how many counts or what length of time is required to count a sample with a certain standard deviation as seen in Table 7.2. If a 5.0 ml/min flow rate and 500 μl cell were used, the radioactivity would be observed for 6 sec, thus a %SD of 2% could be obtained with 25,000 cpm, or for 1000 cpm, a %SD of 10. If the cell size is increased to 2.5 ml and a 4.0 ml/min flow rate, the residence time would be 37.5 sec. Therefore, the residence time of 37.5 sec would result in 640 cpm

TABLE 7.2. Count Times Required to Obtain Various Counting Precisions (% Standard Deviations)

Counts cpm	*%SD(10) Time (sec)*	*%SD(5) Time (sec)*	*%SD(2) Time (sec)*
50,000	0.12	0.48	3
10,000	0.6	2.4	15
5,000	1.2	4.8	30
1,000	6.0	24.0	150
250	24.0	96.0	600

with a %SD of 5, and 160 cpm would give a %SD of 10. The results agree with those obtained earlier using multiple injections of various samples of ^3H and ^{14}C containing 10 to 60,000 cpm (%SD = 2.5), and for samples of ^3H of 350 dpm [14], a %SD = 13 was obtained. This agrees with the theoretical result obtained using the above count-time equation.

One other factor greatly influences the sensitivity of counting, that is the type of cell used in the flow counter. A sample of 150,000 dpm of ^3H and 8800 dpm of ^{14}C are injected on a solid scintillator (calcium fluoride) 0.250 ml cell with an HPLC flow rate of 1.0 ml/min. The same samples were injected into a 0.5 ml liquid (homogeneous) cell with 1.0 ml/min HPLC and 1.0 ml scintillation solution. These conditions provide both samples with the same residence times, thus, the total cpm should be the same if the sensitivity of the two cell types is the same. Figure 7.8 shows the results of these injections. For the solid cell, 1500 cpm resulted from the 150,000 dpm ^3H (% efficiency = 1.00), 6000 cpm for ^{14}C, or 68% efficiency. For the liquid flow cell, the ^3H efficiency has increased to 38,000 cpm of the total 150,000 dpm, or 25% efficiency, and 7600 cpm for ^{14}C (86% efficiency). Therefore, as can be seen from the previous results and Figure 7.8, the solid cell is very inefficient for quantitation of low energy beta emitters (^3H).

APPLICATIONS

After investigating both resolution and sensitivity parameters, it is important for the researcher to decide how and what parameters must be set to obtain the proper resolution and sensitivity for each particular application.

Next, the subjects of applications and what isotopes can be quantitated and the analysis of double-labeled samples are considered. First, the flow-through radioactivity detector has been used to quantitate a large number of different isotopes used in biochemical research. A listing of the isotopes most commonly used in research is shown in Table 7.3. The two isotopes, ^{14}C and ^3H, represent over 75% of all isotopes quantitated in the flow-through detection system. In general, all radioisotopes that can be quantitated in the liquid scintillation counter can be quantitated in the flow-through radioactivity detector.

Second, a comprehensive list of the types of applications using the flow-through radioactivity detector is shown in Table 7.4. As can be ascertained from the table, almost all areas of chemical, biochemical, pharmaceutical, and research areas use radioisotopes and HPLC for the separation and quantitation of radiolabeled products and precursors.

Third, two isotopes have been used extensively in many metabolic studies using radionuclides. One isotope (usually ^3H) is used as a probe to monitor the fate of a labeled precursor in the metabolic or catabolic process. The other isotope (usually ^{14}C) is utilized to monitor the effectiveness of the extraction procedure or a separation method (a recovery tracer). Because

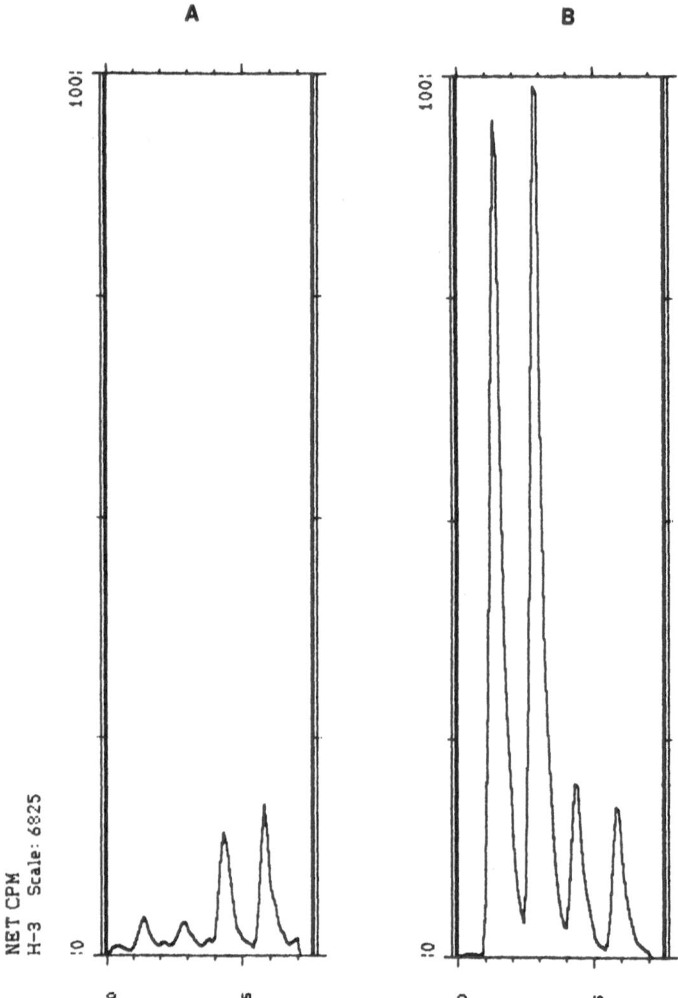

FIG. 7.8. A comparison of the sensitivities of the heterogeneous (A) solid scintillator (0.250 ml XE-calcium fluoride cell) and the homogeneous (B) flow cell (0.5 ml) mixing with 1.0 ml/min of liquid scintillation solution (FLO-SCINT II). The chromatographic conditions are the same as in Figure 7.6. Two injections of ^{3}H-testosterone (150,000 dpm) followed by two injections of ^{14}C (8800 dpm) were injected into the ODS II column.

two isotopes are being simultaneously monitored, it is important to be able to accurately quantitate the amount of each isotope present in each radioactive peak. To do this, one must obtain the energy spectrum of each of the two isotopes under the conditions used for quantitation. This can easily be accomplished with the spectrum analysis program incorporated into the radioactivity flow detector. Spectra of both ^{14}C and ^{3}H are shown in Figure 7.9. Therefore, if a double-labeled sample were analyzed under the same condi-

TABLE 7.3. Radioisotopes Used in Biochemical Experiments

Isotope	*Particle Type*	*Energy (MeV)*
^{3}H	β^-	0.018
^{11}C	β^+	1.98
^{14}C	β^-	0.155
^{18}F	β^+	1.65
^{22}Na	β^+	0.54
	γ	1.277
	γ^+	0.51
^{32}P	β^-	1.71
^{35}S	β^-	0.167
^{36}Cl	β^-	0.714
^{45}Ca	β^-	0.254
^{51}Cr	γ	0.325
	X rays	0.0055
^{55}Fe	X rays	0.0065
^{59}Fe	β^-	0.27, 0.46
	γ	1.1, 1.3
^{63}Ni	β^-	0.067
^{90}Sr	β^-	0.545
^{99}Tc	β^-	0.292
^{125}I	γ	0.035
	X rays	0.033
^{129}I	β^-	0.15
	γ	0.038
^{131}I	β	0.335, 0.608
	γ	0.284, 0.364 0.637

TABLE 7.4. Present Applications of Flow-Through Radioactivity Detectors

Types of Radiolabeled Biological Molecules
Proteins and peptides [15,16]
Carbohydrates [17]
Steroids [18–22]
Lipids and phospholipids [23,24]
Catecholamines [25]
Prostaglandins [26]
Nucleotides, DNA, RNA [27–33]
Polynuclear aromatic hydrocarbons [34,35]
Pharmaceuticals and metabolism studies [36–41]

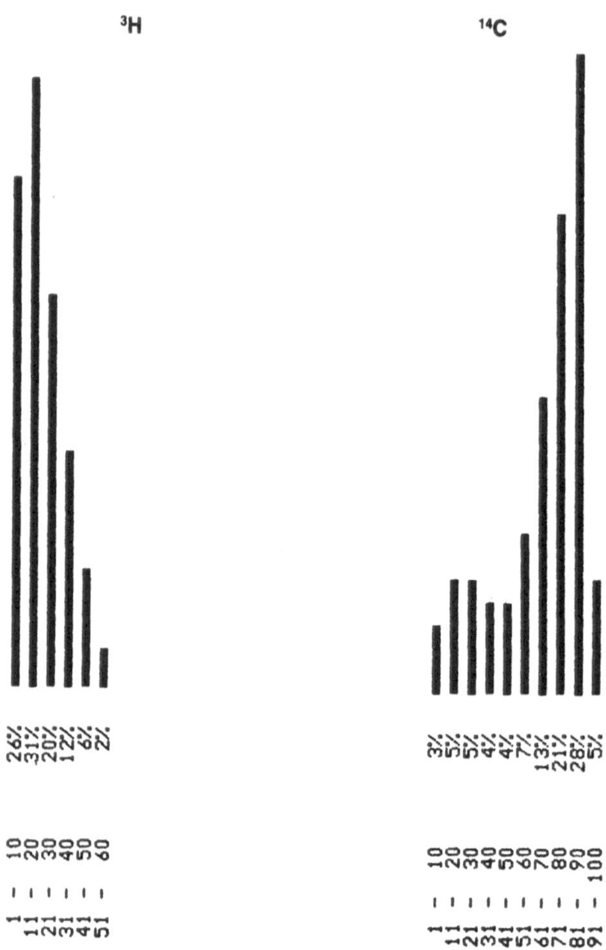

FIG. 7.9. An energy spectrum of two individual samples of ^{3}H and ^{14}C-testosterone mixed with liquid scintillation solution (FLO-SCINT II), pumped into a 2500-μl flow cell, and counted using the spectrum analysis program of the radioactivity detector. The relative position of each spectrum is dependent on the high voltage on the photomultiplier tubes and the amplification on the pulse-height analysis board.

tions used in the spectrum analysis, window settings of 1 to 50 would be set for ^{3}H. These settings are chosen because no ^{3}H occurs above the 50 window setting, and only 25% of the ^{14}C occurs in this window setting (crossover). The ^{14}C window would be set at 50 to 100 to obtain most of the ^{14}C counts in the sample. The counting efficiency for the ^{3}H (1–50) was 28% and for ^{14}C, 70% in the 50 to 100 windows. Therefore, by using these window settings, efficiencies, and crossover, the researcher is able to obtain the number of dpm of both ^{3}H and ^{14}C in each peak eluting from the HPLC. This method of double-label analysis has been used extensively in the metabolic study of steroids [42–44].

CONCLUSION

Two basic techniques exist for the quantitation of radioactive molecules eluting from the HPLC system. The first technique involves the collection of fractions of the HPLC eluent, aliquoting fractions, adding scintillation solution, and quantitating in a liquid scintillation counter.

The second technique is the flow-through radioactivity detection system, which involves the direct, on-line quantitation of radioactive peaks as they elute from the HPLC. Both techniques use the method of scintillation counting (homogeneous or heterogeneous) for the final quantitation of the radioactive peaks.

The flow-through radioactivity detector has several advantages over the fraction collector/liquid scintillation counter method. First, the flow-through radioactivity detector enables the researcher to obtain quantitation of each radioactive peak in real time as it elutes from the HPLC system. This includes on-line graphics, data reduction, and integrated peaks for each radioactive molecule. Second, the time of analysis is substantially reduced from approximately 10 hr to only the time required for the HPLC run. Third, the cost of analysis is substantially reduced using the flow detector technique to about one-tenth of the fraction collector/scintillation technique. Therefore, it is obvious that the flow-through radioactivity detector can substantially assist the user to obtain quantitative results of radioactive molecules eluting from the HPLC in about 5–10% of the time, and at about 10% of the cost of the alternative method of fraction collector/scintillation counting.

The overall sensitivity, resolution, and reproducibility of the flow detector was investigated. The minimum detectable limit for the flow detector was found to be 40 dpm for ^{14}C and 80 dpm for ^{3}H using a 2.5-ml homogeneous flow cell. The reproducibility of the results (peak area) from repetitive injections was found to be about 2–5% for counts above 2500 cpm and 5 to 10% for counts from 2500 to 500 dpm. These results are excellent, and in many cases, better than the results obtained by the fraction collector/scintillation counter technique.

The flow-through radioactivity detector has been used to quantitate various radioisotopes (^{14}C, ^{3}H, ^{35}S, ^{32}P, etc.) as they elute from the HPLC. Finally, the use of radioisotopes in all areas of research has increased dramatically and with the increasing popularity of HPLC, the flow-through radioactivity detector has been used to quantitate almost all types of radioactive molecules.

REFERENCES

1. Scott, R.P.W.: *Liquid Chromatography Detectors*. Elsevier Publishing, New York, 1977.
2. Kissinger, P.T.: Amperometric and coulometric detector for HPLC. *Anal. Chem.* **49**:477–485, 1977.

3. Swartzfager, D.C.: Ultraviolet detectors for HPLC. *Anal. Chem.* **48:**2189–2194, 1971.
4. Maggs, R.J.: *Practical High Performance Liquid Chromatography,* Simpson, C.F., ed. Heyden, London, 1976.
5. Byrne, S.H., Jr.: In *Modern Practices of Liquid Chromatography,* Kirkland, J.J., ed., Wiley Interscience, New York, 1971, Chapter 3.
6. Kessler, M.J.: A sensitive radioactivity detector for HPLC. *Amer. Lab.* **8:**52–63, 1982.
7. Rapkin, E. and Gibbs, J.: A system for continuous measurement of radioactivity in flowing streams. *Nature* **194:**34–49, 1962.
8. Schram, E. and Lombaert, R.: Determination of tritium and carbon-14 in aqueous solution with anthracene powder. *Anal. Biochem.* **3:**68–74, 1962.
9. Hunt, J.: Continuous flow monitor system for detection of UV absorbance, ^{14}C, and ^{3}H in effluent of a column chromatogram. *Anal. Biochem.* **23:**289–300, 1968.
10. Kessler, M.J.: A rapid method of quantitating steroids resulting from the incubation of gonadal tissues with radioactive precursors, *J. Liquid Chromatogr.* **5:**313–325, 1982.
11. Rapkin, E.: In *The Current Status of Liquid Scintillation Counting,* Bransome, E.D. Jr., ed., Grune and Stratton, New York, 1970.
12. Birks, J.B.: *The Theory and Practice of Scintillation Counting.* Pergamon Press, Oxford, 1964.
13. Laney, B.: In *Liquid Scintillation Counting, Current Developments,* Stanley, P.E. and Scoggins, B., eds., Academic Press, New York, 1974.
14. Kessler, M.J.: Quantitation of radiolabeled biological molecules separated by high-performance liquid chromatography. *J. Chromatogr.* **255:**209–215, 1983.
15. Bucala, R., Fishman, J. and Cerami, A.: Formation of covalent adducts between cortisl and 16-hydroxyestrone and protein: possible role in the pathogenesis of cortisol toxicity and systemic lupus erythematosus. *Proc. Natl. Acad. Sci. USA* **79:**3320–3324, 1982.
16. Abruzzini, L.N., Zacheis, M.P. and Schwartz, B.D.: Tryptic peptide disparity between serologically indistinguishable guinea pig 1a.3,5 antigens. *J. Immunol.* **131:**845–850, 1983.
17. Warner, T.G., Mock, A.K., Nyhan, W.L. and O'Brien, J.S.: Alphamannosidosis: analysis of urinary oligosaccharides with high performance liquid chromatography and diagnosis of a case with unusually mild presentation. *Clinical Genetics* **25:**248–255, 1984.
18. Browning, J.Y., Tcholakian, R.K., Kessler, M.J. and Grotjan, H.E. Jr.: Steroid metabolism by purified adult rat Leydig cells in primary culture. *Steroids* **40**(5):535–549, 1982.
19. Kessler, M.J.: Quantitation of radiolabeled compounds eluting from the HPLC system. *J. Chromatogr. Sci.* **20:**523–527, 1982.
20. Kessler, M.J.: High performance liquid chromatography of steroid metabolites in the pregnenolone and progesterone pathways. *Steroids* **39**(1):21–32, 1982.
21. Kessler, M.J.: A rapid method of quantitating steroids resulting from the incubation of gonadal tissues with radioactive precursors. *J. Liquid Chromatogr.* **5**(2):313–325, 1982.
22. Lafont, R., Pennetier, J-L., Adrianjafintrimo, M., Claret, J., Modde, J-F. and

Blais, C.: Sample processing for high performance liquid chromatography of ecdysteroids. *J. Liquid Chromatogr.* **236:**137–149, 1982.

23. Alam, I., Smith, J.B. and Silver, M.J.: Metabolism of platelets and plasma. *Lipids* **18**(8):534–538, 1983.
24. Koch, T.K.: Phospholipid methylation in myogenic cells. *Biochem. Biophys. Res. Commun.* **114**(1):339–347, 1983.
25. Ben-Jonathan, N. Arbogast, L.A., Rhoades, T.A., Schillo, K.K., Pau, K.F. and Jackson, G.L.: Plasma catecholamines in the chronically cannulated sheep fetus: predominance of L-dihydroxyphenylalanine. *Endocrinology* **43**(1):216–221, 1982.
26. Nolan, J.C., Pickett, W., English, K., Oronsky, A.L. and Kerwar, S.S.: Stimulation of prostaglandin E_2 synthesis in chondrocytes by a factor derived from activated macrophages. *Prostaglandins* **24**(4): 443–449, 1982.
27. Wang, C.C., Verham, R., Tzeng, S.F., Aldritt, S. and Cheng, H.: Pyrimidine metabolism in *Tritrichomonas* foetus. *Proc. Natl. Acad. Sci. USA* **80:**2564–2568, 1983.
28. Allaben, W.T., Weis, C.C., Fullerton, N.F. and Beland, F.A.: Formation and persistence of DNA adducts from the carcinogen N-hydroxy-2-acetylaminofluorene in rat mammary gland in vivo. *Carcinogenesis* (summer–fall), 1067, 1983.
29. Berman, J.D. and Webster, H.K.: In vitro effects of mycophenolic acid and allopurinol against *Leishmania tropica* in human macrophages. *Antimicrobial Agents and Chemotherapy* **21**(6):887–891, 1982.
30. Webster, H.K. and Whaun, J.M.: Purine metabolism during continuous erythrocyte culture of human malaria parasites (*P. falciparum*). In *The Red Cell: Fifth Ann Arbor Conference,* Alan R. Liss, Inc., New York, 1981, pp. 557–570.
31. Kessler, M.J.: Reverse phase high performance liquid chromatography of the 2′- and 3′-nucleotide monophosphates. *J. Liquid Chromatogr.* **5**(1):111–123, 1982.
32. Ozawa, N. and Guengerich, F.P.: Evidence for formation of an S-[2-(N^7-guanyl)ethyllgluthathione adduct in glutathione-mediated binding of the carcinogen 1,2-dibromoethane to DNA. *Proc. Natl. Acad. Sci. USA* **80:**5266–5270, 1983.
33. Boulieu, R., Bory, C., Baltassat, P. and Gonnet, C.: The application of UV-radioactivity high performance liquid chromatography to the study of hypoxanthine transport in human erythrocytes. *J. Liquid Chromatogr.* **7:**1013–1021, 1984.
34. Marshall, M.V. and Rogers, W.G.: Carcinogen metabolism in the cigarette smoking baboon. In *Polynuclear Aromatic Hydrocarbons: Seventh International Symposium on Formation, Metabolism and Measurement,* Cooke, M.W. and Dennis, A.J., eds., Battelle Press, Columbus, Ohio, 1982.
35. Cundy, K.C. and Crooks, P.A.: Unexpected phenomenon in the high-performance liquid chromatographic analysis of racemic ^{14}C-labelled nicotine: Separation of enantiomers in a totally achiral system. *J. Chromatogr.* **281:**17–33, 1983.
36. Fong, K-L.L. and Hwang, B.Y-H.: Aromatization of 7,8-dichloro-1,2,3,4-tetradydroisoquinoline by rat liver microsome. *Drug Metabolism and Disposition* **12:**14–19, 1984.

37. Fong, K-L.L. and Hwang, B.Y-H: Dog liver N-methyltransferase: a drug-metabolizing enzyme. *Biochem. Pharm.* **32**:2781–2786, 1983.
38. Minchin, R.F. and Boyd, M.R.: Uptake and metabolism of doxorubicin in isolated rat lung. *Biochem. Pharm.* **32**:2829–2832, 1983.
39. Webster, H.K. and Whaun, J.M.: Antimalarial properties of bredinin. *Clin. Invest.* **70**:461–469, 1982.
40. Zryd, J-P., Bauer, J., Wyler, H. and Lavanchy, P.: Pigment biosynthesis and precursor metabolism in red beet semi-continuous cell suspension cultures. Proc. 5th Intl. Cong. Plant Tissue and Cell Culture, 1982, pp. 387–388.
41. Webster, H.K. and Whaun, J.M.: Application of simultaneous UV-radioactivity high-performance liquid chromatography to the study of intermediary metabolism. *J. Chromatogr.* **209**:283–292, 1981.
42. Steinberger, E., Steinberger, A. and Ficher, M.: Reproductive steroid metabolism. *Recent Prog. Horm. Res.* **26**:547–553, 1970.
43. Steinberger, E., Tcholakian, R.K. and Steinberger, A.: Hormonal control of steroidogenesis. *J. Steroid Biochem.* **11**:185–194, 1979.
44. Mather, J.P., Saez, J.M. and Haour, F.: Primary culture of Leydig cells from rat, mouse, and pig: advantages of porcine Leydig cell cultures on an experimental system. *Steroids* **38**:35–48, 1981.

CHAPTER 8

Flow Detector Designs: Build Your Own or Buy?

RICHARD D. HICHWA

INTRODUCTION

Numerous techniques exist to detect and measure trace amounts of radioactivity in the laboratory. With the expanding field of chromatography and particularly high pressure liquid chromatography (HPLC) for the assay and preparation of radiopharmaceuticals, it becomes necessary to develop highly specific radiation detection systems for identifying the radiochemical components of a pharmaceutical. Choosing the radiodetector system for HPLC is dependent upon the types of radiopharmaceuticals being manufactured and the environment in which the measurement of the radioactive components is performed. The type of chromatography governs the choice of detectors to be used as well as the sensitivity and resolution needed to adequately resolve radiopeaks.

REQUIREMENTS

There are several requirements that must be met by flow-through HPLC radioactivity detectors. In many cases, improvements in the response of the detector system are at the expense of one of the other attributes that characterize detection of radioactive disintegrations in flow detectors.

Sensitivity vs. Resolution

In the optimal situation, sensitivity is balanced against resolution [1]. Maximum sensitivity is most important for low activity and low yield experi-

Abbreviations: Beta particle, β^-; Direct current, DC; Disintegration/min, dpm; Gas chromatography, GS; High pressure liquid chromatography, HPLC; Photomultiplier tube, PMT; Positron, β^+; Positron emission tomography, PET: Transistor-transistor logic, TTL.

ments. Sensitivity is defined as the response of the detector to a known amount of radioactivity. Sensitivity of typical radioactivity detectors used for flow cells is enhanced by increasing the scintillator size, by matching the output wavelength of the scintillator emissions to the optimal PMT response [2], by matching the scintillator to the type and energy of radiation to be detected, and by increasing the solid angle of detection. This last factor is accomplished by simply adding multiple PMTs to the scintillator or by increasing the scintillator size. More light is collected by the PMT or a higher percentage of emitted radiation is detected with this technique. Various scintillators for a given thickness will have different detection efficiencies for various energies of incident radiation. Sensitivity is often used interchangeably with efficiency, which may be stated as the ratio of detected radiation to the number of disintegrations emitted by the radioactive material [3]. These disintegrations may be either photons or particles. The particles emitted from typical nuclear medicine radiopharmaceuticals are, for the most part, either β^- (beta) or β^+ (positron) particles.

The dynamic range of the detector system becomes very important for high sensitivity devices [4]. For any detector system, counts are defined as those disintegrations that have been registered as events within the detector and associated counting electronics. High count rates will exceed the system's capacity to register events, thereby degrading the response of the detector. A system that has a wide dynamic range implies that the detector's response is measurable for count rates that span several orders of magnitude, i.e., from low radioactivity concentrations to high radioactivity concentrations within the field-of-view of the detector.

Over the dynamic range of the system, linearity is a measure of how well the detector responds to different levels of radiation. A system is linear if, for example, by doubling the amount of radiation impinging upon the detector, the observed count rate is doubled. A linear system allows for accurate and precise measurements with a calibration scheme that requires a minimal number of correction factors. High count rate corrections can be implemented to yield a linear response if the entire counting system can be mathematically characterized, but the error associated with these measurements increases.

Elevated detector sensitivity may also be achieved by altering the characteristics of the radioactivity concentration that passes through the detector. Low flow rates of radioactive materials passing through the detector will increase sensitivity, while enhanced flow rates reduce sensitivity of the flow-through detector.

Resolution is defined as the system's response to abrupt changes in radioactivity presented to the detector. A sharp or drastic change in activity is measured as a smooth transition from low count rates to high count rates. Flow rate and detector cell volume affect both resolution and sensitivity. Uniquely sharp separation of peaks by HPLC techniques will be lost if the residence time in the flow-through cell is too long. Count rates increase as flow is diminished and/or the sampling volume of the flow cell is increased. But too large a flow cell or too slow a flow rate significantly degrades resolution. Hence, the detector system will be unable to resolve the separation of two closely spaced radiopeaks. Changes in cell size by a factor of 5 or from a realistic size of 100 μl to 500 μl result in a resolution change of more than a factor of 15, which is an appreciable degradation in peak determination.

Energy and Particle Discrimination

No single detector system can be matched to every nuclide currently used for labeling radiopharmaceuticals. Table 8.1 compares several characteristics of conventional nuclides used for medical imaging and tracer development. The user must choose the correct scintillator to detect either particulate or electromagnetic radiation or contend with a reduced detection efficiency for both.

Beta detection (for ^{3}H, ^{14}C, ^{32}P, etc.) is best measured by constraining the radioactive material to flow through the actual scintillator material. This could be simply scintillator tubing encased in a light guide and attached to a PMT or a machined channel in a block of scintillator coupled directly to a PMT. The disadvantages of these techniques are simply that the scintillator is in contact with the solvent system used for the HPLC procedure. For very low energies, liquid scintillator may be added to the solvent to permit direct interaction of the beta particles with the liquid scintillator. PMTs, in this case, are coupled to light guides through which pass optically coupled trans-

TABLE 8.1. Characteristics of Several Useful Nuclides

Nuclide	*Decay*	*Energy (MeV)*	*Half-life*
^{3}H	β^-	0.0186	12.33 yr
^{11}C[a]	β^+	0.961	20.4 min
^{14}C	β^-	0.156	5730 yr
^{13}N[a]	β^+	1.19	9.97 min
^{15}O[a]	β^+	1.73	122 sec
^{18}F[a]	β^+	0.635	109.8 min
^{32}P	β^-	1.710	14.28 d
^{123}I	γ	0.1589	13.2 hr
^{131}I	β^-	0.606	8.041 d
	γ	0.364	

[a] Positron annihilation also creates two 0.511 MeV diametrically opposed photons.

parent tubes carrying the liquid scintillator and radiopharmaceutical. High counting efficiency is required for low energy beta emitters such as tritium. The efficiency must be in the range of 25–50%. A minimum detectable limit is approximately 200 dpm for correctable data. This implies that flow-rate and cell-size corrections must be applied to the raw data to yield the 200 dpm limit.

Gamma detection is much easier to accomplish because most gamma-emitting nuclides used in nuclear medicine [5] (^{99m}Tc, ^{123}I, ^{131}I, etc.) have sufficient energy to penetrate thin-walled tubing (of metal, teflon, glass, etc.) to directly interact with the scintillator. The HPLC preparation has no opportunity to contaminate the detector system. Residual activity may be flushed from the tubing that passes through the detector to lower background count rates or may be completely replaced if necessary. Additional shielding is required, however, to reduce the interaction of activity from other nearby chemistry experiments that may affect the desired radiochromatography measurement. This shielding should be designed to reduce the response from outside activity by approximately 10^3 for the highest energy gamma ray to be used in the laboratory. Several tables have been compiled to aid in determining the adequate shielding for detector systems [6,7].

The efficiency of most scintillators is inversely proportional to gamma energy. Thicker scintillators are necessary for equal detection efficiency of higher gamma ray energies. For example, the ^{99m}Tc efficiency is 80% for a 5×5 cm NaI well counter (assume a loop passes through the well). However, for 0.511 MeV positron annihilation radiation, the efficiency is only 31%.

Positron emission tomography or PET has stimulated the development of an entirely new field of fast, high specific-activity radiochemistry. The radioactive pharmaceuticals produced are compatible with the body's own biochemistry. The detection systems used by the radiochemist that measure radioactivity from the conventional PET nuclides of ^{11}C, ^{13}N, ^{15}O, and ^{18}F must be designed for maximum shielding. The half-value layer of lead for 0.511 MeV gamma rays is 5.0 mm. A standard 5 cm thick lead brick reduces the unwanted background by only 10^3. High background activity in the laboratory will affect a detector that does not have adequate shielding. Use of electronic collimation or coincidence electronics permits detection systems to operate in areas of high background count rates [8]. Only the sample in the field-of-view of the scintillators registers a valid response to activity in the flow cell. This desirable type of system is achieved at increased expense resulting from the necessary electronics and added complexity of coincidence counting.

Positrons can be detected instead of the annihilation photons. Systems similar to β^- detection can be used. Positrons from low Z neutron deficient nuclides are typically higher in energy and easier to detect. Liquid scintillator need not be used because the positron can penetrate the walls of thin tubing and can be detected by external scintillators. Detector contamination is therefore kept to a minimum.

Corrections

The data must have several corrections applied to the observed radioactivity plots before true count rates are obtained.

Quench corrections are necessary to compensate for varying concentrations of solvents and buffers if liquid scintillator counting is employed.

Delays between the radioactivity detector and other detectors within the HPLC system should be as small as possible. Laminar streaming within the tubing that connects the HPLC to external detectors causes a parabolic radioactivity wavefront to pass through the scintillation detector. An abrupt radioactivity change at the HPLC manifests itself as a smooth transition at the detector, which, in turn, distorts the measured peak width. Hence, resolution is degraded. Sensitivity, however, is virtually unaffected and the total integration of counts in a peak is conserved.

Very high count rates cannot typically be measured by the system. As the count rates increase, the detection electronics cannot handle every incoming count. The system is said to be busy. Some counts are lost. These deadtime losses are corrected by several techniques [4]. Injection of a known specific activity into the system with a nuclide of short half-life enables measurement of the detector count rate response over several decades of radioactive decay. A model of deadtime losses can be established and parameters characterizing the deadtime response of the system can be determined. The correction factors can be folded into the data acquisition scheme to yield online true detector count rates.

TYPES OF FLOW CELLS

There are two basic types of flow cells that are used for HPLC or similar types of radiochemical analyses, solid and liquid scintillator cells.

The solid scintillator cell may be composed of several types of materials as shown in Table 8.2 [9]. In nearly all cases, the efficiencies are relatively poor for ^{14}C and only 1–5% for ^{3}H. If strictly gamma radiation is to be measured, the scintillator must be 10^2 to 10^4 times thicker to increase the detection efficiency. Of course, the β^+ from short-lived positron-emitting nuclides can be detected with equal or better efficiencies than ^{14}C and far more easily than high energy gamma rays. The annihilation radiation from the positron is not realistically detected with thin scintillators.

Several advantages are achieved by using a liquid scintillator flow system. High salt or buffer concentrations do not significantly affect the counting efficiency of the flow detector. Problems that arise are due to the binding,

TABLE 8.2. Flow Cells

Type	*Size*	*Efficiency*
Solid Scintillator Cell		
Plastic	200 μm	20–25% for ^{14}C
Glass	70–90 μm	20–25% for ^{14}C
		1% for ^{3}H
CaF(Eu)	100–200 μm	75–80% for ^{14}C
		5% for ^{3}H
NaI(Tl)	2 cm	50% for ^{199m}Tc
		15% for β^{+} annihilation
Liquid Scintillator Cell		
PTFE tubing	10–2000	
	μl	75–80% for ^{14}C
		20–60% for ^{3}H

irreversibly in some cases, of the radiopharmaceuticals to the solid support of the scintillator. Some of the scintillators are water soluble (CaF) or hydroscopic (NaI). In these cases, the materials are sealed in thin-walled casings.

Thin scintillator materials are easily fashioned into tubes that serve as detectors. The NaI scintillator for gamma detectors must be machined under special conditions and is usually made to order by several companies that manufacture scintillators [10–13]. Light pipes guide visible or near visible (UV) photons emitted from the scintillator to the photomultiplier tube for conversion into electrical signals.

Liquid scintillator cells are primarily constructed of PTFE tubing. The HPLC sample is mixed with nongelling liquid scintillation fluid [14]. The beta particles interact with the fluid to create photons, which, in turn, are converted into electrical signals by PMTs. The liquid scintillators have much higher detection efficiencies than other types of scintillators for ^{14}C and ^{3}H. There is little or no adsorption on the PTFE lines and therefore little contamination. The greatest problem that arises with the use of liquid scintillation fluid is that no further chemical or biological studies of the original radiopharmaceutical can be performed.

A third type of detector currently being investigated for use as a flow cell is a solid state photodiode [15,16]. The photodiode may be optically coupled to scintillator materials previously discussed or used independently. Beta particles can interact directly with the P–N junction of the photodiode to create a small current pulse (direct measurement). This pulse, in turn, is amplified and counted by the data acquisition system. The photodiode detector has the advantage that it may also be coupled to any type of scintillator in which the scintillator response is matched to the photodiode response (optical measurement). The photodiode does not need a PMT nor its associated high voltage connection. Hence, the detector package is small, uses low voltage and has variable sensitivity based on the desired scintillator material or on direct interaction of radiation with the photodiode.

DATA ACQUISITION AND ANALYSIS CONSIDERATIONS

If an HPLC or GC flow-through detector system is constructed in the laboratory, it is of relatively little use without some means for acquiring, analyzing, and storing the radioactivity data. This is in addition to IR, UV, and thermoconductivity detector information that must also be recorded. Many conventional chromatographic systems provide upgrade capabilities for other detector systems but require specific types of signal inputs (low level analog, 10–100 mV; high level analog, 1–10 V; digital, 5 V TTL). The signals from homemade detector systems will have to adhere to the manufacturer's design specifications for a given HPLC or GC system in order for them to be coupled to commercial chromatographs.

Conventional microcomputers or personal computers are extremely useful and easily adapted for real-time data acquisition of chromatographic data. The Apple II [17], IBM PC [18], and DEC LSI 11 [19] systems are the best choices. These computers offer a wide range of interfaces for real-time data acquisition, high level language compilers for programmability, and numerous graphics capabilities for display of data.

For those detector systems that incorporate an amplifier with a single channel analyzer, simple TTL pulses can be scaled as a function of time in the computer [8]. The pulse heights from the detector are proportional to the energy of the incident particle or photon. If the amplitude exceeds some lower threshold, a single TTL pulse of 1–2 μsec is created. The radioactivity curve is described by the number of counts registered in each small time bin along the time axis.

For detector systems that integrate a number of photons or particles per unit time and convert this to a varying DC voltage level, which is essentially the job performed by a ratemeter, simple analog-to-digital (A/D) conversion is performed at a temporal sampling rate to sufficiently define the radiochromatographic curve. The conventional method of displaying this type of data is with a strip chart recorder. However, the use of a computer to store and display the data permits magnification of small peaks, minification of large peaks, peak integration, etc.

Most microcomputers are capable of handling several channels of data (IR, UV, radioactivity) simultaneously. The microprocessor operational speed is critical, however. Slow processors will not execute acquisition software fast enough to adequately sample narrow, rapidly varying peaks. Numerous additional features that can be programmed into the computer are calibration information, multiple displays of data, data recall, and overlaying of several spectra.

TABLE 8.3. Software Routines for Chromatography

1. Scale data (magnify or minify spectra)
2. Integrate peaks
3. Peak strip
4. Fit peaks to defined mathematical functions
5. Overlay spectra (in color)
6. Display log of data
7. Identify peak centroid
8. Window spectrum for magnified temporal peak display
9. Auto identification of peaks (storage of calibration data)
10. Provide calculator mode
11. Full programmability in FORTRAN, BASIC, PASCAL, and/or FORTH
12. Provide timer mode for laboratory operations
13. Maintain a database system for cataloging and storing spectra

A list of features that should be implemented in the acquisition and analysis software is given in Table 8.3. A more difficult acquisition process that is easily resolved by using a microcomputer is the simultaneous measurement of radioactivity from two or more nuclides. Dual-labeling techniques are becoming more important for nuclear medicine studies. Also, nuclides that decay by emission of multiple energy photons are best handled with fast nuclear A/D converters. A complete nuclear spectrum can be acquired if this type of interface is installed in the microcomputer, essentially creating a low cost multichannel analyzer. A general tool of this type in the radiochemistry laboratory is becoming more important as single photon experiments are run in conjunction with PET experiments. Radionuclide impurities as well as radiochemical impurities can be easily measured using this scheme.

CONCLUSIONS

Solid scintillator flow cells are good for the detection of high energy beta particles. Liquid scintillator cells offer increased detection efficiency for low energy beta particles, but no further analysis of the sample is possible. Detection of high energy photons is achieved with thick organic scintillators, but this approach requires sufficient lead shielding to minimize unwanted counts. Electronic coincidence detection schemes further reduce background and noise. Solid state detectors achieve results similar to those of the solid scintillator methods, but employ a much more compact design.

Commercially available flow-through radioactivity detectors offer reasonable detection hardware at moderate costs. Systems built in the user's laboratory better suit the individual needs and requirements for radiopharmaceutical production at that site. Incorporation of a computer as part of the system is essential for a radiation detector to be useful. However, the development of software tools necessary for data analysis is often too large a time investment for the user. Hardware development times average only 10–20% of the time required to build a total system.

The actual choice between building your own system or buying a commercially available system depends upon the resolution and sensitivity requirements of the system as well as the type and energy of the nuclides to be measured. Careful thought should be given to the evaluation of available systems. Only when the manufacturers cannot meet the design specifications dictated by an individual user should the user embark upon a path of building his own system.

REFERENCES

1. Knoll, G.F.: *Radiation Detection and Measurement.* John Wiley and Sons, New York, 1979.
2. Engstrom, R.W.: *Photomultiplier Handbook.* RCA, Solid State Division, Lancaster, Pennsylvania, 1980.
3. Price, W.J.: *Nuclear Radiation Detection.* McGraw-Hill Book Company, New York, 1964.
4. Evans, R.D.: *The Atomic Nucleus.* McGraw-Hill Book Company, New York, 1955, pp. 785–818.
5. Sorenson, J.A. and Phelps, M.E.: *Physics in Nuclear Medicine.* Grune and Stratton, New York, 1980.
6. Marion, J.B. and Young, F.C.: *Nuclear Reaction Analysis.* North-Holland Publishing Company, Amsterdam, 1968.
7. *Radiological Health Handbook.* U.S. Department of Health, Education and Welfare, Rockville, Maryland, (No. 017-011-00043-0), 1970, pp. 129–218.
8. *Experiments in Nuclear Science (AN34).* ORTEC, Oak Ridge, Tennessee, 1971.
9. Kessler, M.J.: A sensitive radioactivity detector for HPLC. *Am. Lab.* **14**:52–63, 1982.
10. The Harshaw Chemical Co., Crystal and Electronics Products, 6801 Cochran Road, Solon, Ohio 44139.
11. Nuclear Enterprises Ltd., 935 Terminal Way, San Carlos, California 94070.
12. Bicron Corporation, 12345 Kinsman Road, Newbury, Ohio 44065.
13. Teledyne Isotopes, 50 VanBuren Avenue, Westwood, New Jersey 07675.
14. Beckman Instruments, Inc., Scientific Instruments Division, P.O. Box C-19600, Irvine, California 92713.
15. Hamamatsu Corporation, 420 South Avenue, Middlesex, New Jersey 08846.
16. DEVAR, Inc., Control Products Division, 706 Bostwick Avenue, Bridgeport, Connecticut 06605.
17. Apple Computer, Inc., 20525 Mariani Avenue, Cupertino, California 95014.
18. IBM Instruments, Inc., Orchard Park, P.O. Box 332, Danbury, Connecticut 06810.
19. Digital Equipment Corporation, 146 Main Street, Maynard, Massachusetts 01754.

Applications

CHAPTER 9

Radio-HPLC: Application to Organics and Metal Chelate Chemistry

ALAN R. FRITZBERG and ADRIAN D. NUNN

INTRODUCTION

This chapter discusses applications of HPLC to radiopharmaceuticals. Because detailed discussions of theory and general applications are provided in books and in the literature, this chapter provides a means to indicate the scope of applications at this time, the utility of HPLC in solving radiochemical problems, and the considerations that go into developing assay conditions.

Table 9.1 lists applications of HPLC to radiochemicals of radiopharmaceutical interest that have been described in the literature over the last few years [1–53]. The list is not intended to be exhaustive, but inclusive of representative assays and conditions for different classes of radiochemicals. Covalently bound radiolabels in organic compounds represent isotopic or isosteric replacement, and characterization of their chemistry with nonradioactive forms is straightforward. In many cases, biochemical or pharmaceutical literature exists that provides either assay conditions for very closely related compounds. We have chosen to discuss the applications of ^{99m}Tc chelates because application of HPLC to such inorganics is relatively new. The illustrations used represent the combined experiences of the two authors. We hope that these experiences will provide a broad basis from which to develop HPLC separations of radioinorganics.

Abbreviations: High pressure liquid chromatography, HPLC; No carrier added, nca; Diamido disulfur ligand, DADS; Carbon chain length of bonded phases, C_8, C_{18}, etc.; Hepatobiliary-iminodiacetic acids, HIDAs; Structure distribution relationship, SDR; Bisdimethylphosphinoethane, DMPE; Hydroxyethylidenediphosphonic acid, HEDP; Methylenediphosphonic acid, MDP; Monocarboxylate derivative of DADS, CO_2DADS; C_{18} column packing from Altex-Beckman, ODS; Relative hydrophobicity of a fragment, π; Phosphate buffer, Pi.

TABLE 9.1. HPLC Assays for Radiolabeled Compounds of Interest to Nuclear Medicine

Compound	*Column*	*Conditions*	*Reference*
	^{99m}Tc Complexes		
^{99m}Tc-acetanilideiminodiacetates	μ Bondapak C_{18}, 5 μm	0.025 *M* Pi, pH 6.8/CH_3OH (1 : 1)	1
^{99m}Tc-disofenin	Ultrasphere ODS, 5 μm	0.01 *M* Pi, pH 6.5/CH_3CN (7 : 3)	Fritzberg and Johnson unpublished
^{99m}Tc-etiofenin	Ultrasphere ODS, 5 μm	0.01 *M* Pi, pH 6.5/CH_3CN (7 : 3)	Fritzberg and Johnson unpublished
^{99m}Tc-diphosphonates	Aminex A-27	0.85 *M* Na acetate	2
^{99m}Tc-N,N′bis(mercaptoacetyl)-ethylenediamine	μ Bondapak C_{18}	0.005 *M* (*n*-butyl)$_4$Pi/CH_3OH (8 : 3—2.8 : 10 min)	3
^{99m}Tc-N,N′bis(mercaptoacetyl)-ethylenediamine	Ultrasphere ODS, 5 μm	0.01 Pi/CH_3CN (1 : 9—4 : 6/10 min)	4
^{99m}Tc-N,N′bis(mercaptoacetyl)-ethylenediamine	Ultrasphere ODS, 5 μm	0.01 *M* Pi/ethanol (9 : 1)	Fritzberg and Johnson unpublished
^{99m}Tc-N,N′bis(mercaptoacetyl)-2,3-diaminopropanoate	Ultrasphere ODS, 5 μm	0.01 *M* Pi/CH_3CN (95 : 5—70 : 30/10 min)	5
^{99m}Tc-N,N′-bis(mercaptoacetyl)-2,3-diaminopropanoate	Ultrasphere ODS, 5 μm	0.01 *M* Pi/ethanol (95 : 5)	Fritzberg and Johnson unpublished
^{99m}Tc-acetanilideiminodiacetates	Ultrasphere ODS, 5 μm	0.01 *M* Pi/CH_3CN (75 : 25—40 : 60/10 min)	6
^{99m}Tc-bis-aminoethanethiols	PRP-1 RP	CH_3CN/H_2O (85 : 15)	7
^{99m}Tc-$(DMPE)_2Cl_2^+$ and -$(DMPE)_3^+$	C_8 RP	CH_3OH/H_2O (7 : 3), aq 0.02 *M* heptane SO_3H, 0.003 *M* Pi	8,9
^{99m}Tc-diethyldithiocarbamate	μ Bondapak phenyl	CH_3OH/H_2O (85 : 15)	10
$^{99m}TcO_4^-$, $^{99}TcO_4^-$	Spherisorb Amino bonded	NaOAc, pH 4.5	11
^{99m}Tc-cyclam,-ethylenediamine,1,5,8-12-tetraazadodecane	μ Bondapak CN	0.015 *M* CH_3SO_3H and 0.005 *M* 1-heptanesulfonic acid	13
^{99m}Tc-MDP	Aminex	0.85 *M* NaOAc	14

$^{99m}TcO_4^-$, $^{99}TcO_4^-$	Waters Fatty Acid Analysis Sepralyte C_{18}	Bu_4N^+ in 40% CH_3OH	15
^{99m}Tc-human serum albumin	Spherogel-TSK 3000	0.1 *M* Pi, or MOPS	12
	Radiolabeled Sugars		
2-deoxy-2-[^{18}F]fluoro-D-glucose (FDG)	Bio-Rad carbohydrate analysis	Water	16
[1-^{11}C]2-deoxy-D-glucose-2-deoxy-	Ultrasil-NH_2 HPX-87C Ca^{2+}	0.1 *M* Pi/CH_3OH (55 : 45)	17
2-[^{18}F]fluoro-2-deoxy-D-glucose	exchanged polystyrene sulfonate resin	H_2O	18
2-deoxy-2-[^{18}F]fluoro-D-glucopyranosyl-[^{18}F] fluoride	Bio-Rad carbohydrate analysis column	H_2O	19
3-[^{11}C]-methyl-D-glucose	Lichrosorb NH_2, 10 μm	CH_3CN/H_2O (4 : 1)	20
2-deoxy-2-[^{18}F]fluoro-3-*o*-methyl-D-glucose	ODS	H_2O/CH_3CN (5 : 95)	21
2-deoxy-D-[1-^{11}C]glucose	Partisil 10 PAC M 9/50 600 CH Carbohydrate	CH_3CN/H_2O (80 : 20)	22
[^{11}C]-2-deoxy-D-glucose	HPX-87P Aminex	H_2O	23
	Radiohalogenated Drug Analogues		
16α-[^{77}Br]bromo-11-β methoxyestradiol-17β	Silica gel-10 μm	Hexane/CH_2Cl_2/2-propanol (85 : 12 : 3)	24
[^{77}Br]-17-β-bromoethynyl-estradiol	Lichrosorb RP 18, 10 μm	Ethanol/H_2O (7 : 3)	25
16α-[^{77}Br]-bromoestradiol-17β	Whatman Partisil M-9 silica gel	Hexane/CH_2Cl_2/2-propanol (92 : 6.4 : 1.6)	26
benzodiazepines	Lichrosorb RP-18, 7 μm	CH_3OH/H_2O (3 : 2)	27
N′-(4-[^{11}C]methyl)-imipramine	Si-NH_2, 10 μm	*n*-Hexane/2-propanol/CH_3OH dimethylamine (98 : 2.2 : 0.05)	28
^{11}C-BCNU	Partisil PAC	ether/hexane (1 : 1)	29
[^{131}I]iodobenzylguanidine	μ Bondapak C_{18}	THF/0.1 *M* Pi (12 : 88)	30
4-[^{82}Br]bromomisonidazole	RP, C_{18}	H_2O/CH_2OH (9 : 1)	31

TABLE 9.1. *Continued*

Compound	*Column*	*Conditions*	*Reference*
^{131}I-*N*-isopropyl-*p*-iodo-amphetamine	Lichrosorb RP 18, 7 μm	CH_2OH/0.05 *M* NH_4Cl (4 : 1)	32
[^{18}F]haloperidol	PAX 5-1025 ODS-3	0.1 mM NH_4OAc/CH_3OH (65 : 35)	33
[2-^{11}C]5,5-dimethyloxazolidine-2,4-dione	Whatman Partisil M9 10/25 ODS-2	Ethanol/saline (1 : 9)	34
[^{14}C]butoprozine	Lichrosorb RP 8, 5 μm	$CH_3/CH_2/H_2O$ (100 : 35 : 15) +0.5% dimethylamine	35
[^{11}C]antipyrine	R Sil C18 HL 10 μm	0.05 *M* Pi/ethanol (77.5 : 22.5)	36
[^{11}C]practolol	μ Bondapak C_{18} Magnum 9 Partisil 1050	$CHCl_3$/ethanol (4 : 1) with 1.5% $C_2H_5NH_2$ and 2.5% H_2O	
^{77}Br-*p*-bromospiroperidol	Micropak CH-10 prep.	Ethanol/H_2O/4M NH_4OAc/HoAc (530 : 42 : 23 : 10)	37
	R Sil C_{19} HL anal.	Ethanol/H_2O/4M NH_4OAc/HoAc (390 : 595 : 15 : 10)	
[*N*-methyl-^{11}C]erythromycin	μ Partisil	$CH_2Cl_2/C_2H_5OH/NH_4OAc$ (97.5 : 2.5 : 0.25)	38
^{123}I-*m*-iodobenzylguanidine	Bonded-phase NH_2	Ethanol/Pi (1 : 4)	39
3-quinuclidinyl benzilate (QNB)	RP C_{18}	$CH_3CN/CH_3OH/H_2O$ (36 : 30 : 40) 5 *mM* octane SO_3H	40
	Amino Acids		
L-[1-^{11}C]leucine	Ultrasphere ODS, 5 μm	8 *mM* $Cu(OAc)_2$, 17 *mM* L-proline 30 *mM* NaOAc, pH 5.0/CH_3OH (9 : 1)	41
L- and D-[^{11}C]-DOPA	(L-proline)$_2$-Cu stationary phase	Pi, pH 4.5	42

	Radiolabeled Fatty Acids/Lipids and Analogues		
α-CH_3[1-^{11}C]heptadecanoic acid	ODS, 5 μm	Tetrahydrofuran/CH_3CN/H_2O (4 : 4 : 2)	43
15-(*p*-[[75]Br]bromphenyl)pentadecanoic acid	Lichrosorb RP-18, 10 μm	CH_3OH/H_2O/HOAc (86 : 9 : 5)	44
17-[^{123}I]heptadecanoic acid	Lichrosorb RP-18, 7 μm	Ethanol/H_2O/HOAc (83 : 5 : 16.5 : 0.1)	45
[^{131}I]-11-*N*(*p*-iodophenyl sulfonamideundecanoic acid	Ultrasphere ODS, 5 μm	0.01 *M* Pi/CH_3CN (30 : 70)	46
[1-^{11}C]palmitic acid	Ultrasphere ODS, 5 μm	CH_3OH	47
ω-(*p*-^{123}I-iodophenyl)-pentadecanoic acid	Lichrosorb RP-18, 10 μm	CH_3OH/HOAc/H_2O (95 : 4 : 1)	48
C-11 valproate	Partisil 10-ODS-3	CH_3CN/H_2O HOAc (35 : 64 : 1)	49
C-11 palmitate	μ Bondapak fatty acid	THF/CH_5CN/H_2O (2 : 6 : 2)	49
[11]C-ethers, alcohols	RP 5 ODS-3 RAC	H_2O/ethanol, 45–73% ethanol depending on compound	50
	Miscellaneous		
[[57]Co]bleomycins	Nucleosil 10 C_{18} Radial Pak A	1% NH_4OAc/CH_3OH (6 : 4)	51
[18]F-fluoride	HP X-87 carbohydrate or SAX (Cl-form)	H_2O or 0.9% NaCl respectively	52
5-iodo-2′-deoxycytidine triphosphate	Lichrosorb RP 18, 10 μm	0.05 *M* NH_4Pi pH 3.2	53

STABILITY UNDER HPLC CONDITIONS

A most important concern in the analysis of radiochemicals/radiopharmaceuticals by HPLC is their stability during HPLC procedures. Covalently bound organic radiolabels are unlikely to present problems, although the well-known problems of weak carbon iodine bonds require concern in the analysis and use of radioiodinated materials. It is well known that some technetium complexes based on ligands such as glucoheptonate, citrate, pyrophosphate, phosphonates, etc., are kinetically labile and undergo exchange in the presence of complexing agents with higher avidity for technetium. The loss of radioactivity on silica-backed columns if nca ^{99m}Tc phosphonate solutions are chromatographed without ligand in the eluate is a case in point.

Stability should be evaluated by collecting radioactivity corresponding to eluted peaks and rechromatographing them. This is particularly true if HPLC cuts are being taken to enable testing of a single component of a multicomponent system. Such cuts are invariably substantially free of unbound ligand or reducing agent and are prone to change. HPLC analysis should be performed on the cuts at the time of injection if, for instance, animal studies are being done. Another of the conditions of interest is likely to be stability in plasma. It is preferable not to inject proteins into reversed-phase columns in quantity, especially if eluants are high in organic solvent content that will clearly denature the proteins. If proteins are removed by, for example, the addition of acetonitrile to 50% (v/v), care should be taken that the injectate volume does not alter the mobile phase properties of the HPLC system. An alternative is to use column switching techniques, i.e., a TSK column followed by a reversed-phase column. The stability of radioactive species in bile has already been described in the case of the ^{99m}Tc HIDAs.

An example of a surprising source of exchange problems may be useful. Studies of ^{99m}Tc pyridoxyl-5-methyltryptophan in animals required HPLC purification because the heating step for isomerization of chelate forms often produced some unbound ^{99m}Tc pertechnetate [54]. The addition of ascorbate resulted in up to 50% of a new radiochemical form. This was found to be due to a few micrograms of thioglycerol added to pharmaceutical ascorbate solution. The additive was described on the package insert but not on the ampoules. Use of "unstabilized" solid ascorbic acid eliminated the problem. This is similar to the unexpected synthesis of Tc-oxo-dithiolate complexes reported by DePamphilis et al. [55].

Tc-N_2S_2 CHELATES

Over the past several years, a number of N_2S_2, diamide disulfur (DADS), chelating agents that contain a variety of functional groups have been studied [4,5]. A consistent structural pattern for this series has been observed by

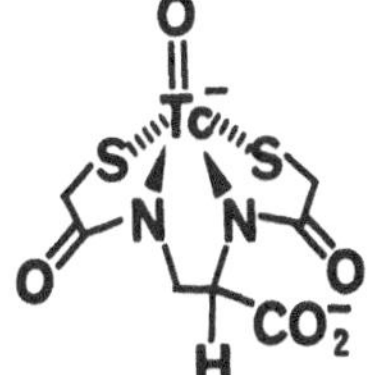

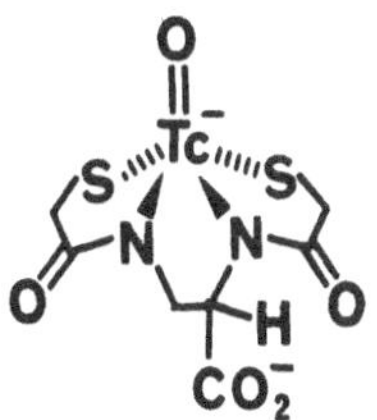

FIG. 9.1. Tc-CO_2 DADS-A and -B stereochemical isomers.

crystallographic studies of the parent compound, oxo[N,N′-ethylene bis(2-mercaptoacetamido)technetate (V) (Tc-DADS)] [3] and mass spectral studies on the carboxylate derivative Tc-CO_2DADS [56], as shown in Figure 9.1. HPLC has been especially useful in the separation of chelate ring stereoisomers and in the purification of radiochemical components so that biodistribution studies could be carried out on single radiochemical forms.

Comparison of C_8 and C_{18} Bonded Phases

The retention of ^{99m}Tc CO_2DADS epimers on C_8 and C_{18} bonded phases was compared using aqueous phosphate buffer or 5% ethanol/95% buffer, as shown in Table 9.2. Differences between columns were very small, with only an indication of slightly greater retention of ^{99m}Tc CO_2DADS-A and -B on the C_8 column in 5% ethanol. These observations are consistent with results that showed a rapid increase in retention up to a critical chain length, C_6 to C_{10}, and gradual changes at greater chain lengths [57]. Other data in Table 9.2 show the increase in separation between chelate ring epimers as the ethanol solvent fraction is decreased to 0. However, the retention volumes (hence times) are inconveniently long.

TABLE 9.2. Comparison of Retention of ^{99m}Tc N,N′-bis(mercaptoacetyl)-2,3-diaminopropanoate on Different Alkyl Chain Length Bonded Phases

	Retention Volumes	
Mobile Phase	Octyl	Octadecyl
0.05 *M* PO_4, pH 6	12.3 ml, 21.3 ml	14.2 ml, 22.2 ml
0.05 *M* PO_4, pH 6/ethanol (95 : 5)	4.2 ml, 6.3 ml	4.0 ml, 5.4 ml

Comparison of Reversed Phase and Ion Exchange

The use of anion exchange HPLC has proved superior to reversed-phase conditions in a few instances. Table 9.3 shows comparative retention volumes for selected ^{99m}Tc chelates. The chelate ring carboxylate, ^{99m}Tc CO_2DADS, is separated to a comparable extent on either, except that the order of elution is reversed, as shown in Figure 9.2.

This is as expected because the most polar or least retained compound under reversed-phase conditions would be expected to interact more strongly with the adsorbent cationic groups. In the case of the acetate derivative, complete separation was not possible using 5% ethanol and an ODS column, however, the anion exchange column resulted in complete separation, as shown in Figure 9.3. Comparison of the carboxylate derivatives with the parent DADS chelates and pertechnetate show retention differences between dianion and monoanion forms. Thus, in these relatively small, compact chelates, retention volumes of about 5 ml for monoanions and 10 ml for dianions result, with 0.010 *M* sulfate/0.010 *M* phosphate as the mobile phase.

A series of ^{99m}Tc N_2S_2 chelates with changes both in the position of amide carbonyl groups and in the addition of different groups to the center chelate ring and their HPLC behavior on an ODS column is shown in Table 9.4. The parent chelate compound, Tc-DADS, chromatographs as a single compo-

TABLE 9.3. Comparison of Reversed-Phase and Anion Exchange of Tc-N_2S_2 Chelates

	Retention Volumes	
Chelate	C_{18} *(ml)*[a]	*AX (ml)*[b]
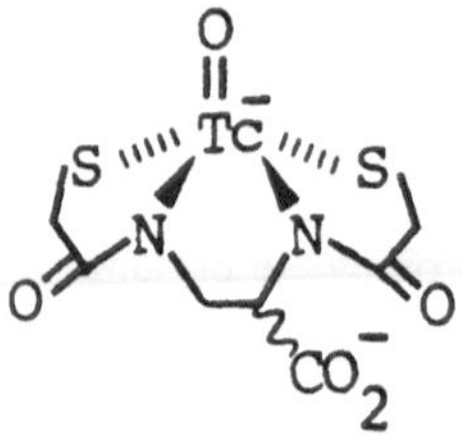 Tc(=O) N_2S_2 chelate with CO_2^- on center ring	4.0, 5.4	11.8, 10.2
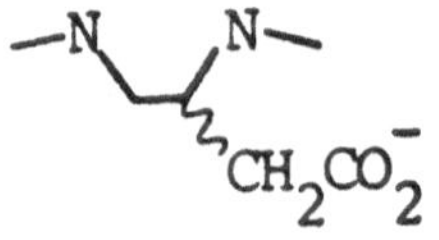 —N N— with $CH_2CO_2^-$	6.1, 6.4	10.2, 8.8
 —N N—	6.2	5.2
TcO_4^-	2.8	6.6

[a] Ethanol/0.010 *M* Pi, pH 6.0 (5 : 95), 5 μm ODS, 4.6 × 250 mm column.
[b] 0.010 *M* sulfate/0.010 *M* Pi, 10 μm Ultrasil-AX.

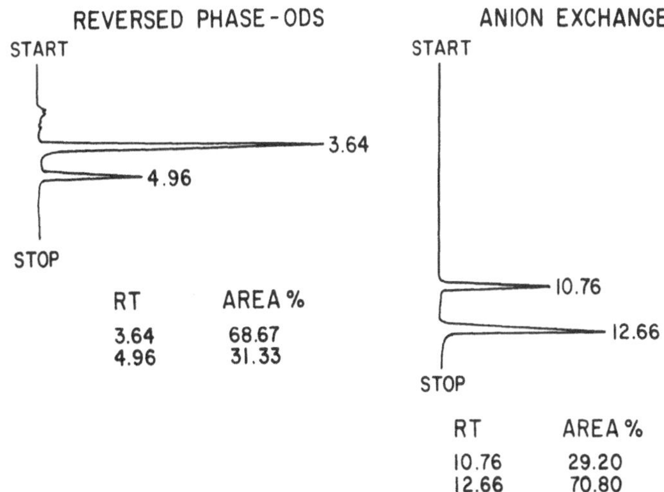

FIG. 9.2. Comparison of ^{99m}Tc CO_2DADS-A and -B epimer resolution on reversed-phase and anion exchange HPLC columns. Reversed-phase column was 4.6 × 250 mm with 5-μm ultrasphere ODS packing, ethanol/0.010 *M* Pi, pH 6.0 (5 : 95), flow rate 1.0 ml/min. Anion exchange column was 10-μm Ultrasil AX packing, 0.010 *M* sulfate/0.010 *M* phosphate (1 : 1), pH 6.0, flow rate 1.5 ml/min. Order of peak elution is reversed on the two types of columns.

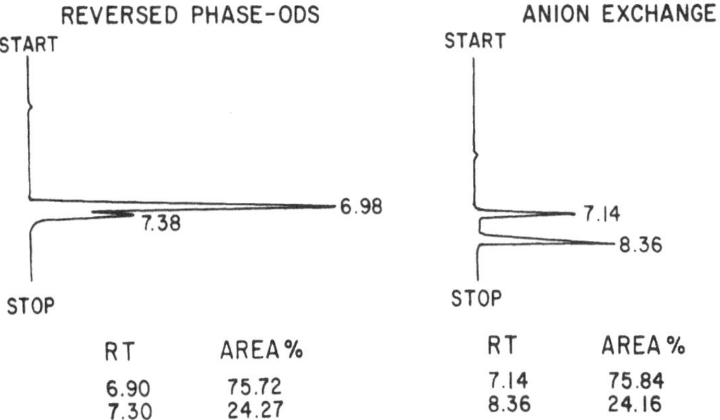

FIG. 9.3. Comparison of ^{99m}Tc N,N′-bis(mercaptoacetyl)-3,4-diaminobutanoate (CH_2CO_2DADS) on reversed-phase and anion exchange columns. Conditions were as described for Fig. 9.2. Note incomplete resolution on reversed phase while anion exchange provided base-line separation.

TABLE 9.4. Reversed-Phase Retention of Various Tc-N_2S_2 Derivatives[a]

	Retention Volumes	
Chelate	*20% Ethanol*	*5% Ethanol*
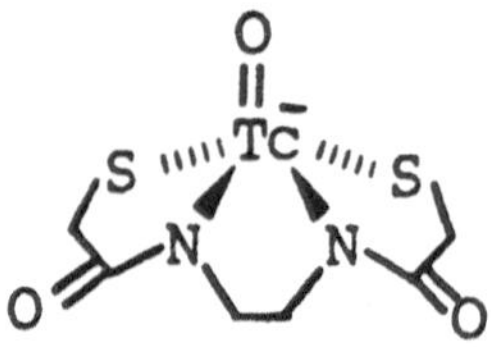	5.47 ml	60 ml
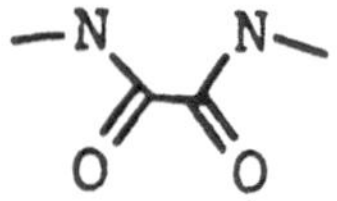	7.31	—
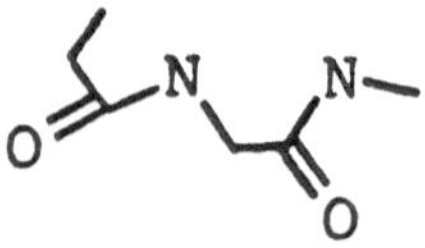	7.59	—
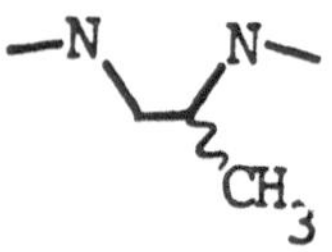	6.12, 7.54	—
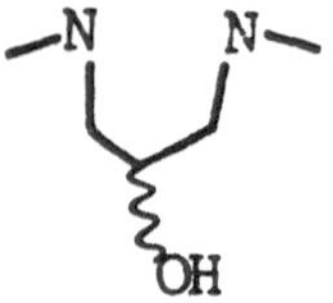	3.30, 4.03	—
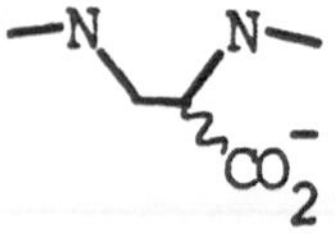	—	3.98, 5.40
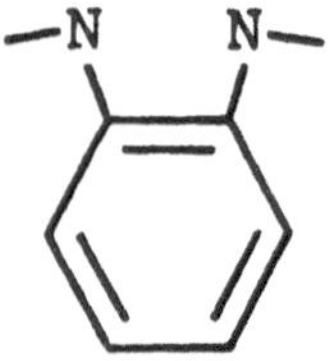	51.98	

[a] Conditions were ethanol/0.010 *M* Pi, pH 6.0, 5 μm ODS, 4.6 × 250 μm, 1.0 ml/min.

nent and requires 60 ml to elute at 5% ethanol and only 5.5 ml at 20% ethanol. Oxalyl and asymmetric amide carbonyl ligands resulted in the HPLC behavior shown for the second and third chelates. The 20% ethanol HPLC system was able to discriminate between them with unsubstituted ethylene groups between nitrogen and sulfur donor atoms apparently resulting in greater retention (lipophilicity). Addition of a methyl group to the center chelate ring increased retention as would be expected, and also gave two peaks assigned to chelate ring methyl epimers syn or anti to the Tc=0 group. In contrast to the increased retention caused by the methyl group, a polar hydroxy group decreased retention, as would be expected. Addition of the highly polar carboxylate group shortens retention such that reduction in organic solvent percentage to 5% is required for significant retention and separation of chelate ring stereoisomers in this reversed-phase system. The phenyl derivative gave, as expected, a more strongly retained single radioactive peak.

Determination of pK_a Values

The results of chromatographing ^{99m}Tc CO_2DADS epimers under ion pair conditions are shown in Table 9.5. The increased retention is clear, since comparable retention volumes are found at 5% ethanol for the sodium counter ion and 50% methanol for the tetrabutylammonium cation as counter ion. Surprisingly, the separation was lost as both epimers (preparatively separated under the sodium counter ion conditions and separately run with

TABLE 9.5. Ion-Pair Separation of ^{99m}Tc CO_2DADS Epimers (A & B)

	Retention Volumes	
Epimers	*Na$^+$*	*(Butyl)$_4$N$^+$*
A	3.98 ml[a]	5.21 ml[b]
B	5.40	5.19
% Organic modifier	5	50

[a] Ethanol/0.01 *M* Pi, pH 6 (5:95).
[b] 0.005 *M* (Butyl)$_4$NOH, 0.01 *M* NaH_2PO_4, pH 7. Both systems run on 0.5-μm ODS, 4.6 × 250 mm column, 1.0 ml/min.

injected radioactivity accounted for) had essentially identical retention volumes in the ion pair system. These results contrast to the excellent resolution obtained with the anion exchange column and indicate the complexity of mechanisms potentially operative under the ion pair conditions.

Compounds that contain pH-dependent ionizable groups will also have pH-dependent retention in all other forms of HPLC. This dependency has been applied to the determination of pK_a values of carboxylic acids with values obtained in good agreement with those obtained by titration [58]. We have applied anion exchange HPLC of ^{99m}Tc CO_2DADS to the determination of the pK_a of the chelate ring carboxylic acid group (Johnson and Fritzberg, unpublished data). Because the components separable by HPLC were determined to be epimers with significantly different biodistribution properties [5], it was of interest to know whether or not the carboxylate group of the anti (axial) epimer was bound to the technetium as described for the D-penicillamine complex of technetium [59]. A pH dependency was observed for both epimers as shown in Figure 9.4. Importantly, both show an inflection at pH 3.1, which indicates that they both have pK_a values in a range expected for peptide carboxylic acids and that neither is bound to the metal center. This methodology is particularly important for radioactive compounds since it is easily applied at tracer levels. Absolute pK_as are harder to determine by HPLC because organic modifiers affect the hydrogen ion activ-

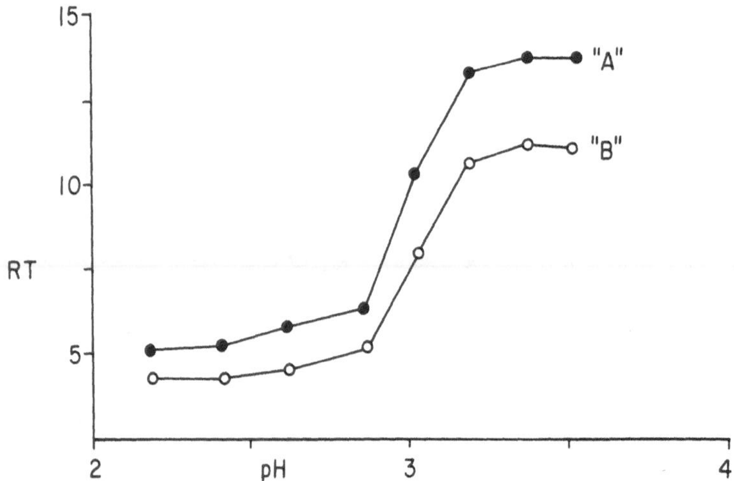

FIG. 9.4. Titration curves for ^{99m}Tc CO_2DADS-A and -B epimers on anion exchange column. Solvent conditions were 0.010 *M* sulfate/0.010 *M* phosphate at pH values shown. Control studies showed little elution power due to phosphate in this pH range. At higher pH values, retention times decreased slowly owing to increased amounts of dianionic phosphate. Complexes are monoanionic at low pH and dianionic at upper range shown. Inflection point at pH 3.1 indicates both epimers have p*K* values of about 3.1.

ity. A number of papers have appeared describing the changes and giving graphs of how to assess the true pH of an organic/aqueous mixture starting with a value for the pH measured with a pH meter [60,61].

Separation of Isomers and Homologous Series

The observations of renal secretion differences attributable to carboxylate stereoisomers prompted the synthesis of isomers in which the carboxylate and amide carbonyl groups occupy a variety of side chelate ring positions. Table 9.6 shows the reversed-phase HPLC behavior of these positional isomers. In each case, separation of epimers was possible with differences of 0.7 to 1.4 ml. However, the A and B peak retentions are similar and there is insufficient separation for chelate identification. An increase in separation can be achieved by using large columns, as shown in Figure 9.5. The 4.6-mm diameter column resulted in about 2000 theoretical plates for the peaks, while the 10-mm column resulted in about 10,000 plates. The ratio approxi-

TABLE 9.6. Reversed-Phase Retention of Positional Isomers[a]

	Retention Volumes	
^{99m}*Tc chelate*	*Peak A (ml)*	*Peak B (ml)*
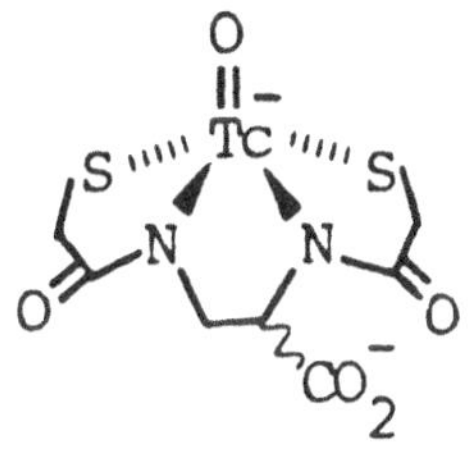	3.98	5.40
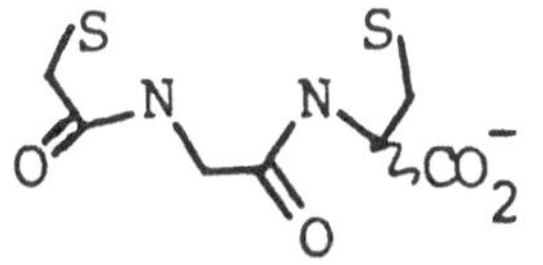	3.98	4.83
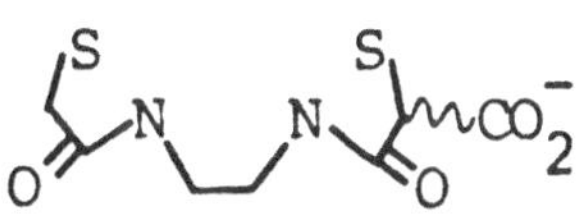	3.43	4.21
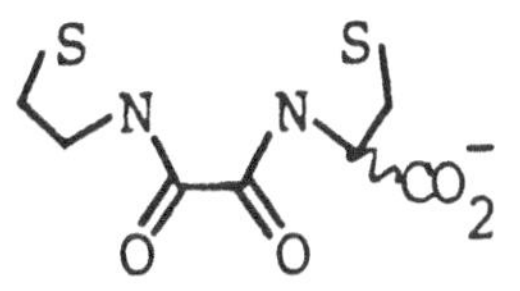	3.54	4.99

[a] Conditions were ethanol/0.01 *M* Pi, pH 6 (5 : 95), 5 μm ODS, 4.6 × 250 mm, flow rate 1.0 ml/min.

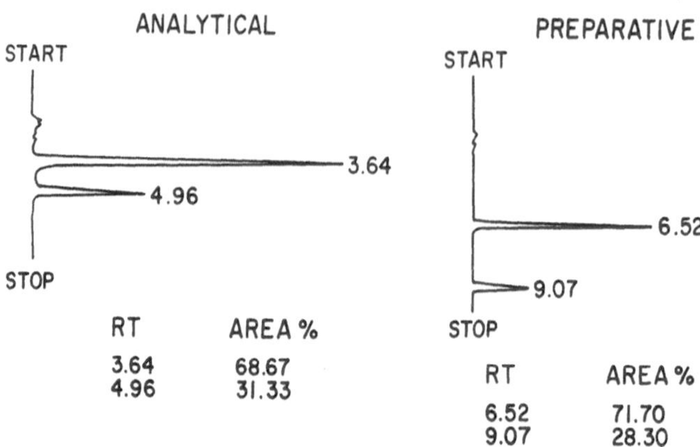

FIG. 9.5. Comparison of ^{99m}Tc CO_2DADS-A and -B epimer resolution on a 4.6 × 250 mm analytical column and a 10 × 250 mm preparative column. Packing was 5-μm ultrasphere ODS and solvent system was ethanol/0.010 *M* phosphate, pH 6 (5 : 95) at 1.0 ml/min for the analytical column and 3.0 ml/min for the preparative column.

mates the difference in column packing volumes. (The values are lower than specified by the manufacturer. Since runs on test samples using high organic content mobile phases give results that are similar to specifications, the difference is probably due to the low fraction of organic modifier present.)

Table 9.7 shows a set of ^{99m}Tc N_2S_2 carboxylate chelates that differ by addition of methylene or methyl groups. The vicinal methyl derivative resulted in four major products, as expected based on stereochemical possibilities. The first peak contained two components that were resolved on anion exchange HPLC. Interestingly, carboxylate on a propyl bridge rather than ethylene did not increase retention. However, the acetate homolog clearly resulted in increased retention, although separation of isomers was reduced. This is consistent with the observation that reversed-phase HPLC is especially good at separations where compounds differ by methylene units [62]. The addition of a second carboxylate group decreases retention to approximately column void volume. All of these chelates are quite polar, as they are dianions at pH 6. Retention could be increased significantly by incorporating alkyl quaternary amines as ion pairing agents in the mobile phase or by lowering the pH of the mobile phase to protonate the carboxylate groups. We have not done this for three reasons. First, we have sufficient resolution except for the vincinal dicarboxylate. Second, we have found it difficult to remove all traces of ion pairing agents from the columns. These traces could affect the subsequent use of the column in a non-ion pairing mode with nca ^{99m}Tc. (This can be obviated by dedicated columns.) Third, we wish to minimize organic solvent content of the mobile phase so that effects on biological studies are negligible.

TABLE 9.7. Reversed-Phase Separation of Carboxylate Derivatives[a]

	Retention Volumes	
	A (ml)	*B (ml)*
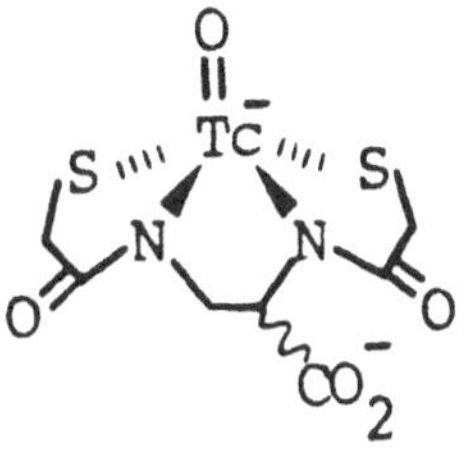	3.98	5.40
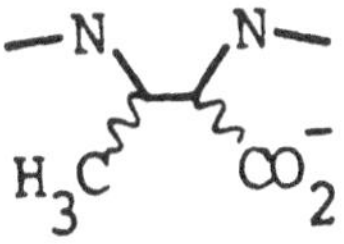	4.24	4.86, 8.86
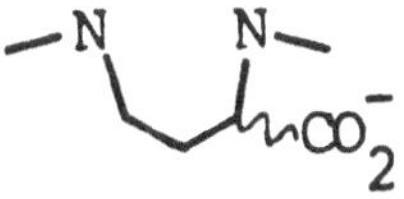	3.81	5.14
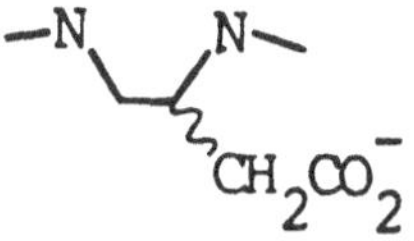	6.09	6.42
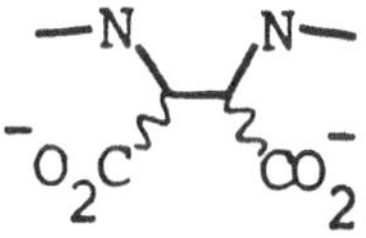	2.58	—

[a] Conditions were ethanol/0.01 *M* Pi, pH 6.0 (5 : 95), 4.6 × 250 mm ODS, 5-μm column, 1.0 ml/min. First peak contained two components that were separable by anion exchange HPLC. Fourth component.

Separation of Isomers with Different Biological Activity

Metabolism and excretion are known to be stereospecific in many cases. The differences in biological behavior that have been observed for stereoisomers separable by these HPLC conditions indicate the importance of using such a powerful chromatographic method for radiopharmaceutical development. As an example, the early peak of ^{99m}Tc CO_2DADS was shown to be more efficiently transported by the renal tubular transport system in animals [5]. The percent of the injected dose in human urine at 30 min is 58% for the earlier (A) component and 19% for the later (B) component [63]. Interestingly, the methyl derivative gave kidney retention of 29% of the injected dose for the A peak and 14% for the B at 10 min in mice with almost quantitative excretion of each in the urine by 2 hr [64]. As a result, we routinely purify preparations by HPLC before biodistribution studies so that the evaluation of radiopharmaceutical potential and hence design is not based on conclusions resulting from the study of radiochemical mixtures.

The use of HPLC in aiding the animal study portion of the development program has been a factor in the choice of solvent system. As discussed in Chapter 5, methanol, acetonitrile, and tetrahydrofuran are the most popular reversed-phase solvents. However, the pharmacologic effects of these solvents are not well known, and we have chosen to use ethanol as our organic solvent because there is some understanding of its pharmacologic dose dependency [65]. In order to minimize effects, mouse and rat studies are performed with at least 100-fold dilutions of the HPLC eluate so that the dose of ethanol injected into the animal is less than 1% in the mouse, and 3% in rats, of the dose reported to cause minimal effects, e.g., 250 mg/kg. Injection of

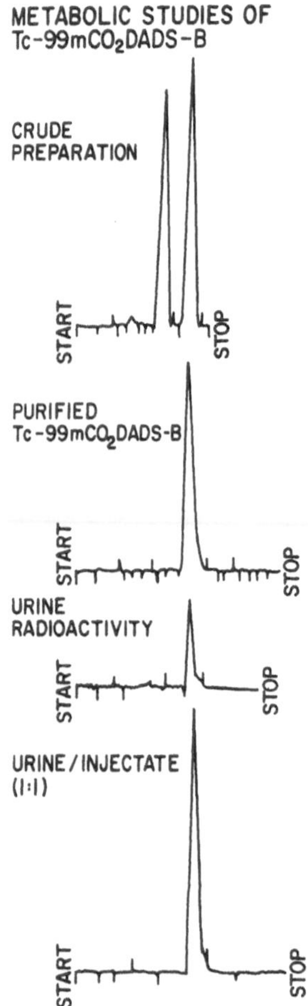

FIG. 9.6. Metabolic studies of ^{99m}Tc CO_2DADS-B by HPLC analysis of rat urine. Chromatograms represent crude preparation, purified ^{99m}Tc CO_2DADS-B, urine radioactivity, and coinjection of purified epimer B and urine radioactivity. Although no shoulder or extra component is indicated in the coinjection comparison, the fact that area-to-height ratios for the urine and coinjection peaks were 0.265 and 0.268 provides further nonsubjective support for the conclusion that the urine radioactivity was unchanged. Conditions were as described in Fig. 9.5 for the analytical column.

up to 2 ml without dilution in humans of the 5% ethanol-95% buffer eluate is also less than 1% of the 250 mg/kg minimal effect dose. The concentration of ethanol in the injection solution should be less than 25% v/v so as to avoid hemolysis.

Metabolic Studies

The efficient separation potential of HPLC makes it a very useful method for studying the metabolism of radiochemicals. We have evaluated urine and bile samples for changes in ^{99m}Tc CO_2DADS preparations, as shown in Figures 9.6 and 9.7. Both urine and bile are low in proteins and thus can be injected directly without removal of these proteins. For technetium compounds that are stable in the absence of excess ligand, the only concerns are reproducible retention times and column operating efficiencies. The retention times may be variable or different because of the surfactant properties

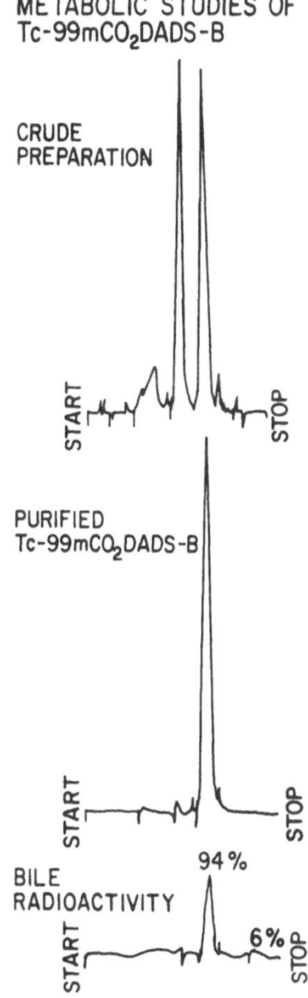

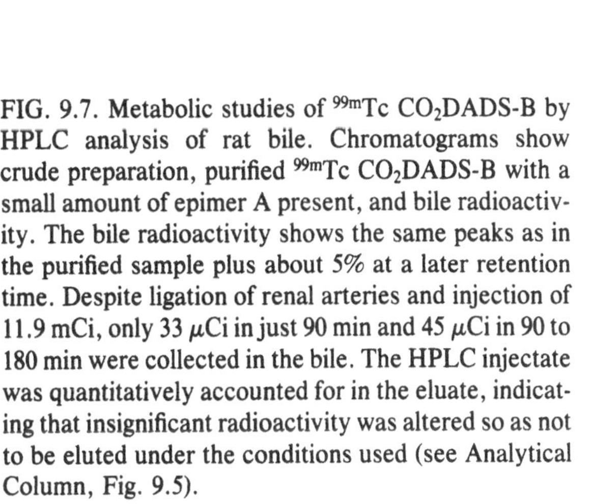
FIG. 9.7. Metabolic studies of ^{99m}Tc CO_2DADS-B by HPLC analysis of rat bile. Chromatograms show crude preparation, purified ^{99m}Tc CO_2DADS-B with a small amount of epimer A present, and bile radioactivity. The bile radioactivity shows the same peaks as in the purified sample plus about 5% at a later retention time. Despite ligation of renal arteries and injection of 11.9 mCi, only 33 μCi in just 90 min and 45 μCi in 90 to 180 min were collected in the bile. The HPLC injectate was quantitatively accounted for in the eluate, indicating that insignificant radioactivity was altered so as not to be eluted under the conditions used (see Analytical Column, Fig. 9.5).

of bile. If the retention times are different or variable, it is advisable to mix approximately equal radioactivity portions of starting preparation with comparable urine or bile samples. A resulting single peak will bolster confidence that differences in retention time are due to HPLC variables and not to radiochemical differences.

USE OF HPLC IN STEREOCHEMICAL STUDIES OF Tc and Re PENICILLAMINE COMPLEXES

Penicillamine has been bound to ^{99m}Tc and used as a renal agent [66] or hepatobiliary agent [67] depending on conditions of preparation. Little attention was given to stereochemistry, and in some cases conditions were probably harsh enough to racemize the penicillamine. Johnson et al. prepared ^{99m}Tc penicillamine [68] under conditions similar to those described for O=Tc-(D-penicillamine)$_2$ [59]. If either D-penicillamine or L-penicillamine was used, a single product of the same HPLC properties was observed. If D,L-penicillamine was used, two products separable by HPLC, as shown in Figure 9.8, were formed. Characterization of ^{99m}Tc products by NMR and mass spectroscopy indicated that the new product resulted from a mixed complex, O=Tc-(D-penicillamine)(L-penicillamine), shown in Figure 9.9. Interestingly, the mixed complex has both carboxylate groups anti to the technetium oxo bond and undergoes intramolecular racemization. The rhenium penicillamine complexes were also prepared and behaved similarly, except for displaying somewhat shorter retention times on reversed-phase HPLC, as shown in Figure 9.10. NMR analysis of the intramolecular racemization allowed a comparison of Tc (13 kcal) and Re (18 kcal) activation energies for the carboxylate exchange process.

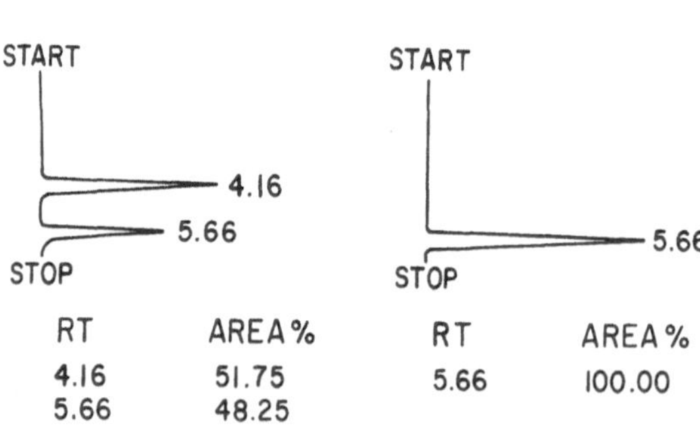

FIG. 9.8. HPLC chromatograms of penicillamine complexes of technetium indicating stereochemical fate. The single peak results from either L-penicillamine or D-penicillamine and the metal, the two peaks from racemic penicillamine. The additional peak represents the mixed ligand complex, TcO(L-penicillamine)(D-penicillamine). HPLC conditions were ethanol/0.01 *M* phosphate, pH 6 (15:85) on 5-μm ultrasphere (ODS, 1.0 ml/min flow rate. Radiometric and 423 nm detection were equivalent.

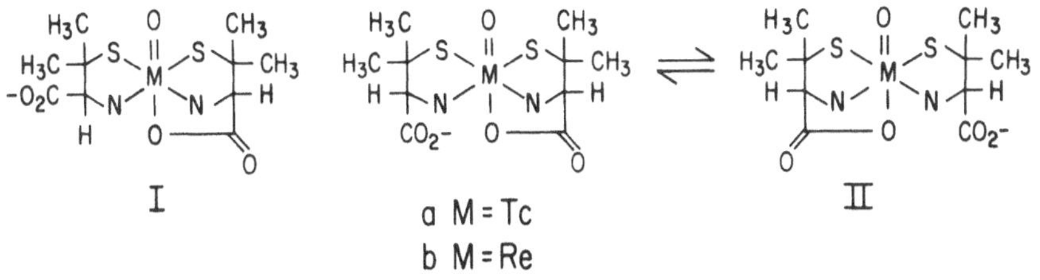

I

II

a M = Tc
b M = Re

FIG. 9.9. Structures of technetium or rhenium complexes resulting from stereochemically pure or racemic penicillamine. Complexes with two D- or two L-penicillamines are indistinguishable by HPLC. The additional peak resulting from an L- and a D-penicillamine interestingly has cis sulfur and nitrogen donor atoms and undergoes intramolecular racemization.

HPLC IN THE DEVELOPMENT AND ANALYSIS OF HIDAs

Determination of the Stoichiometry

One of the earliest group of radiotracers that benefited from HPLC is the substituted iminodiacetic acid complexes of technetium (HIDAs). HPLC has been used to determine the structure of the complexes, to analyze reaction mixtures and develop formulations, to investigate metabolism, and to develop better complexes using an SDR approach. In all cases, classical reversed-phase HPLC systems were employed.

Analysis of Reaction Mixtures

Loberg and Fields [69] used a reversed-phase system consisting of a μ Bondapak C_{18} column eluted with a 0.005 M phosphate buffer, (pH 6.8)/acetonitrile (90 : 10) mobile phase to determine the stoichiometry of the technetium complexes. They were particularly interested to find out whether tin was an integral part of the complex. By using ^{99}Tc, ^{99m}Tc, ^{113}Sn and ^{14}C HIDA they were able to show that the HIDA forms a 2 : 1 complex with the technetium and that no tin is included in the complex. They showed equivalence be-

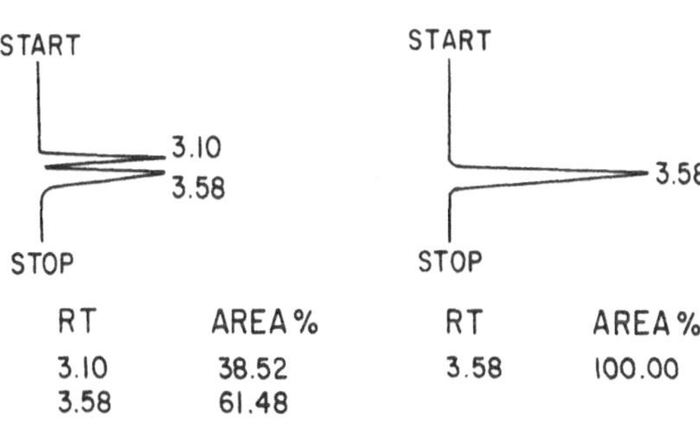

FIG. 9.10. HPLC chromatograms of penicillamine complexes of rhenium. Conditions were as described in Fig. 9.8. Stereochemical results are the same as those of technetium. Note shorter retention times, however. Detector was set at 340 nm.

tween the ^{99}Tc and ^{99m}Tc preparations. Fields et al. [70] used the same HPLC system to examine the effect of the pK_a of the imino nitrogen on the radiochemical purity and were able to show that in aqueous solution the pK_a had to be about 6 to achieve stable, rapid labeling.

A number of groups were also examining HIDA complexes by gel chromatography and found that they could see two major radioactive peaks, one of which converted to the other within 2 hr after reconstitution [71,72]. There was some speculation that 1:1 and 2:1, HIDA:Tc complexes were being separated. The peaks were not reported by others using HPLC [70]. Fritzberg and Lewis [6] used HPLC to examine this question in more detail and found that the peak proportions were pH dependent and were present in a number of different HIDA derivative preparations. They concluded that equilibrium was reestablished upon intravenous injection as indicated by studies with the diethyl derivative. Nunn and coworkers [73,74] studied different formulations and different HIDA derivatives using HPLC and confirmed the observations of Fritzberg and Lewis [6]. They were able to cycle the pH of the reaction solution and show that the radioactivity cycled between the two major peaks. They also showed that the different HIDA derivatives took different times to reach equilibrium and investigated the effects of ligand concentration on the labeling rates. There is still no published proof as to the existence of the postulated 1:1 complex or for its postulated greater *in vivo* efficacy. Objections to its existence and relevance have been raised [6,72,75], and it seems that HPLC is giving a better representation of the clinical significance than the gel chromatography.

Of some note is that all of these studies on the effect of the pH of the reaction mixture on the radiochemical purity were examined using eluants of one pH. The fact that different results were obtained shows that the separation rate of the HPLC systems used is fast compared with the rate of attainment of equilibrium of the reaction mixtures. This is true for the HIDAs, but may not be so for other faster reacting complexes.

Metabolism

The excreted form of the HIDAs was determined by HPLC analysis of the bile. The bile duct was cannulated in a rat and the bile collected at 5-min intervals. About 100 μl were excreted and collected for each time point. This was diluted with 0.5 ml of saline to reduce the natural surfactant concentration prior to injection on the HPLC. This protects the column and helps prevent changes in the retention time of the relevant peaks. The collection of samples is faster than their analysis, so checks have to be made to see if the drug is being degraded in the bile after collection, as shown in Figure 9.11. This was done by analyzing one of the more radioactive early samples at a number of times after collection. Some degradation was seen, and so all other results were adjusted for this degradation. The results showed that unchanged drug was excreted into the bile.

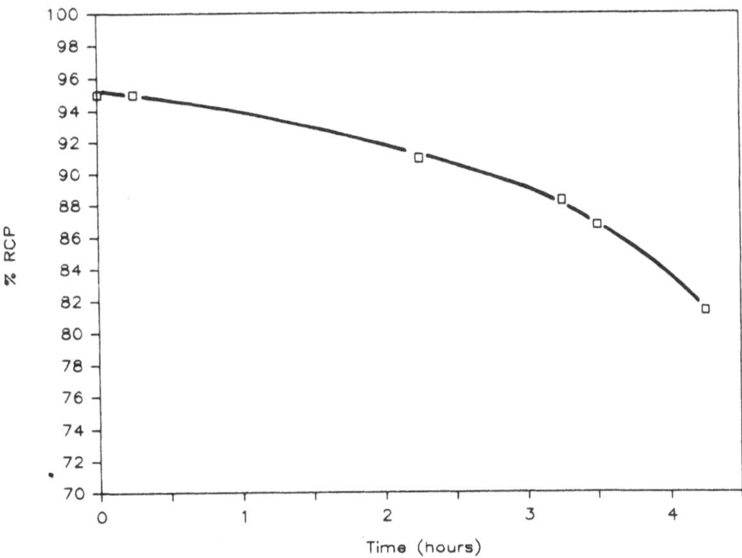

FIG. 9.11. Degradation of Tc-mebrofenin in bile.

Development of SDRs Leading to New Compounds

Radiopharmaceuticals suffer the same developmental problems as therapeutic drugs, i.e., how to optimize the characteristics of the class in the most efficient manner. A stochastic approach is the normal initial response until enough data are available to start constructing SDRs. There are normally two sides to the problem as described by Topliss [76]. One is where synthesis is rapid but testing is slow; the other is where synthesis is slow but testing is rapid. (The Topliss and the Hansch approaches.)

The development of the HIDAs was reduced to measuring the lipophilicity of each complex [1,74]. To do this by traditional shake-flask methods on the complexes would be time consuming and would involve handling undesirable amounts of air-sensitive radioactive material. Instead, they assumed that the complexes all had the same core and that the relative lipophilicity of the ligands could be used to rank the complexes. The shake-flask method was not used to rank the ligands because of the wide range in lipophilicity of the ligands and because there were so many of them. An HPLC method was used to measure the relative lipophilicity of the ligands. This was reported as the capacity factor (K') where:

$$K' = \frac{t_r - t_m}{t_m} \tag{1}$$

t_m is the elution point of unretained compounds, and t_r is the elution point of a retained compound. Each should be in the same units, e.g., time, volume, distance (from a chart recorder trace). The capacity factor is the preferred means of reporting HPLC retention times, especially for correlative work, because it is directly applicable from system to system. All other reporting methods are sensitive to fitting dead volumes, flow rate inaccuracies, column packing efficiencies, detector volumes, etc. The reproducibility of capacity factors from system to system depends upon the accurate measurement of t_m. Wells and Clark [77] have addressed this problem and recommend that inorganic ionic solutes be used to determine t_m.

The capacity factors of the ligands were then correlated with the sum of the π values of the substituents in the aromatic ring of the ligand. The relationship between the theoretical and measured lipophilicity of the ligands showed that some of the substituents were shielded and did not contribute the expected amount to the lipophilicity of the ligand (specifically the ortho substituents). This situation was also true for a limited number of technetium complexes [1], which validated the assumption that the contribution of the core was constant. In this way, the search for the ligand necessary to produce a technetium complex of the required lipophilicity was considerably shorter and easier.

PHOSPHINES

Cationic technetium complexes of diphosphines have been investigated as potential myocardial imaging agents because they may act as potassium analogues [78]. Bisdimethylphosphinoethane (DMPE) is the ligand most used. No reducing agent is required because the ligand itself acts as reducing agent. The reduction of pertechnetate by DMPE is a multistep process and produces three major cationic species: the dioxo-, the dichloro-, and the tris-DMPE.

$$\underset{\text{Tc(VII)}}{TcO_4^-} \rightarrow \underset{\text{Tc(V)}}{[Tc(DMPE)_2O_2]^+} \rightarrow \underset{\text{Tc(III)}}{[Tc(DMPE)_2Cl_2]^+} \rightarrow \underset{\text{Tc(I)}}{[Tc(DMPE)_3]^+} \quad (2)$$

In animals, their ability to concentrate in the myocardium increases as the oxidation state of the technetium is reduced; it is, therefore, very important that each can be separated from the other. A number of different analyses of this cascade have been developed, aided by the fact that ^{99}Tc complexes have been characterized by HPLC. Vanderheyden et al. [8] have described a reversed-phase column eluted with an ion pairing solvent system. The system consists of a C_8 column, 250 × 4.5 mm I.D. (Jones Chromatography) eluted with 70/30 methanol : 0.003 *M* phosphate buffer, pH 7, (containing 0.02 *M* sodium hepansulfonate), at a flow rate of 1.4 ml/min. They report that with all the systems they tried, the retention time of the technetium complexes was dependent on the number of moles injected; this is probably because theirs is an ion-pairing system. However, the normal nca masses were too low to show any variation in retention time. In all cases, the UV-absorbing peaks derived from ^{99}Tc material were coincident with the radioactive peaks derived from the ^{99m}Tc material. Nunn (unpublished) used a variety of reversed-phase columns with no ion-pairing agents in the eluate and did not find any great concentration dependence of the retention times. The separation did change from reversed to normal phase or some form of ion chromatography, depending upon the amount of nonaqueous modifier in the solvent, as shown in Figure 9.12. The separations were done using a μ Bondapak C_{18} column, 250 × 4.1 mm I.D., eluted with 0.25 *M* phosphate buffer, pH 6.8, together with varying proportions of THF at a flow rate of 1 ml/min. The samples were fully characterized ^{99m}Tc material with the detector set at 235 nm. The elution order of the dioxo vs. the dichloro species could be reversed by replacing tetrahydrofuran with methanol. This is probably a result of binding of tetrahydrofuran to the oxygens in a similar manner to the binding of tetrahydrofuran to ether oxygens as reported by Roggendorf and Spatz [79].

PHOSPHONATES

Unlike the phosphines, which have not made the transition from the research environment to routine clinical use, technetium complexes of diphosphonates have been shown to be essential for the management of cancer involving the osseous part of the skeleton. Their use started before the advent of HPLC in technetium chemistry.

Pinkerton [2] performed ion exchange chromatography using an Aminex A-27 column eluted with 0.85 *M* acetate, pH 8.4, at 0.2 ml/min. The stable technetium was reduced with borohydride at pH 8.5 in the presence of

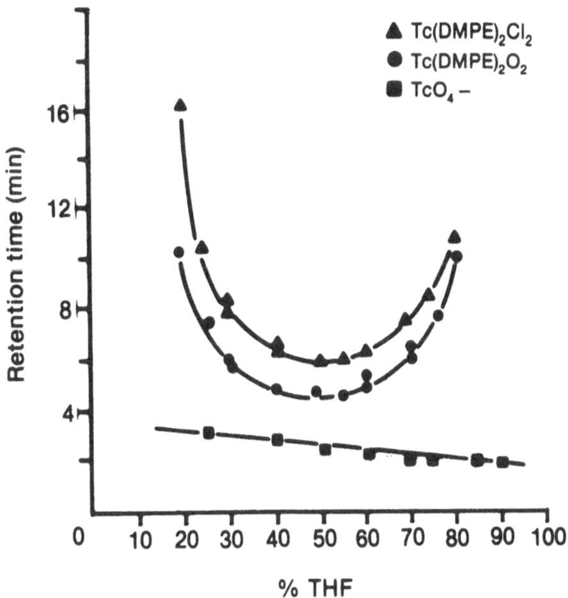

FIG. 9.12. Chromatography of the Tc-DMPE cascade.

HEDP. Over a dozen peaks were identified by their absorbance at 405 nm as containing ^{99}Tc. The number and relative proportions of the peaks could be changed by preparing the samples in the presence or absence of oxygen or by oxidizing or rereducing the samples. When ^{99m}Tc was added to the stable technetium just prior to reduction, the distribution of ^{99m}Tc as measured by its radioactivity could be determined. Note that extinction coefficients must be known to determine the percent abundances by UV (because different species will have different extinction coefficients at a given wavelength), but the detection efficiency is equal for the radioactivity. More recently, Tanabe et al. [13] from the same laboratories have shown that the pH of the reaction mixture affects the composition of technetium MDP complexes formed by reduction of the technetium with borohydride. Using the same HPLC system, Aminex A-27/acetone eluant, they measured the in vivo distribution of HPLC cuts and found that the major component made at pH 8.2 in carrier-containing solutions had the best in vivo characteristics. Unfortunately, this species was not made in the nca solutions [80].

Srivastava et al. [81] examined clinical solutions with a number of HPLC systems including anion exchange, C_2-, C_8-, C_{18}-, NH_2-, and CN-bonded columns. They added ascorbic acid and the phosphonate ligand to the eluant to stabilize the Tc component. (Without added ligand, little or no technetium is eluted from silica-backed columns when examining nca solutions). In their hands Pinkerton's modified system separated HEDP complexes but not MDP complexes. In contrast, C_{18} columns with an eluant consisting of 10% dioxane, 0.0025 *M* *n*-butylammonium hydroxide, 0.0025 *M* MDP, and 0.001 *M* sodium acetate, pH 7.8 at 2 ml/min gave good results. They found multiple peaks with relative proportions that changed with time, temperature, oxidation, etc., and noted that the different fractions had different in vivo characteristics in animals.

To date, no fully characterized ^{99}Tc phosphonate has been unequivocally identified with a species in a clinically used solution. Some of these separation systems are quite complex and could be producing artifacts. As with the phosphines, a number of different technetium oxidation states could be involved. In addition, the phosphonate ligands are not as strong as the phosphines, so a number of mixed ligand complexes containing water, halogen, or hydroxide ions as well as the phosphonates may form [82]. The normal reducing agent in clinical formulations is Sn(II). This cannot be used in carrier preparations because it and the Sn(IV) produced compete with the technetium for the ligands and generally interfere. Borohydride or dithionite are used instead, but the equivalence of the products has not been proven.

CONCLUSIONS

We feel that there are four major areas where radio-HPLC is almost essential for the development of radiopharmaceuticals. They are, 1) the determination of radiochemical purity, 2) the determination of stoichiometry and structure, 3) the development of SDRs, and 4) the fractionation of multicomponent mixtures.

We have given practical examples of technetium chelates to illustrate how these four operations can be achieved. In doing so, we hope to have given our idea of where radio-HPLC has helped radiopharmaceutical chemistry in the past and the methods by which it will advance it in the future.

ACKNOWLEDGMENTS

The support of Department of Energy Contract DE-AC02-83ER 60140 to A.R. Fritzberg is acknowledged.

REFERENCES

1. Nunn, A.D.: Structure-distribution relationships of radiopharmaceuticals. Correlation between the reversed-phase capacity factors for Tc-99m phenylcarbamoylmethyliminodiacetic acids and their renal elimination. *J. Chromatogr.* **255:**91–100, 1983.
2. Pinkerton, T.C., Ferguson, D.L., Deutsch, E., Heineman, W.R. and Libson, K.: In vivo distributions of some component fractions of Tc($NaBH_4$)-HEDP mixtures separated by anion exchange high performance liquid chromatography. *Int. J. Appl. Radiat. Isot.* **33:**907–915, 1982.
3. Jones, A.G., Davison, A., LaTegola, M.R., Brodack, J.W., Orvig, C., Sohn, M., Toothaker, A.K., Lock, C.J.L., Franklin, K.J., Costello, C.E., Carr, S.A., Biemann, K. and Kaplan, M.L.: Chemical and in vivo studies of the anion oxo[N,N′-ethylenebis(2-mercaptoacetimido]-technetate (V). *J. Nucl. Med.* **23:**801–809, 1982.
4. Fritzberg, A.R., Klingensmith, W.C. III, Whitney, W.P. and Kuni, C.C.: Chemical and biological studies of Tc-99m N,N′bis(mercaptoacetamido)-ethylenediamine: a potential replacement for I-131 hippurate. *J. Nucl. Med.* **22:**258–263, 1981.
5. Fritzberg, A.R., Kuni, C.C., Klingensmith, W.C., Stevens, J. and Whitney, W.P.: Synthesis and biological evaluation of Tc-99m N,N′-bis(mercaptoacetyl)-2,3-diaminopropanoate: a potential replacement for [^{131}I]*o*-iodohippurate. *J. Nucl. Med.* **23:**592–598, 1982.
6. Fritzberg, A.R. and Lewis, D.: HPLC analysis of Tc-99m iminodiacetate hepatobiliary agents and a question of multiple peaks. *J. Nucl. Med.* **21:**1180–1184, 1980.
7. Kung, H.F., Molnar, M., Billings, J., Wicks, R. and Blau, M.: Synthesis and biodistribution of neutral lipid-soluble Tc-99m complexes that cross the blood–brain barrier. *J. Nucl. Med.* **25:**326–332, 1984.
8. Vanderheyden, J-L., Deutsch, E., Libson, K. and Ketring, A.R.: Synthesis and characterization of [^{99m}Tc(dmpe)$_3$]$^+$, a potential myocardial imaging agent. *J. Nucl. Med.* **24:**9, 1983.
9. Vanderheyden, J-L., Libson, K., Nosco, D.L., Ketring, A.R. and Deutsch, E.: Preparation and characterization of [^{99m}Tc(DMPE)$_2$X$_2$], x = Cl, Br (DMPE) =

1,2-bis(dimethylphosphino)ethanol. *Int. J. Appl. Radiat. Isot.* **34:**1611–1615, 1983.

10. Baldas, J., Bonnyman, J., Pajer, P.M., Williams, G.A.: The influence of reducing agents on the composition of ^{99m}Tc-complexes: implications for ^{99m}Tc-radiopharmaceutical preparation. *Eur. J. Nucl. Med.* **7:**187–189, 1982.
11. Deutsch, E., Heineman, W.R., Zodda, J.P., Gilbert, T.W. and Williams, C.C.: Preparation of "no-carrier added" technetium-99m complexes: determination of the total technetium content of generator eluants. *Int. J. Appl. Radiat. Isot.* **33:**843–848, 1982.
12. Vallabhajosula, S., Goldsmith, S.J., Pollina, R. and Lipszyc, H.: Radiochemical analysis of Tc-99m human serum albumin with high pressure liquid chromatography. *J. Nucl. Med.* **23:**326–329, 1982.
13. Volkert, W.A., Troutner, D.E. and Holmes, R.A.: Labeling of amine ligands with ^{99m}Tc in aqueous solutions by ligand exchange. *Int. J. Appl. Radiat. Isot.* **33:**891–896.
14. Tanabe, S., Zodda, J.P., Deutsch, E. and Heineman, W.R.: Effect of pH on the formation of Tc($NABH_4$)-MDP radiopharmaceutical analogs. *Int. J. Appl. Radiat. Isot.* **34:**1577–1584, 1983.
15. Bonnyman, J.: Effect of milking efficiency on ^{99}Tc content of ^{99m}Tc derived from ^{99m}Tc generators. *Int. J. Appl. Radiat. Isot.* **34:**901–906, 1983.
16. Shiue, C-Y., Arnett, C.D. and Wolf, A.P.: Synthesis and biodistribution of 2-deoxy-2-[^{18}F]fluoro-D-glucopyranosyl[^{18}F]fluoride. *Eur. J. Nucl. Med.* **9:**77–80, 1984.
17. Padgett, H.C., Barrio, J.R., MacDonald, N.S. and Phelps, M.E.: The unit operations approach applied to the synthesis of [1-^{11}C]2-deoxy-D-glucose for routine clinical applications. *J. Nucl. Med.* **23:**739–744, 1982.
18. Ehrenkaufer, R.E., Potocki, J.F. and Jewett, D.M.: Simple synthesis of F-18-labeled 2-fluoro-2-deoxy-D-glucose. *J. Nucl. Med.* **25:**333–337, 1984.
19. Shiue, C-Y., Salvadori, P.A., Wolf, A.P., Fowler, J.S. and MacGregor, R.R.: A new improved synthesis of 2-deoxy-2-[^{18}F]fluoro-D-glucose from ^{18}F labeled acetyl hypofluorite. *J. Nucl. Med.* **23:**899–903, 1982.
20. Laufer, P. and Kloster, G.: Remote control synthesis of 3-[^{11}C]-methyl-D-glucose. *Int. J. Appl. Radiat. Isot.* **33:**775–776, 1982.
21. Levy, S., Livni, E., Elmaleh, D.R., Varnum, D.A. and Brownell, G.L: 2-Deoxy-2-[^{18}F]fluoro-3-O-methyl-D-glucose synthesis and animal biodistribution studies. *Int. J. Appl. Radiat. Isot.* **34:**1560–1562, 1983.
22. Vora, M.M., Boothe, T.E., Finn, R.D., Smith, P.M. and Gilson, A.J.: Quality control procedures in the preparation of 2-deoxy-D-[1-^{11}C]-glucose radiopharmaceuticals. *J. Labelled Compd. Radiopharm.* **XX:**417–427, 1983.
23. Stone-Elander, S.A., Ehrin, E. and Nilsson, J.L.G.: A semiautomatic remote controlled synthesis of [^{11}C]-2-deoxy-D-glucose. *J. Labelled Compd. Radiopharm.* **XIX:**1372, 1982.
24. Katzenellenbogen, J.A., McElvaney, K.D., Senderoff, S.G., Carlson, K.E., Landvatter, S.E., Welch, M.J. and the Los Alamos Group: 16α-[^{77}Br]-Bromo-11-β-methoxyestradiol-17β a gamma-emitting estrogen imaging agent with high uptake and retention by target organs. *J. Nucl. Med.* **23:**411–419, 1982.
25. Gibson, R.E., Eckelman, W.C., and Francis, B., O'Brien, H.A., Mazaitis, J.K., Wilbur, S. and Reba, R.C.: [^{77}Br]-17-β-Bromoethynylestradiol: in vivo and in vitro characterization of an estrogen receptor radiotracer. *Int. J. Nucl. Med. Biol.* **9:**245–250, 1982.

26. Senderoff, S.G., McElvany, K.D., Carlsen, K.E., Heiman, D.F., Katzenellenbogen, J.A. and Welch, M.J.: Methodology for the synthesis and specific activity determination of 16α-[^{77}Br]-bromoestradiol-17β and 16α-[^{77}Br]-11B-methoxyestradiol-17β two estrogen receptor-binding radiopharmaceuticals. *Int. J. Appl. Radiat. Isot.* **33**:545–551, 1982.
27. Scholl, H., Kloser, G. and Stocklin, G.: Bromine-75-labeled 1,4-benzodiazepines: potential agents for the mapping of benzodiazepine receptors in vivo. *J. Nucl. Med.* **24**:417–422, 1983.
28. Denutte, H., Goethals, P., Cattori, H., Bogaert, M., Vanderwalle, T., Vandecasteele, C., Jonckheere, J. and De Leenheer, A.: The production in high yield of N′-(4-[^{11}C]methyl)-imipramine. *J. Nucl. Med.* **24**:1185–1187, 1983.
29. Diksic, M., Farrokhazad, S., Yamamoto, L. and Fernidel, W.: Synthesis of "no carrier added" 1,3-bis-(2-chloroethyl)nitrosourea (BCNU). *J. Nucl. Med.* **23**:895–898, 1982.
30. Wieland, D.M., Wu, J., Brown, L.E., Mangner, T.J., Swanson, D.P. and Beierwaltes, W.H.: Radiolabeled adrenergic neuron-blocking agents: Adrenomedullary imaging with [^{131}I]iodobenzylguanidine. *J. Nucl. Med.* **21**:349–353, 1980.
31. Jette, D.C., Wiebe, L.I. and Chapman, J.D.: Synthesis and in vivo studies of the radiosensitizer 4-[^{82}Br]bromomisonidazole. *Int. J. Nucl. Med. Biol.* **10**:205–210, 1983.
32. Carlson, L. and Andersen, K.: ^{131}I-labelled N-isopropyl-*p*-iodoamphetamine. *Eur. J. Nucl. Med.* **7**:280–281, 1982.
33. Zanzonico, P.B., Bigler, R.E. and Schmall, B.: Neuroleptic binding sites: specific labeling in mice with [^{18}F]haloperidol, a potential tracer for positron emission tomography. *J. Nucl. Med.* **24**:408–416, 1983.
34. Berridge, M., Comar, D., Roeda, D. and Syrota, A.: Synthesis and *in vivo* characteristics of [2-^{11}C]5,5-dimethyloxazolidine-2,4-dione (DMO). *Int. J. Appl. Radiat. Isot.* **33**:647–665, 1982.
35. Drenth, B.F.H. and De Zeeuw, R.A.: Radiochemical purity control of radiolabeled drugs. *Int. J. Appl. Radiat. Isot.* **33**:681–683, 1982.
36. Van Haver, D., DeClercq, P., Venderwalle, T. and Vandecasteele, C.: Remote-controlled production of [^{11}C] antipyrine. *Int. J. Appl. Radiat. Isot.* **33**:751–754, 1982.
37. DeJesus, O.T., Friedman, A.M., Prasad, A. and Revenaugh, J.R.: Preparation and purification of ^{77}Br-labelled *p*-bromospiroperidol. *J. Labelled Compd. Radiopharm.* **XX**:745–756, 1983.
38. Pike, V.W., Palmer, A.J., Horlock, P.L., Perun, T.J., Freiberg, L.A., Dunnigan, D.A. and Liss, R.H.: Semi-automated preparation of C-11 labelled antibiotic—[N-methyl-C-11]erythromycin A lactobionate. *Int. J. Appl. Radiat. Isot.* **35**:103–109, 1984.
39. Angelberger, P., Wagner-Loffler, M. and Hruby, R.: I-123 (131)-*m*-iodobenzylguanidine: optimized preparation and tissue distribution. *J. Labelled. Compd. Radiopharm.* **XIX**:1479–1481, 1982.
40. Francis, B.F., Rzeszotarski, W.J., Eckelman, W.C. and Reba, R.C.: Nucleophilic iodination of 3-quinuclidinyl benzilate (QNB). *J. Labelled Compd. Radiopharm.* **XXX**:1499–1500, 1982.
41. Barrio, J.R., Keen, R.E., Ropchan, J.R., MacDonald, N.S., Baumgartner, F.J., Padgeth, H.C. and Phelps, M.E.: L-[1-^{11}C]Leucine: routine synthesis by enzymatic resolution. *J. Nucl. Med.* **24**:515–521, 1983.

42. Bolster, J.M., Vaalburg, W., and Van Veen, W., Van Dijk, T.H., Vander Molen, H.D., Wynberg, H. and Woldring, M.G.: Synthesis of no-carrier-added L- and D-[1-^{11}C]-DOPA. *Int. J. Appl. Radiat. Isot.* **34:**1650–1652, 1983.
43. Livni, E., Elmaleh, D.R., Levy, S., Brownell, G.L. and Strauss, W.H.: Beta-methyl[1-^{11}C]heptadecanoic acid: a new myocardial metabolic tracer for positron emission tomography. *J. Nucl. Med.* **23:**169–175, 1982.
44. Coenen, H.H., Harmand, M.F., Kloster, G. and Stocklin, G.: 15-(*p*-[^{75}Br]-Bromophenyl-pentadecanoic acid: pharmacokinetics and potential as a heart agent. *J. Nucl. Med.* **22:**891–896, 1981.
45. Dudczak, R., Klelter, K., Frischauf, H., Losert, U., Angelberger, P. and Schmoliner, R.: The use of ^{123}I-labeled heptadecanoic acid (HDA) as metabolic tracer. *Eur. J. Nucl. Med.* **9:**81–85, 1984.
46. Fritzberg, A.R. and Eshima, D.: Iodophenylsulfonamide fatty acid analogs as potential myocardial imaging agents. *Int. J. Appl. Radiat. Isot.* **33:**451–453, 1982.
47. Padgett, H.C., Robinson, G.D. and Barrio, J.R.: [1-^{11}C]palmitic acid: improved radiopharmaceutical preparation. *Int. J. Appl. Radiat. Isot.* **33:**1471–1472, 1982.
48. Reidel, G., Bauer, R. and Pabst, H.W.: Preparation and purification of ω-(*p*-^{123}I-iodophenyl)-pentadecanoic acid. *Int. J. Appl. Radiat. Isot.* **34:**1642–1644, 1983.
49. Schmall, B., Conti, P.S., Sundoro-Wu, B., Dahl, J.R., Jacobson, J.K. and Lee, R.: Synthesis and high pressure liquid chromatography of carbon-11 labeled carboxylic acids: valproic acid and palmitic acid. *J. Labelled Compd. Radiopharm.* **XIX:**1278–1279, 1982.
50. Dischino, D.D., Welch, M.J., Kilbourn, M.R. and Raichle, M.E.: Relationship between lipophilicity and brain extraction of C-11 radiopharmaceuticals. *J. Nucl. Med.* **24:**1030–1038, 1983.
51. Vos, C.M. and Westera, G.: Labeled bleomycin as a tumor localizing agent III. Selectivity of tumor tissue uptake of the different forms of [^{57}Co] bleomycin A_2 and B_2. *Int. J. Nucl. Med. Biol.* **9:**237–243, 1982.
52. Gatley, S.J. and Shaughnessy, W.J.: Production of ^{18}F-labeled compounds with ^{18}F-produced with a 1-MW research reactor. *Int. J. Appl. Radiat. Isot.* **33:**1325–1330, 1982.
53. Linz, U.: Synthesis and quality control of ^{125}I- and ^{14}C-labelled 5-iodo-2′-deoxycytidine triphosphate. *J. Labelled Compd. Radiopharm.* **XIX:**1151–1160, 1982.
54. Fritzberg, A.R., Bloedow, D.C., Eshima, D. and Johnson, D.L.: Comparison of Tc-99m N-pyridoxyl-5-methyl-tryptophan and Tc-99m N-(3-bromo-2,4,6-trimethylacetanilide)iminodiacetate as hepatobiliary radiopharmaceuticals in rats. *J. Pharm. Sci.* **73:**1861–1863, 1984.
55. DePamphilis, B.V., Jones, A.G., Davis, M.A. and Davidson, A.: Preparation and crystal structure of oxotechnetium bis(thiomercaptoacetate) and its relationship to radiopharmaceuticals labelled with Tc-99m. *J.A.C.S.* **100:**5570–5571, 1978.
56. Costello, C.E., Brodack, J.W., Jones, A.G., Davison, A., Johnson, D.L., Kashina, S. and Fritzberg, A.R.: The investigation of radiopharmaceutical components by fast atom bombardment mass spectrometry. The identification of Tc-HIDA and the epimers of Tc-CO_2DADS. *J. Nucl. Med.* **24:**353–355, 1983.
57. Berendsen, G.E. and de Galan, L.: Role of the chain length of chemically

bonded phases and the retention mechanism in reversed-phase liquid chromatography. *J. Chromatogr.* **196**:21–37, 1980.

58. Horvath, Cs., Melander, W. and Molnar, I.: Liquid chromatography of inorganic substances with nonpolar stationary phases. *Anal. Chem.* **49**:142–154, 1977.
59. Franklin, K.J., Howard-Lock, H.E. and Lock, C.J.L.: Preparation, spectroscopic properties, and structure of 1-oxo-2,3,6-D-penicillaminato-N,S,O)-4,5-(D-penicillaminato-N,S)technetium (V). *Inorg. Chem.* **21**:1941–1946, 1982.
60. Leffold, A. and Vigh, C.Y.: pK* Values of phosphate buffer components in methanol–water reversed-phase eluants. *J. Chromatogr.* **257**:384–386, 1983.
61. Carlson, L. and Andersen, K.: ^{131}I-labelled N-isopropyl-*p*-iodoamphetamine. *Eur. J. Nucl. Med.* **7**:280–281, 1982.
62. Snyder, L.R. and Kirkland, J.J.: *Introduction to Modern Liquid Chromatography*. John Wiley and Sons, New York, 1979, pp. 270–281.
63. Klingensmith, W.C. III, Fritzberg, A.R., Spitzer, V.M., Johnson, D.L., Kuni, C.C., Williamson, M.R., Washer, G. and Weill, R. III: Clinical evaluation of Tc-99m N,N′-bis(mercaptoacetyl)-2,3-diaminopropanoate as a replacement for I-131 hippuran. *J. Nucl. Med.* **25**:42–48, 1984.
64. Fritzberg, A.R., Kasina, S., Eshima, D., Johnson, D.L., Jones, A.G., Lister-James, J., Davison, A. and Brodack, J.W.: Synthesis and evaluation of N_2S_2 complexes of Tc-99m as renal function agents. *J. Nucl. Med.* **25**: P16, 1984.
65. Ritchie, J.M.: The aliphatic alcohols. In *The Pharmacological Basis of Therapeutics,* Gilman, A.G., Goodman, L.S. and Gilman, A., eds., Macmillan Publishing Co., New York, 1980, pp. 377–385.
66. Robinson, R.G., Bradshaw, D., Rhodes, B.A., Spicer, J.A., Visentin, R.J. and Gobuty, A.H.: A kit for the preparation of basic ^{99m}Tc penicillamine for renal scanning, *Int. J. Appl. Radiat. Isot.* **28**:105–112, 1977.
67. Tubis, M., Krishnamurthy, G.T. and Endow, J.S.: ^{99m}Tc-penicillamine: a new cholescintigraphic agent. *J. Nucl. Med.* **13**:652–654, 1972.
68. Johnson, D.L., Fritzberg, A.R., Hawkins, B.L., Kasina, S. and Eshima, D.: Stereochemical studies in the development of technetium radiopharmaceuticals I: fluxional racemization of technetium and rhenium penicillamine complexes. *Inorg. Chem.* **23**:4204–4207, 1984.
69. Loberg, M.D. and Field, A.T.: Stability of Tc-99m labelled N-substituted iminodiacetic acids: ligand exchange reaction between Tc-99m-HIDA and EDTA. *Int. J. Appl. Radiat. Isot.* **28**:687–692, 1977.
70. Fields, A.T., Porter, D.W., Callery, P.S., Harvey, E. and Loberg, M.D.: Synthesis and radiolabelling of technetium radiopharmaceuticals based on N-substituted iminodiacetic acid: effect of radiolabelling conditions on radiochemical purity. *J. Labelled Compd. Radiopharm.* **15**:387–399, 1978.
71. Fonda, U. and Pedersen, B.: Tc-99m-diethyl-HIDA: a contribution to the study of its structure. *Eur. J. Nucl. Med.* **3**:87–89, 1978.
72. Nicholson, R.W., Herman, K.J., Shields, R.A. and Testa, H.J.: The preparation and composition of HIDA. *Eur. J. Nucl. Med.* **5**:313–317, 1980.
73. Nunn, A.D. and Schramm, E.: Analysis of Tc-HIDAs and factors affecting their labeling rate, purity and stability. *J. Nucl. Med.* **22**:52, 1981.
74. Nunn, A.D., Loberg, M.D. and Conley, R.A.: A structure-distribution-relationship approach leading to the development of Tc-99m mebrofenin: an improved cholescintigraphic agent. *J. Nucl. Med.* **24**:423–430, 1983.

75. Pauwels, E.K.J., Feitsma, R.I.J. and Verney, D.: Composition of ^{99m}Tc-HIDA as a function of time after kit preparation. *Eur. J. Nucl. Med.* **6**:433–434, 1981.
76. Topliss, J.G.: Utilization of operational schemes for analogue synthesis in drug design. *J. Med. Chem.* **15**:1006–1011, 1972.
77. Wells, M.J.M. and Clark, C.R.: Liquid chromatographic elution characteristics of some solutes used to measure column void volume on C_{18} bonded phases. *Anal. Chem.* **53**:1341–1345, 1981.
78. Deutsch, E., Bushong, W., Glavan, K.A., Elder, R.C., Sodd, V.J., Scholz, K.L., Fortman, D.L. and Lukes, S.J.: Heart imaging with cationic complexes of technetium. *Science* **214**:85–86, 1981.
79. Roggendorf, E. and Spatz, R.: Systemic use of tetrahydrofuran in reversed-phase high performance liquid chromatography. An example of the selective benefits of ternary mobile phases. *J. Chromatogr.* **204**:263–268, 1981.
80. Tanabe, S., Zodda, J.P., Libson, K., Deutsch, E. and Heineman, W.R.: The biological distribution of some technetium-MDP components isolated by anion exchange high performance liquid chromatography. *Int. J. Appl. Radiat. Isot.* **34**:1585–1592, 1983.
81. Srivastava, S.E., Meinken, G.E., Richards, P., Ford, L.A. and Benson, W.R.: HPLC characterization of clinically used Tc-99m bone agents. Relative tissue distribution of fractionated components in mice. In *Proc. Third World Congress of Nuclear Medicine and Biology*, Raynaud, C., ed., Vol. II, Pergamon Press, Paris, 1982, pp. 1631–1634.
82. Munze, R.: Complexes of technetium IV with hydroxyethylidene diphosphonic acid. *J. Labelled Compd. Radiopharm.* **XV**:215–225, 1978.

CHAPTER 10

Concepts and Techniques Used in Metabolic Tracer Studies

JORGE R. BARRIO, RANDY E. KEEN, DIANE C. CHUGANI, GERALD BIDA, NAGICHETTIAR SATYAMURTHY, and MICHAEL E. PHELPS

INTRODUCTION

Studies of ammonia, amino acid, fatty acid, and glucose metabolism have relied primarily on the use of ^{15}N, ^{14}C, ^{13}C, and ^{3}H-labeled compounds. However, tissue sampling and analysis of enriched metabolites, i.e., ^{15}N and ^{13}C by mass spectrometry [1] and NMR [2–4], and ^{14}C and ^{3}H by liquid scintillation counting, limit the application of these techniques for the noninvasive determination of in vivo tissue tracer kinetics.

More recently carbon-13 and phosphorus-31 NMR have shown applicability to metabolic analyses in living organisms (i.e., bacteria), cells, or isolated organ preparations [5]. The application of NMR to large intact animals is presently restricted to ^{1}H NMR imaging, which can offer excellent delineation of anatomic tissue compartments, but has very limited utility in the study of metabolic events [6]. On the other hand, PET has permitted for the first time the noninvasive application of tracer kinetic principles to humans for the assessment of local physiologic processes [7]. In vivo tracer kinetics and tissue tracer concentration may be determined by use of positron-emitting labeled substrates. Cyclotron-produced positron-emitting radioisotopes of carbon (^{11}C, 20.38 min half-life), nitrogen, (^{13}N, 9.96 min), oxygen (^{15}O,

---

Abbreviations: Nuclear magnetic resonance, NMR; Positron emission tomography, PET; Gas-liquid chromatography, GLC; High pressure liquid chromatography, HPLC; Thin-layer liquid chromatography, TLC; 2-Deoxy-2-fluoro-D-glucose, 2-FDG; 2-Deoxy-2-fluoro-D-mannose, 2-FDM; Glutamine, gln; *o*-phthaldialdehyde, OPT; Spiperone, SP; Transfer RNA, tRNA.

2.03 min), and fluorine (^{18}F, 109.72 min) allow labeling of a wide variety of agents that trace a biochemical process in a predictable manner [8,9].

The advantages of positron-emitting radiolabeled compounds over stable isotope labeling of substrates in biochemical studies are multiple. First, they can be prepared with high specific activity, so that the process to be measured is not perturbed by the tracer introduced. For example, ^{15}N-labeling has been and is useful in biochemistry, but its value is limited by the requirements for large amounts of carrier. Second, the gamma rays produced after positron-annihilation allow for rapid external detection without disturbing the system under study. As a consequence, the time course of radioactive emission can be readily quantitated, permitting the application of tracer kinetic techniques to the study of various processes of interest (i.e., membrane diffusion, metabolism). Lastly, their short half-lives permit extension of these studies to humans.

In order to lend significance to the in vivo tracer kinetic data gathered by using external detection of radioactivity, researchers must know the chemical identity, concentration, and metabolism of the radiotracer. This chemical information can only be obtained by direct serial tissue and plasma (or perfusate) analysis, either in animal models [10–12], isolated organ preparations [13,14], or in man, when possible [15], using methods that are able to provide direct information about tissue biochemistry within the timeframe imposed by the short-lived radionuclides. In this work, the application of various biochemical procedures to positron-emitting-labeling radiotracers is reviewed and a few examples are discussed to illustrate the essential points, particularly in conjunction with tracer kinetic methods.

REQUIREMENTS OF ANALYTICAL TECHNIQUES FOR METABOLIC STUDIES

Identity Considerations

The selection of the analytical methodology to identify and quantitate positron-emitting labeled radiotracers should be a matter of careful consideration. Chromatographic techniques (GLC, HPLC, and TLC) have become major analytical tools in the radiopharmaceutical field. When the chromatographic systems in use exhibit large separation factors for the radiolabeled compounds of interest, and the analytical conditions have been extensively validated, these techniques become irreplaceable tools for radiotracer analysis.

The use of chromatographic retention data as a means of characterizing unknown bands is now common practice; however, caution should be exercised because this approach is simply not applicable in many instances. It should be remembered that chromatographic identification provides only a tentative assignation of chemical structure, with final confirmation requiring the use of ancillary (supplemental) procedures. Of course this is not possible in all cases, e.g., with *no carrier added* chemistry, in which case the degree of confidence of qualitative identity can be increased by either using other chromatographic conditions (i.e., different columns, solvents, etc.), or techniques (i.e., GLC, TLC, HPLC).

Positive evidence for the identity of an unknown can be accomplished by using spectroscopic and spectrometric techniques. NMR spectroscopy is a very powerful tool for this purpose, but its sensitivity may not be sufficient in many instances. Fourier-transform NMR techniques partially overcome this difficulty and produce excellent signal-to-noise ratios, even with samples containing as little as a few micrograms. An interesting example of the usefulness of NMR techniques is the high resolution (470.56 MHz) Fourier-transform ^{19}F-NMR analysis of the product of the reaction of D-glucal with F-18-labeled fluorine [16]. It should be noted that application of ^{19}F-NMR in this case is possible because of the relatively low specific activity of the F-18 samples (1–10 Ci/mmol) having sufficient fluorine-19 mass to permit Fourier-transform NMR spectroscopy after F-18 decay. In the above mentioned reaction, where TLC techniques (silica gel plate, system I, acetonitrile : water, 85 : 15, system II, *n*-butanol : acetic acid : water, 5 : 1 : 1) have shown only one band with properties identical to [^{18}F]2-deoxy-2-fluoro-D-glucose ([^{18}F]2-FDG), ^{19}F-NMR analysis of the reaction product shows four peaks, corresponding to the two anomers of 2-FDG (65.4%, α = 32.85 ppm; β = 32.70 ppm) and 2-deoxy-2-fluoro-D-mannose (2-FDM, 34.6%, α = 38.21 ppm; β = 56.59 ppm) [17]. Similarly, ^{19}F-NMR analysis of the same reaction using F-18-labeled acetyl hypofluorite (prepared in the gas phase) [18] as a reagent conclusively demonstrates that in aqueous media acetyl hypofluorite displays no stereoselectivity for D-glucal (product composition: 2-FDG, 45%; 2-FDM, 55%), an observation that is difficult to corroborate by chromatography [19]. Interestingly, using the same analysis method, it is possible to demonstrate that acetyl hypofluorite shows more marked stereospecificity with 3,4,6-tri-O-acetyl-D-glucal in solvents of low polarity like freon-11 (2-FDG, ~95%) than in acetic acid (2-FDG, 80%; 2-FDM, 20%) [16].

Selection of Chromatographic Conditions

Analysis of ^{13}N-labeled metabolites is technically one of the most difficult because of the short half-life (9.96 min) of ^{13}N. The use of this radionuclide requires a high starting activity and rapid processing of the biological sample in order to obtain statistically significant counts for various metabolites present. These requirements are met by on-site cyclotron production of the radionuclide, remote rapid semiautomated synthesis of positron-labeled molecules, and the use of chromatographic conditions, preferentially HPLC, suitable for resolution of metabolites within a few minutes. Considering the severe technical constraints, ^{13}N metabolic studies are used in this section as a description of worst-case limitations. Obviously, C-11 and F-18 procedures have relaxed time constraints as compared with N-13 procedures.

Since its discovery in 1934 by Joliot and Curie [20], ^{13}N has been initially used as a biological tracer to quantitate the assimilation of ^{13}N-labeled N_2 by plants [21], bacteria [22], and algae [23], and in denitrification studies using ^{13}N-labeled nitrate [24]. These investigations, elegantly reviewed by Krohn and Mathis [25], were more recently followed by the work of Cooper et al., who used ^{13}N-ammonia to evaluate the cerebral permeability and metabolism of ammonia in conscious rats [11]. These ^{13}N-labeled ammonia studies have demonstrated that cerebral ammonia metabolism occurs in at least two compartments, with the majority of ^{13}N-ammonia being metabolized to ^{13}N-glutamine (^{13}N-gln) in a small, rapid turnover glutamate pool that is distinct from a larger, slow turnover glutamate pool [11]. Similar work with ^{13}N-ammonia using the arterially perfused rabbit interventricular septum permitted the generation of tissue time-activity curves, and identification of glutamine synthetase, an enzyme that has been erroneously reported to be absent from myocardium [26], as responsible for incorporating ^{13}N-label from ^{13}N-ammonia into ^{13}N-gln [13].

High performance liquid chromatography using strong anion-exchange resin (Partisil SAX) with 50 *m*M phosphate buffer, pH 3.0, as a mobile phase, permits excellent separation of radiolabeled ammonia, nitrite, and nitrate in less than 5 min. Cooper *et al.* [11], using similar chromatographic conditions, reported the separation of glutamate, glutamine, aspartate, and ammonia. Neutral amino acids are difficult to resolve under those conditions, and electrophoretic procedures do not offer distinct advantages, since many of the amino acids are not completely resolved either. For complete resolution, a bidimensional electrophoresis is needed, which adds significantly to the analysis time (25–40 min).

Quantitation of ^{13}N-ammonia, or L-[^{13}N]amino acids and their metabolites, as shown in Tables 10.1 and 10.2, is best accomplished with reversed-phase, pre-column *o*-phthaldialdehyde (OPT) derivatization techniques [27,28]. Figure 10.1 depicts this process. The elution order of OPT-primary amino acid complexes in reversed-phase HPLC (hydrophobic or lipophilic stationary phase) is governed by the balance between the hydrophobic and hydrophilic groups of these derivatives, as shown in Figure 10.2. Introduction of a more hydrophobic mobile phase lowers the hydrophobic interaction between stationary phase and OPT-primary amine complexes, thus reducing retention times and, consequently, analysis time [27].

In contrast to other popular HPLC methods for detection of primary amines such as 1) ion exchange, post-column ninhydrin [29–31]; 2) ion exchange, post-column OPT [31–34]; 3) reversed-phase, pre-column dansylation [35,36]; and 4) reversed-phase, pre-column phenylthiocarbamylation [37,38]; reversed-phase precolumn OPT derivatization offers distinctive advantages. Among them we find: 1) a rapid reaction (1–2 min pre-column reaction at room temperature with OPT reagent); 2) precision (i.e., fluores-

TABLE 10.1. Nitrogen-13 Distribution in Myocardial Tissue 6 Min after Bolus Injection of ^{13}N-Ammonia[a,b]

	Relative Distribution of Counts[c]	
^{13}N-Metabolite	*Normal Septa*[d]	*Methionine Sulfoximine Treated Septa*[e]
Ammonia ion	27 ± 19	64 ± 4
Glutamine, ω	59.2 ± 1.0[f]	24.6 ± 0.2
Glutamine, α	<0.1[f]	
Glutamate	2.1 ± 0.06	0.4 ± 0.3
Aspartate	1.5 ± 0.4	0.3 ± 0.1
Alanine	<0.1	<0.1
Unidentified compounds	4.2 ± 0.9	1.6 ± 0.6
Acid insoluble precipitate	6 ± 1	9 ± 3

[a] Modified from [13].
[b] Isolated arterially perfused intraventricular septa of male white New Zealand rabbits (1.5–2.0 kg body wt) were used. The perfusate contained 130 *mM* NaCl, 1.5 *mM* $CaCl_2$, 5 *mM* KCl, 1 *mM* $MgCl_2$, 0.435 *mM* NaH_2PO_4, 9 *mM* $NaHCO_3$, 5.56 *mM* dextrose, 51 μM NH_4HCO_3, and 5 U/1 insulin. One-ml bolus containing 30 to 100 mCi ^{13}N-ammonia in isotonic saline was delivered arterially. A 0.10-g tissue sample was cut from septa 6 min after the bolus injection and analyzed for metabolites as described.
[c] All values are % ± SD.
[d] (n = 3)
[e] Standard perfusate was switched to standard perfusate plus 1 *mM* L-methionine-RS-sulfoximine 15 min before delivery of 1 ml bolus of ^{13}N-ammonia; (n = 3).
[f] Position of nitrogen-13 in glutamine was determined by enzymatic procedure as described.

TABLE 10.2 Nitrogen-13 Distribution in Myocardial Tissue 6 Min after Bolus Injection of L-[ω-^{13}N]Glutamine[a,b]

	Relative Distribution of Counts[c]
^{13}N-Metabolite	*Normal Septa*[d]
Ammonia ion	7 ± 2
Glutamine, ω	75.1 ± 2.4
Glutamine, α	
Glutamate	2.5 ± 2.8
Aspartate	1.3 ± 0.5
Alanine	<0.1
Unidentified compounds	8.2 ± 1.6
Acid insoluble precipitate	6 ± 0.6

[a] Modified from [13].
[b] Same as Table 10.1, footnote (b) except a 1-ml bolus containing 10–15 mCi L-[ω-^{13}N]glutamine in isotonic saline was delivered arterially.
[c] All values are % ± SD.
[d] (n = 3)

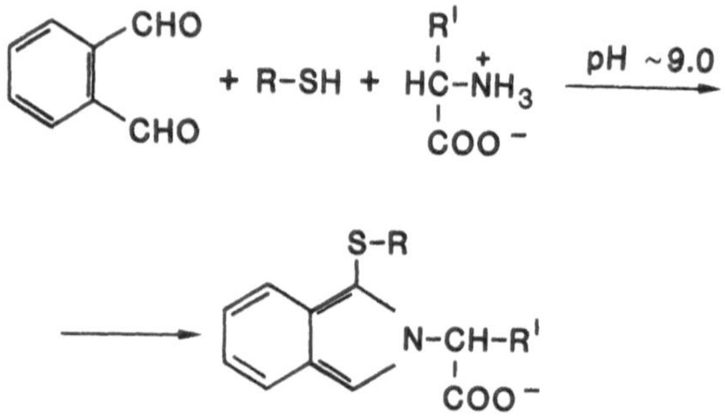

FIG. 10.1. Scheme depicting the *o*-phthaldialdehyde (OPT) derivatization reaction of amino acids. The amino acid-OPT complexes (isoindigo derivatives) present excellent fluorescent properties that permit detection at subpicomole levels.

cence response ±0.5%, retention times are reproducible within ±0.3%); 3) detection to subpicomole levels (50 fmol) as a result of the high sensitivity of fluorescence detection; 4) accuracy (intercalibration with amino acid analyzer (>95%); 5) recovery of standards, which averaged close to 100%; and 6) rapid analysis (<15 min for major tissue amino acids) [27,28].

Ion-pair reversed-phase post-column OPT offers some of the advantages of the reversed-phase pre-column OPT procedure in terms of analysis time, resolution, and sensitivity [39–41]. Good chromatographic resolution can be achieved by using sodium decyl sulfate [39] or sodium dodecyl sulfate (SDS) [40] as the ion pair reagent, with solvent gradients. Using shorter carbon chain ion pair reagents (i.e., hexane sulfonate) results in poor resolution of polar amino acids (e.g., aspartic acid, glutamic acid, glycine, serine, threonine, and alanine). The need of solvent gradients to achieve adequate resolution of amino acids makes column re-equilibration necessary. Excessive re-equilibration time in this case (greater than 30 min) [40] limits the usefulness of this technique for application to short-lived radionuclides.

More extensive limitations are found in other analytical techniques. For example, ion exchange HPLC requires several mobile phase buffers, a total analysis time of more than 60 min to resolve all physiological amino acids, and additional time (approximately 30 min) to regenerate the column for the next analysis [11,29–34]. Both ion exchange, post-column ninhydrin, and post-column OPT require a post-column reactor to derivatize the primary and secondary amines so they may be detected via UV or fluorescence detectors [29–34].

Reversed-phase HPLC of pre-column dansylated derivatives has the drawback of long pre-column reaction times, high reaction temperatures, and formation of multiple derivatives for basic amino acids [35,36]. The long derivatization time and multireaction make reversed-phase, pre-column dansylation unsuitable for fast quantitative analysis of nitrogen-13-labeled amino acids and metabolites. Similarly, reversed-phase HPLC of phenylthiocarbamyl amino acid derivatives requires long (up to 40 min) pre-column

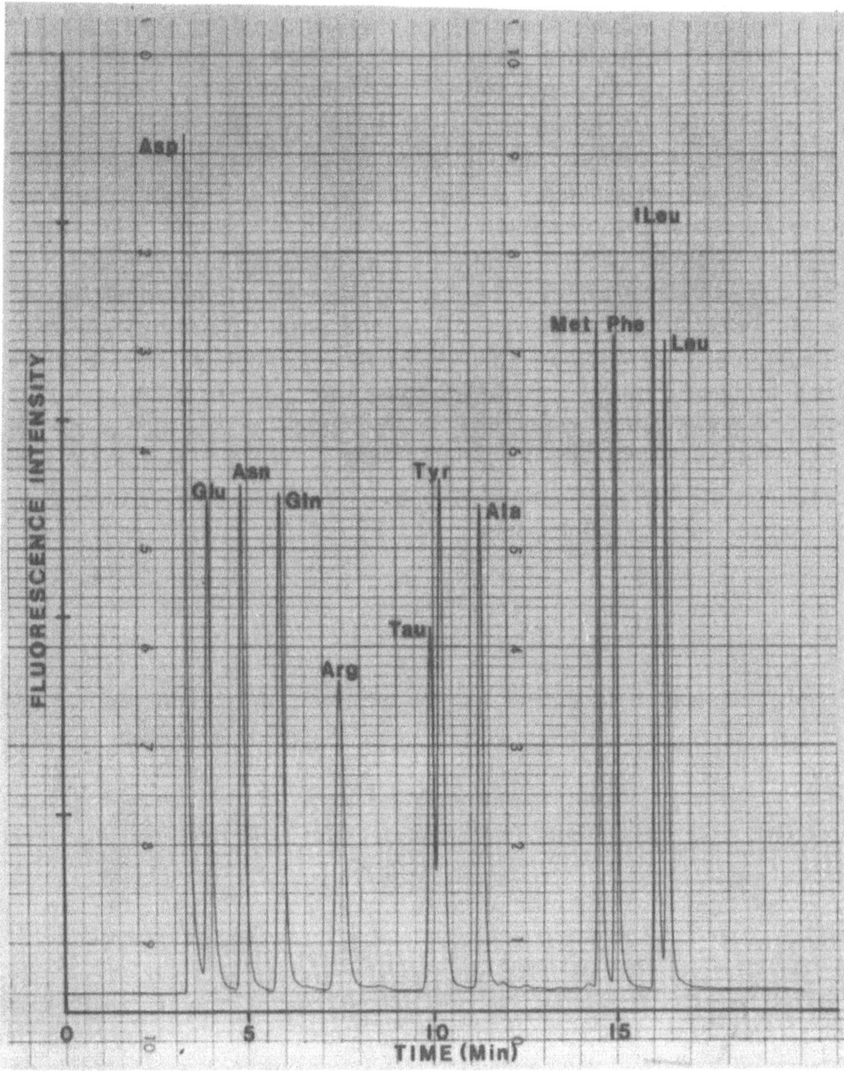

FIG. 10.2. Gradient run (55% 100 *mM* potassium phosphate, pH 6.2, 45% MeOH to 100% MeOH in 15 min) using an Ultrasphere ODS 5 μm, 4.6 $\times$ 250 mm; Spheri-5 RP-18, 5 μm 4.6 $\times$ 30 mm pre-column and fluorescence detection. Amino acids are identified by common abbreviations: aspartic acid, *Asp*; glutamic acid, *Glu*; asparagine, *Asn*; glutamine, *Gln*; arginine, *Arg*; taurine, *Tau*; tyrosine, *Tyr*; alanine, *Ala*; methionine, *Met*; phenylalanine, *Phe*; isoleucine, *Ileu*; and leucine, *Leu*.

sample preparation times, which includes phenylthiocarbamylation of amino acids, evaporation of excess reagent and byproducts [37,38], making this otherwise excellent technique unsuitable for nitrogen-13 metabolite analysis.

By comparison with the above mentioned techniques, reversed-phase, pre-column OPT affords maximum speed, reproducibility, and resolution.

For example, serial analyses every 15 min can be made for major myocardial tissue amino acids (i.e., aspartic acid, glutamic acid, glutamine, taurine, and alanine) using isocratic conditions (Ultrasphere ODS 5 μm, 4.6 × 250 mm; solvent 55% 100 *mM* potassium phosphate, pH 6.2, 45% methanol), obviating the need for column re-equilibration associated with solvent gradients [27]. Using solvent gradients, as shown in Figure 10.2, more hydrophobic amino acids can be resolved within 20 min. Column re-equilibration to initial solvent conditions requires 12 min. Re-equilibration times are reduced to 2 min using a shorter (75 mm) Ultrasphere ODS column with a smaller particle size (3 μm) [28]. In addition, OPT derivatization of amino acids affords strongly fluorescent amino acid-OPT complexes (isoindigo derivative) with excellent fluorescent properties (λ_{em} = 445 nm upon excitation at 340 nm), adding high sensitivity to the analytical method. The presence of the long wavelength absorption is important because it permits excitation of the fluorophore without interference from most other ultraviolet-absorbing moieties present in biological samples.

Sample Preparation and Analysis

Sample preparation in HPLC requires removal of components that may interfere with the chromatographic separation or ruin the analytical columns. Positron-emitting labeled metabolites in tissue can be conveniently divided into protein and nonprotein fractions by homogenization in 0.4 *N* perchloric acid, precipitation of the protein, and removal of the acidic supernatant. The supernatant may be neutralized with KOH, and the insoluble $KClO_4$ precipitate may be removed by filtration or centrifugation. The neutralized (pH 6.0–7.0) filtered supernatant is thus ready for direct HPLC analysis or derivatization prior to injection.

In some instances, removal of labeled or unlabeled components that may interfere with the chromatographic system is necessary. This may be carried out with commercially available solid phase extraction columns containing a variety of polar and nonpolar phases (i.e., Waters Sep-Pak™; Analytichem International Bond Elut™, Harbor City, California, etc.), again followed by direct HPLC analysis of the column effluent.

Fractions from the HPLC column (as small as 0.1 ml each) are collected with the aid of an automatic fraction collector, and radioactivity measured in a NaI(Tl) gamma well counter when radioactivity levels are low (subnanocurie levels). For higher radioactivity levels, a flow-through NaI(Tl) gamma detector, in line with a fluorescence detector, permits simultaneous determination of mass (down to subpicomole levels) and radioactivity, and, consequently, a direct measurement of the specific activity of the radiolabeled compounds.

Regardless of the procedure, it is essential to keep in mind that quantitation of positron-emitting labeled metabolites is only possible when 100% of the activity is accounted for during sample preparation and analysis. In addition, the chromatographic system should permit resolution of all components as well as their initial identification.

Verification Procedures

As indicated above, in radiotracer biochemistry, metabolite identity should be verified by the use of multiple procedures. Results from these various tests should reveal and validate all chemical forms of the tracer present. One of the preferred ancillary experiments involves the introduction of inhibitors of metabolite production. In the above mentioned studies by Cooper et al. [11] and Krivokapich et al. [13], the addition of the glutamine synthetase inhibitor L-methionine-SR-sulfoximine [42] significantly decreases the amount of [^{13}N]-gln synthesized in tissue, consistent with the glutamine synthetase reaction being the primary mechanism of ^{13}N-ammonia trapping. In addition, the isolated radiolabeled fractions should be subjected to specific enzymatic treatments to elicit specific transformations. For example, verification of the position of the label in ^{13}N-gln formed in brain [11] and heart [13] from ^{13}N-ammonia was performed in vitro by the use of *E. coli* glutaminase, which demonstrated that most of the radioactivity was present as amide-labeled ^{13}N-gln. In addition, jack bean urease was used to detect the presence of ^{13}N-urea in tissue myocardium. Any radiolabeled ^{13}N-ammonia formed may be trapped on a cation exchange column and quantitated [43].

APPLICATIONS IN TRACER KINETIC MODELS

Glucose Metabolism

The most frequently used radiolabeled compound for measuring glucose utilization with PET is [^{18}F]2-FDG. The approach developed by Sokoloff and colleagues [44] and initially applied to autoradiography with [^{14}C]2-deoxy-D-glucose, ensures that the radioactivity, at all times during measurements, will be contained only in the deoxysugar precursor or in the product of the hexokinase reaction, 2-deoxyhexose-6-phosphate. This permits the formulation of a tracer kinetic model that has been evaluated in cerebral [45–48] and myocardial [7,49–52] tissues in humans under various conditions, so that the utilization rate of glucose is calculated from the kinetics of [^{18}F]2-FDG.

The ability of a model to fit the measured data is a necessary but not a sufficient condition, because only direct chemical assays can test the validity of the model. Obviously the most appropriate, and direct, chemical validation procedure in tracer kinetics involves performing a simultaneous tracer kinetic and chemical assay study. The invasiveness of the chemical assay (i.e., tissue analysis) precludes its utilization in humans and, necessarily, limits its use to animals or isolated preparations.

A workable 2-FDG model has been found to fit the tracer kinetic data in myocardium [14]. In these experiments, [^{18}F]2-FDG was administered as a step-function to an interventricular rabbit septum, and F-18 activity was monitored as a function of time by a pair of coincidence detectors. At the end of the experiment (30, 60, and 90 min after start of infusion of the tracer), tissue analysis revealed only the radiolabeled 2-FDG and its 6-phosphate, in proportions that compared very favorably with the predictions obtained from the tracer kinetic data and the use of the model. Details are shown in Table 10.3.

Technically, the longer half-life of F-18 (109.8 min) allows the most flexibility in selection of chromatographic conditions for analysis. Using the same buffered solvent as in the reversed-phase, pre-column OPT system, an amino-bonded phase column resolves carbohydrates ([^{18}F]2-FDG or [1-^{11}C]2-DG) and their phosphate metabolites by weak anion exchange mode without derivatization. Analysis of hexoses and metabolites is complete in less than 18 min. Absence of a derivatization procedure allows direct analysis of hexoses from a neutralized supernatant, thus shortening the overall analysis time.

The methodology involves the preparation of a tissue homogenate, centrifugation, and separation of the supernatant, an aliquot of which is injected into the HPLC (Ultrasil-NH_2, 55% 100 *mM* potassium phosphate, pH 7.0; 45% methanol) [14]. Controls are prepared by enzymatic transformation of [^{18}F]2-FDG in vitro, in a buffered solution containing ATP, $MgCl_2$, and yeast

TABLE 10.3. Fluorine-18-Distribution in Myocardial Tissue after Constant Infusion of [^{18}F]2-FDG [14][a]

	Relative Distribution of Counts[b]		
^{18}F-Metabolite	*Normal Septa (30-min Sample)*[c]	*Normal Septa (60-min Sample)*[d]	*Normal Septa (90-min Sample)*[c]
2-Fluoro-6-phosphoglucono-δ-lactone	<0.1	<0.1	<0.1
2-Deoxy-2-fluoro-D-glucose	51.8 ± 9.0	37.2 ± 4.3	34.7 ± 2.4
2-Deoxy-2-fluoro-D-glucose-6-phosphate	48.2 ± 9.0	62.8 ± 4.3	65.3 ± 2.4
	(43.8 ± 1.5)	(55.4 ± 5.4)	(62.6 ± 5.1)

[a] Same as Table 10.1, footnote (b) except after 1 hr perfusion with standard perfusate, the media was switched to standard perfusate plus 0.4–1.1 mCi/l [^{18}F]2-FDG. 0.10-g Tissue samples were taken at 30, 60, and 90 min after beginning infusion with [^{18}F]2-FDG and analyzed for metabolites as described in methods. Relative distribution of 2-deoxy-2-fluoro-D-glucose-6-phosphate estimated by model in parenthesis.

[b] All values are % ± SD.

[c] (n = 2)

[d] (n = 4)

hexokinase. Conversion of 2-FDG 6-phosphate to 2-fluoro-6-phosphonoglucono-δ-lactone in vitro is catalyzed by glucose 6-phosphate dehydrogenase [14]. No significant amount of the pentose shunt metabolite gluconolactone or its hydrolysis metabolite (gluconate) was present in myocardium, as shown in Table 10.3.

Amino Acid Metabolism and Protein Biosynthesis

In the development of different types of biochemical assays with PET, it is highly desirable to identify procedures in vivo that follow the basic biochemical principles so successfully applied in the deoxyglucose method for the determination of glucose metabolic rates [44,46–48]. Specifically, the selected method should be such that radioactivity in tissues is confined, at all times during the experiment, to one product or, at least, to a few well-defined chemical species. In vivo determination of protein biosynthesis is a process of primary importance that can, a priori, be modeled following the principles of Sokoloff's procedure for deoxyglucose [53,54].

In addition to their incorporation into proteins, amino acids are substrates for various metabolic pathways. Ideally these metabolic pathways should be traceable, and the radiolabeled metabolites eliminated from tissue, allowing for the calculation of protein synthesis rates from the residual tissue activity. This problem is dealt with by using the L-isomer labeled at carbon-1 as originally proposed by Sokoloff and coworkers [53]. This strategy of selective position labeling illustrated for L-[1-^{11}C]leucine is also effective for other radiolabeled amino acids (e.g., L-[1-^{11}C]phenylalanine). In the metabolic pathway, L-leucine is reversibly transaminated with alpha-ketoglutarate to form alpha-ketoisocaproate, which is then oxidatively decarboxylated in mitochondria [55,56]. Labeled carbonic anhydride is then removed from tissue by diffusion and blood flow.

This labeling strategy is necessary for accurate estimations of protein synthesis rates using PET. Because branched-chain amino acids contribute carbon skeletons to the tricarboxylic acid cycle, other labeling strategies with C-11 may result in delayed oxidation and production of radiolabeled carbonic anhydride, or incorporation in radiolabeled precursors for biosynthetic processes in the cell, all of which would obscure the quantification of protein synthesis rates.

The essential variables that must be determined to structure properly a tracer kinetic model for a quantitative estimation of protein synthesis rates are the temporal sequences of the specific activities of the amino acid precursor pools in plasma and tissue, as well as for the reaction product(s) in tissue [15,54,57].

A major problem is the determination of the specific radioactivity of the labeled precursor amino acid at the site of protein synthesis. Calculation of protein synthesis rates using the specific radioactivity of the amino acids in intracellular spaces as the precursor pool produce erroneous estimations. Indeed, most studies indicate that the general intracellular pool is not the precursor of amino acids for protein synthesis [58–60].

Esterification of L-amino acids to their corresponding transfer-RNAs (tRNA) to form aminoacyl-tRNAs is catalyzed by the highly specific cytosolic aminoacyl-tRNA synthetase. Activation of the amino acids to aminoacyl-tRNAs is the first step in the formation of a polypeptide chain in a process by which the aminoacyl-tRNAs are recognized by the messenger RNA coding system on the ribosome. For this reason, the most direct approach to estimate the specific radioactivity of the amino acid precursor pool is the measurement of this parameter in the aminoacyl-tRNA in vivo [59,61,62].

Significant data have been accumulated suggesting that for most amino acids the aminoacyl-tRNA pool is not in equilibrium with the intracellular free amino acid pool [59,61,63–65]. Actually, the precursor pool specific activity is significantly higher than the intracellular amino acid pool specific activity, but lower (or similar) than the extracellular (plasma) amino acid specific activity [64]. This indicates that both intracellular and, to a larger extent, extracellular amino acids, may contribute to the amino acid precursor pool for tRNA aminoacylation.

It is important to note that the specific radioactivity of the amino acid precursor pool (aminoacyl-tRNA) may be calculated from the measured values of the molar concentrations and specific activities of intra- and extracellular (plamsa) amino acids [65], which can be determined using the chromatographic conditions described above for L-[^{13}N]amino acids. This has great significance when the isolation of aminoacyl-tRNA is not technically or ethically (e.g., in human experiments) possible.

There are also other kinetic and biochemical variables that need to be studied. For example, the rate-limiting steps in the competing metabolic pathways [66], the concentration and turnover rate of the small precursor pool [15], the degree to which this pool communicates with the plasma pool and the intracellular amino acid metabolic pool, and the rates of formation and breakdown of the various proteins are essential variables for the formulation of the model. Once these variables are understood more clearly, more accurate estimates of protein biosynthesis rates will be possible.

Receptor Localization

The development of quantitative receptor-binding assays using PET is one of the most challenging and potentially informative areas of research today. The success of in vivo receptor assays with PET depends upon the availability of ligands, that meet the criteria for imaging of areas containing specific receptor sites. In vivo binding studies in rodents [67–69] have generally shown that specific binding can only be demonstrated with very high affinity ligands ($K_D < 1.0$ nM). With such high affinity compounds, it has been shown that tissues with high concentrations of receptors accumulate radioactivity for at least one and up to several hours after a single intravenous injection of radioactive ligands, whereas in tissues with few or no receptors, the radioactivity washes out with time. The choice of receptor radiotracers, then, may begin with estimating their binding affinity by performing receptor binding assays with in vitro tissue preparations and tritiated ligands. Extensive in vitro binding experiments with [^{3}H]spiperone ([^{3}H]-SP, phenyl-4-(^{3}H)spiperone), a neuroleptic butyrophenone with high affinity for D-2 dopamine receptors have demonstrated that spiperone fills this first criteria ($K_D \sim 0.1\ nM$) and was, in fact, one of the first ligands with which in vivo binding studies were performed [67–69].

Some radioligands that are very useful in vitro, however, may not be appropriate in vivo receptor labels. The ligand must be able to reach the target tissue after systematic administration and be rapidly cleared from nonspecific binding sites. For example, in brain studies, it must cross the blood–brain barrier. Because most ligands would be expected to cross the blood–brain barrier by passive diffusion, a relative estimate of the compound's ability to get into the brain may be obtained by determining its octanol/water partition coefficient. This measurement also gives some indication about the amount of nonspecific binding. A compound that is very hydrophobic readily enters the brain, and thus will have considerable nonspecific binding. The ideal ligand, then, enters the brain in adequate amounts for imaging purposes, but does not have excessive amounts of nonspecific binding that will obscure the measurement of specific binding sites.

In order to determine finally the utility of a radiotracer for PET studies, in vivo studies must be performed. With in vivo autoradiography in animals, much valuable information may be obtained about the utility of a ligand for PET studies. The technique offers several parallels with PET, namely: 1) delivery of the radiotracer by intravenous injection, 2) measurement of the temporal course of arterial blood input function in terms of chemical composition of labeled and unlabeled constituents, 3) imaging of the localized substrate/ligand tissue concentration, and, 4) quantitation of localized substrate/ligand tissue concentration, measurement of the temporal course of radioactivity, and performance of competitive studies. In addition, the animal studies are valuable tools in the planning strategies of the PET studies since they permit 1) evaluation and screening of ligands or substrates prior to development of time-consuming chemical synthetic procedures, 2) the development, evaluation, and analysis of tracer kinetic models, and 3) the potential possibility to investigate mechanistic questions raised by human studies.

As indicated above, in vivo studies using [^{3}H]-SP help to determine the feasibility of using positron-emitting labeled SP as a ligand for PET. In these experiments, [^{3}H]-SP is administered to rats intravenously and arterial blood samples are collected from implanted cannulae. The rats are sacrificed at various times after the administration of the [^{3}H]-SP, and the brains removed are either dissected into regions of interest and processed for biochemical assays or prepared for autoradiographic work. With the brain and blood samples collected, researchers can determine the concentration of tritium using liquid scintillation counting and the chemical composition of the tritiated compounds using HPLC. [^{3}H]-SP is rapidly metabolized peripherally, but most of the tritium in the brain is in the form of [^{3}H]-SP, with only a small amount as tritiated water 2 hr after administration of the ligand [70]. It is therefore not likely that metabolites will interfere with imaging spiperone binding sites in brain [71,72]. Labeled metabolites of other ligands, however, may compete at receptor sites or increase the amount of nonspecific binding. While the metabolites of spiperone in the blood will not affect the imaging of receptor sites in the brain, they will influence quantitation. For this reason, the metabolic time course must be clearly defined in order to correct for metabolism in the blood radioactivity measurements.

From in vivo autoradiography experiments [67,73] it is possible to visualize that radioactivity accumulates in the corpus striatum, nucleus accumbens, and olfactory tubercles, areas containing large numbers of dopamine receptors, and that the radioactivity rapidly clears from the cerebellum, an area devoid of dopamine receptors.

It is interesting to note that [^{3}H]-SP in vivo autoradiographs show very little radioactivity in the frontal cortex (2 hr after injection), whereas in vitro studies have demonstrated a large number of high affinity SP-binding sites in frontal cortex, especially lamina IV, which are serotonergic in nature [73]. In addition, for spirodecanone sites reported in in vitro studies are observable in vivo. Kinetic analyses of rat brain regions following intravenous administration of [^{3}H]-SP show that SP accumulates in striatum (dopamine receptors) up to 3 hr after injection, whereas its concentration declines after 30 min in frontal cortex (serotonergic receptors) and more rapidly in cerebellum (nonspecific binding) [73].

These autoradiographic studies show that there are major differences between in vivo and in vitro ligand binding. They suggest that higher sensitivity for dopamine receptors can be obtained at later imaging times owing to the increasing ratio of specific dopamine to nonspecific binding of SP with time. The serotonin component in frontal cortex, however, may be characterized in vivo with SP at early points in time.

In order to understand what portion of in vivo deposition of ligands in brain or other tissues is determined by receptor number and affinity, other critical variables determining their deposition must be determined. Among those factors that may be investigated are local differences in blood flow, diffusion rates, and nonspecific binding. In addition, competition of the neurotransmitter with the ligand at the receptor site and functional changes in receptor conformation should be studied. The value of in vivo receptor assays using PET is questionable unless the critical variables determining ligand deposition can be identified and measured.

CONCLUSIONS

To get the most out of the biochemical analyses using short-lived radionuclides, care is required in selecting the right analytical techniques and using the proper conditions for a particular analysis problem. Generalizations are difficult because of the variety of analytical procedures that can be used. However, it is apparent that a combination of chromatographic and spectroscopic, enzymatic and autoradiographic techniques should permit the resolution of a large number of problems involving position-emitting labeled substrates.

ACKNOWLEDGMENTS

The authors are grateful to the many colleagues who contributed time and data. This work was supported in part by DOE contract DE-AC03-76-SF00012, USPHS grant 9 R01 NS 20867-08, USPHS grant 1 P01 NS15654, NIMH grant 1 R01 MN 447916-01, NIH 1 R01 HL 30673-01, and donations from the Will's Foundation, Houston, Texas; Fritts Family Foundation, Bakersfield, California; the Hereditary Disease Foundation, Los Angeles, California; and Jennifer Jones-Simon Foundation, Los Angeles, California.

REFERENCES

1. Berl, S., Takagaki, G., Clarke, D.D. and Waelsch, H.: Metabolic compartments in vivo: Ammonia and glutamic acid metabolism in brain and liver. *J. Biol. Chem.* **237**:2562–2569, 1962.
2. Walker, T.E., Han, C.H., Koffman, V.H., London, R.E. and Matwiyoff, N.A.: ^{13}C nuclear magnetic resonance studies of the biosynthesis by microbacterium ammoniaphilum of L-glutamate selectively enriched with carbon-13. *J. Biol. Chem.* **257**:1189–1195, 1982.
3. Herak, J.N. and Adamic, K.G.: *Magnetic Resonance in Chemistry and Biology*. Marcel Dekker, New York, 1971.
4. Gadian, D.G.: *Nuclear Magnetic Resonance and Its Application to Living Systems*. Oxford University Press, New York, 1982.
5. Levy, G.C., Lichter, R.L. and Nelson, G.L.: *Carbon-13 Nuclear Magnetic Resonance Spectroscopy*. 2nd ed., John Wiley and Sons, New York, 1980.
6. Smith, F.W.: Nuclear magnetic resonance in the investigation of cerebral disorders. *J. Cerebral Blood Flow Metab.* **3**:263–269, 1983.
7. Phelps, M.E., Hoffman, E.J. and Kuhl, D.E.: Physiological tomography: a new approach to in vivo measure of metabolism and physiologic function. In *Medical Radionuclide Imaging*, International Atomic Energy Agency, Vienna, 1977, Vol. 1, pp. 233–253.
8. Barrio, J.R.: Biochemical parameters in radiopharmaceutical design. In *Positron Emission Tomography of the Brain*, Heiss, W.-D. and Phelps, M.E., eds., Springer-Verlag, New York, 1983, pp. 65–76.

9. Phelps, M.E.: Emission computed tomography. *Semin. Nucl. Med.* **7**:337–365, 1977.
10. Henze, E., Schelbert, H.R., Barrio, J.R., Egbert, J.E., Hansen, H.W., MacDonald, N.S. and Phelps, M.E.: Evaluation of myocardial metabolism, with N-13- and C-11-labelled amino acids and positron computed tomography. *J. Nucl. Med.* **23**:671–681, 1982.
11. Cooper, A.J.L., McDonald, J.M., Gelbard, A.S., Gledhill, R.F. and Duffy, T.E.: The metabolic fate of ^{13}N-labeled ammonia in rat brain. *J. Biol. Chem.* **254**:4982–4992, 1979.
12. Barrio, J.R., Egbert, J.E., Henze, E., Schelbert, H.R. and Baumgartner, F.J.: L-[4-^{11}C]Aspartic acid: enzymatic synthesis, myocardial uptake, and metabolism. *J. Med. Chem.* **25**:93–96, 1982.
13. Krivokapich, J., Barrio, J.R., Phelps, M.E., Watanabe, C.R., Keen, R.E., Padgett, H.C., Douglas, A. and Shine, K.I.: Kinetic characterization of ^{13}N-ammonia and ^{13}N-glutamine metabolism in rabbit heart. *Am. J. Physiol.* **246** (Heart Circ. Physiol. 12):H267–H273, 1984.
14. Krivokapich, J., Huang, S.-C., Phelps, M.E., Barrio, J.R., Watanabe, C.R., Selin, C.E. and Shine, K.I.: Estimation of rabbit myocardial metabolic rate for glucose using fluorodeoxyglucose. *Am. J. Physiol.* **243** (Heart Circ. Physiol. 12):H884–H895, 1982.
15. Phelps, M.E., Barrio, J.R., Huang, S.-C., Keen, R.E., Chugani, H. and Mazziotta, J.C.: Criteria for the tracer kinetic measurement of cerebral protein synthesis in humans with positron emission tomography. *Ann. Neurol.* **15** (Suppl):S192–S202, 1984.
16. Bida, G., Satyamurthy, N. and Barrio, J.R.: The synthesis of [F-18]labeled 2-deoxy-2-fluoro-D-glucose using glycals: a reexamination. *J. Nucl. Med.* **25**:1327–1334, 1984.
17. Phillips, L. and Wray, V.: Stereospecific electronegative effects. Part I. The ^{19}F nuclear magnetic resonance spectra of deoxyfluoro-D-glucopyranoses. *J. Chem. Soc.* (B)1618–1624, 1971.
18. Jewett, D.M., Potocki, J.F. and Ehrenkaufer, R.E.: A preparative gas–solid phase synthesis of acetylhypofluorite. *Synth. Commun.* **14**:45–51, 1984.
19. Ehrenkaufer, R.E., Potocki, J.F. and Jewett, D.M.: Simple synthesis of F-18 labeled 2-fluoro-2-deoxy-D-glucose. *J. Nucl. Med.*, **25**:333–337, 1984.
20. Joliot, F. and Curie, I.: An artificial production of a new kind of radioelement. *Nature* **133**:201–202, 1934.
21. Ruben, S., Hassid, W.Z. and Kamen, M.D.: Radioactive nitrogen in the study of N_2 fixation by non-leguminous plants. *Science* **91**:578–579, 1940.
22. Nicholas, D.J.D., Silvester, D.J. and Fowler, J.F.: Use of radioactive nitrogen in studying nitrogen fixation in bacterial cells and their extracts. *Nature* **189**:634–636, 1961.
23. Thomas, J., Wolk, C.P., Shaffer, P.W., Austin, S.M. and Galonsky, A.: The initial organic products of fixation of ^{13}N-labeled nitrogen gas by the blue-green alga *Anabaena cylindrica*. *Biochem. Biophys. Res. Commun.* **67**:501–507, 1975.
24. Gersberg, R., Krohn, K., Peek, N. and Goldman, C.R.: Denitrification studies with ^{13}N-labeled nitrate. *Science* **192**:1229–1231, 1976.
25. Krohn, K.E. and Mathis, C.A.: The use of isotopic nitrogen as a biochemical

tracer. In *Short-Lived Radionuclides in Chemistry and Biology*, Root, J.W. and Krohn, K.A., eds., American Chemical Society, 1981, pp. 233–249.

26. Iqbal, K. and Ottaway, J.H.: Glutamine synthetase in muscle and kidney. *Biochem. J.* **119:**145–156, 1970.
27. Lindroth, P. and Mopper, K.: High performance liquid chromatographic determination of subpicomole amounts of amino acid by precolumn fluorescence derivization with *o*-phthaldialdehyde. *Anal. Chem.* **51:**1667–1674, 1979.
28. Jones, B.N. and Gilligan, J.P.: *o*-Phthaldialdehyde precolumn derivization and reversed-phase high performance liquid chromatography of polypeptide hydrolysates and physiological fluids. *J. Chromatogr.* **266:**471–482, 1983.
29. Hamilton, P.B.: Ion exchange chromatography of amino acids: a single column, high resolving, fully automatic procedure. *Anal. Chem.* **35:**2055–2064, 1963.
30. Murayama, K. and Sugawara, T.: Resolution of 52 ninhydrin-positive compounds with a high-speed amino acid analyzer. Determination of carnosine and homocarnosine in biological materials. *J. Chromatogr.* **224:**315–320, 1981.
31. Cunico, R.L. and Schlabach, T.: Comparison of ninhydrin and *o*-phthalaldehyde post-column detection techniques for high performance liquid chromatography of free amino acids. *J. Chromatogr.* **266:**461–470, 1983.
32. Roth, M.: Fluorescence reaction for amino acids. *Anal. Chem.* **43:**880–882, 1971.
33. Roth, M. and Hampai, A.: Column chromatography of amino acids with fluorescence detection. *J. Chromatogr.* **83:**353–356, 1973.
34. Bohlen, P. and Schroeder, R.: High-sensitivity amino acid analysis: methodology for the determination of amino acid compositions with less than 100 picomoles of peptides. *Anal. Biochem.* **126:**144–152, 1982.
35. Wiedmeier, V.T., Porterfield, S.P. and Hendrich, C.E.: Quantitation of Dns-amino acids from body tissues and fluids using high performance liquid chromatography. *J. Chromatogr.* **231:**410–417, 1982.
36. De Jong, C., Hughes, G.J., Van Wieringen, E. and Wilson, K.J.: Amino acid analyses by high-performance liquid chromatography: an evaluation of the usefulness of pre-column Dns derivatization. *J. Chromatogr.* **241:**345–359, 1982.
37. Heinrikson, R.L. and Meredith, S.C.: Amino acid analysis by reverse-phase high performance liquid chromatography: precolumn derivatization with phenylisothiocyanate. *Anal. Biochem.* **136:**65–74, 1984.
38. Granberg, R.R.: High-resolution analysis of PITC-derivatized amino acids with UV and electrochemical detection. *Liq. Chromatogr.* **2:**776–781, 1984.
39. Radjai, M.K. and Hatch, R.T.: Fast determination of free amino acids by ion-pair high performance liquid chromatography using on-line post-column derivatization. *J. Chromatogr.* **196:**319–322, 1980.
40. Dong, M.W. and DiCesare, J.L.: Amino acid analysis by liquid chromatography. An overview of five common methods. *Liq. Chromatogr.* **1:**222–228, 1983.
41. Hayashi, T., Tsuchiya, H. and Naruse, H.: Reversed-phase ion-pair chromatography of amino acids. Application to the determination of amino acids in plasma samples and dried blood on filter papers. *J. Chromatogr.* **274:**318–324, 1983.
42. Gass, J.D. and Meister, A.: Computer analysis of the active site of glutamine synthetase. *Biochemistry* **9:**1380–1390, 1970.
43. Keen, R.E., Krivokapich, J., Barrio, J.R., Douglas, A., Wittmer, S., Shine, K.

and Phelps, M.E.: Metabolic fate of L-[N-13]glutamate in normal isolated myocardium. *J. Nucl. Med.* **25:**P79, 1984.

44. Sokoloff, L., Reivich, M., Kennedy, C., Des Rosiers, M.H., Patlak, C.S., Pettigrew, K.D., Sakurada, O. and Shinohara, M.: The [^{14}C]deoxy-glucose method for the measurement of local cerebral glucose utilization: theory, procedure and normal values in the conscious and anesthetized albino rat. *J. Neurochem.* **28:**897–916, 1977.
45. Phelps, M.E., Hoffman, E.J., Huang, S.C. and Kuhl, D.E.: Positron tomography: an "in vivo" autoradiographic approach to measurement of cerebral hemodynamics and metabolism. In *Cerebral Function, Metabolism and Circulation,* Ingvar, D.H. and Lassen, N.A. eds., Munksgaard, Copenhagen, 1977, pp. 446–447.
46. Phelps, M.E., Huang, S.C., Hoffman, E.J., Selin, C.E. and Kuhl, D.E.: Tomographic measurement of local cerebral glucose metabolic rate in humans with (F-18)2-fluoro-2-deoxy-D-glucose: validation of method. *Ann. Neurol.* **6:**371–388, 1979.
47. Reivich, M., Kuhl, D., Wolf, A., Greenberg, J., Phelps, M.E., Ido T., Casella, V., Fowler, J., Hoffman, E., Alavi, A. and Sokoloff, L.: The [^{18}F]fluorodeoxyglucose method for the measurement of local cerebral glucose utilization in man. *Circ. Res.* **44:**127–137, 1979.
48. Huang, S.C., Phelps, M.E., Hoffman, E.J., Sideris, K., Selin, C.E. and Kuhl, D.E.: Noninvasive determination of local cerebral metabolic rate of glucose in man. *Am. J. Physiol.* **238**(Endocrinol. Metab. 1):E69–E82, 1980.
49. Gallagher, B.M., Ansari, A., Atkins, H., Casella, V., Christman, D.R., Fowler, J.S., Ido, T., MacGregor, R.R., Som, P., Wan, C.N., Wolf, A.P., Kuhl, D.E. and Reivich, M.: Radiopharmaceuticals XXVII. ^{18}F-labeled 2-deoxy-2-fluoro-D-glucose as a radiopharmaceutical for measuring regional myocardial glucose metabolism in vivo: tissue distribution and imaging studies in animals. *J. Nucl. Med.* **18:**990–996, 1977.
50. Gallagher, B.M., Fowler, J.S., Gutterson, N.I., MacGregor, R.R., Wan, C.-N. and Wolf, A.P.: Metabolic trapping as a principle of radiopharmaceutical design: some factors responsible for the biodistribution of [^{18}F]2-deoxy-2-fluoro-D-glucose. *J. Nucl. Med.* **19:**1154–1161, 1978.
51. Phelps, M.E., Hoffman, E.J., Selin, C.E., Huang, S.C., Robinson, G., MacDonald, N., Schelbert, H. and Kuhl, D.E.: Investigation of [^{18}F]2-fluoro-2-deoxyglucose for the measure of myocardial glucose metabolism. *J. Nucl. Med.* **19:**1311–1319, 1978.
52. Phelps, M.E., Schelbert, H.R., Hoffman, E.J., Huang, S.C. and Kuhl, D.E.: Positron tomography of the heart. *Prog. Nucl. Med.* **6:**183–209, 1980.
53. Smith, C.B., Davidsen, L., Deibler, G., Patlak, C., Pettigrew, K. and Sokoloff, L.: A method for the determination of local rates of protein synthesis in brain. *Trans. Am. Soc. Neurochem.* **11:**94, 1980 (abstract).
54. Barrio, J.R., Phelps, M.E., Huang, S.-C., Keen, R.E., MacDonald, N.S., Smith, C. and Sokoloff, L.: Positron emitting labeled L-amino acids for measurement of protein synthesis. *Trans. Am. Nucl. Soc.* **41:**17–18, 1982.
55. Chaplin, E.G., Goldberg, A.L. and Diamond, I.: Leucine oxidation in brain slices and nerve endings. *J. Neurochem.* **26:**701–707, 1976.
56. Williamson, J.R., Walajtys-Roade, E. and Coll, K.E.: Effects of branched

chain alpha-ketoacids on the metabolism of isolated rat liver cells. *J. Biol. Chem.* **254:**11511–11520, 1979.

57. Sokoloff, L. and Smith, C.: Biochemical principles for the measurement of metabolic rates in vivo. In *Positron Emission Tomography of the Brain,* Heiss, W.-D. and Phelps, M.E., eds., Springer-Verlag, New York, 1983, pp. 2–18.
58. Mortimore, G.E., Woodside, K.H. and Henry, J.E.: Compartmentation of free valine and its relation to protein turnover in perfused rat liver. *J. Biol. Chem.* **247:**2776–2784, 1972.
59. McKee, E.E., Cheung, J.Y., Rannels, D.E. and Morgan, H.E.: Measurement of the rate of protein synthesis and compartmentation of heart phenylalanine. *J. Biol. Chem.* **253:**1030–1040, 1978.
60. Schreiber, S.S., Evans, C.D., Ovatz, M. and Rothschild, M.A.: Problems in evaluating cardiac protein synthesis. *J. Mol. Cell. Cardiol.* **14:**307–312, 1982.
61. Airhart, J., Vidrich, A. and Khairallah, E.A.: Compartmentation of free amino acids for protein synthesis in rat liver. *Biochem. J.* **140:**539–548, 1974.
62. Hod, Y. and Hershko, A.: Relationship of the pool of intracellular valine to protein synthesis and degradation in cultured cells. *J. Biol. Chem.* **251:**4458–4467, 1976.
63. Schreiber, S.S., Evans, C.D., Oratz, M. and Rothschild, M.A.: Protein synthesis and degradation in cardiac stress. *Circ. Res.* **48:**601–611, 1981.
64. Martin, A.F., Rabinowitz, M., Blough, R., Prior, G. and Zak, R.: Measurements of half-life of rat cardiac myosin heavy chain with leucyl-tRNA used as precursor pool. *J. Biol. Chem.* **252:**3422–3429, 1977.
65. Vidrich, A., Airhart, J., Bruno, M.K. and Khairallah, E.A.: Compartmentation of free amino acids for protein biosynthesis. *Biochem. J.* **162:**257–266, 1977.
66. Pardridge, W.M.: Brain metabolism: a perspective from the blood–brain barrier. *Physiol. Rev.* **63:**1481–1535, 1983.
67. Kuhar, M.J., Murrin, L.C., Malouf, A.T. and Klemm, N.: Dopamine receptor binding in vivo: the feasibility of autoradiographic studies. *Life Sci.* **22:**203–210, 1978.
68. Laduron, P.M., Janssen, F.M. and Leyson, J.E.: Spiperone: a ligand of choice for neuroleptic receptors. 1. Kinetics and characteristics of in vitro binding. *Biochem. Pharmacol.* **27:**307–316, 1978.
69. Laduron, P.M., Janssen, F.M. and Leyson, J.E.: Spiperone: a ligand of choice for neuroleptic receptors. 2. Regional distribution and in vivo displacement of neuroleptic drugs. *Biochem. Pharmacol.* **27:**317–321, 1978.
70. Chugani, D.C., Barrio, J.R. and Phelps, M.E.: Spiperone metabolism: significance for kinetic modeling and nonspecific binding estimates. *J. Nucl. Med.* **24:**P106, 1983.
71. Fowler, J.S., Arnett, C.D., Wolf, A.P., MacGregor, R.R., Norton, E.F. and Findley, A.M.: [^{11}C]Spiroperidol: synthesis, specific activity determination and biodistribution in mice. *J. Nucl. Med.* **23:**437–445, 1982.
72. Welch, M.J., Kilbourn, M.R., Mathias, C.J., Mintun, M.A. and Raichle, M.E.: Comparison in animal models of ^{18}F-spiroperidol and ^{18}F-haloperidol: potential agents for imaging the dopamine receptor. *Life Sci.* **33:**1687–1693, 1983.
73. Chugani, D.C., Barrio, J.R. and Phelps, M.E.: To what extent does [^{3}H]spiperone label serotonin receptors in vivo? *Soc. Neurosci.* Abstr. 9(Part 1):724, 1983.

CHAPTER 11

Development of No-Carrier-Added Radiopharmaceuticals with the Aid of Radio-HPLC

D. SCOTT WILBUR

INTRODUCTION

The single most important tool for the development of high specific activity or no-carrier-added (nca) radiopharmaceuticals is a radio-HPLC. The reason for making such a strong statement in favor of HPLC rather than other chromatographic methods is that HPLC offers many advantages over other chromatographic methods. For example, the columns employed in HPLC are much more efficient than conventional thin-layer chromatography (TLC) plates, and the isolation of the products is simpler, more convenient, and safer than either TLC or gas chromatography.

Radiolabeling molecules to obtain compounds with very high specific activities is of particular interest for receptor-binding radiopharmaceuticals [1]. Unfortunately, radiolabeling molecules with high specific activity (e.g., no-carrier-added) radionuclides can often be difficult and frustrating. Reactions that are quite facile using stoichiometric quantities of reagents can, and often do, give significantly different results when no-carrier-added levels of a radionuclide are used. Thus, it nearly always takes a large amount of work changing reaction conditions to optimize a no-carrier-added radiolabeling reaction yield. The use of radio-HPLC can shorten the time required for optimizing the reaction conditions by rapid monitoring of the results of a

---

Abbreviations: High performance liquid chromatography, HPLC; Thin-layer chromatography, TLC; Gas-liquid chromatography, GLC; No-carrier-added, nca; Carrier-added, ca; Chloramine-T, CAT; N-chlorosuccinimide, NCS; *tert*-Butyl hypochlorite, TBHC; Curie, Ci; Millimole, mmol.

particular change in reaction conditions. Furthermore, the resolution obtainable on an HPLC gives reasonable assurance that the peak observed on the chromatogram is that of the desired radiolabeled compound.

In the following text, some personal biases in the use of radio-HPLC are described. Also, the steps taken in scaling a reaction from stoichiometric levels of reagents to no-carrier-added levels of radionuclide are presented.

RADIO-HPLC

The radio-HPLC design is not discussed here as it is adequately discussed in other chapters in this book. For developmental work, it is important to have an HPLC setup that can accommodate a variety of different chromatographic separations. A schematic of such an HPLC setup is shown in Scheme I. It is advisable that the setup be modular, because lead shielding may be needed around columns, detectors, injectors, and waste or sample collection areas. It is not necessary to have a setup with the capability of switching between two columns [2]. However, it is advisable to have several columns available so that the radiation exposure can be minimized if a column should become contaminated with radioactivity.

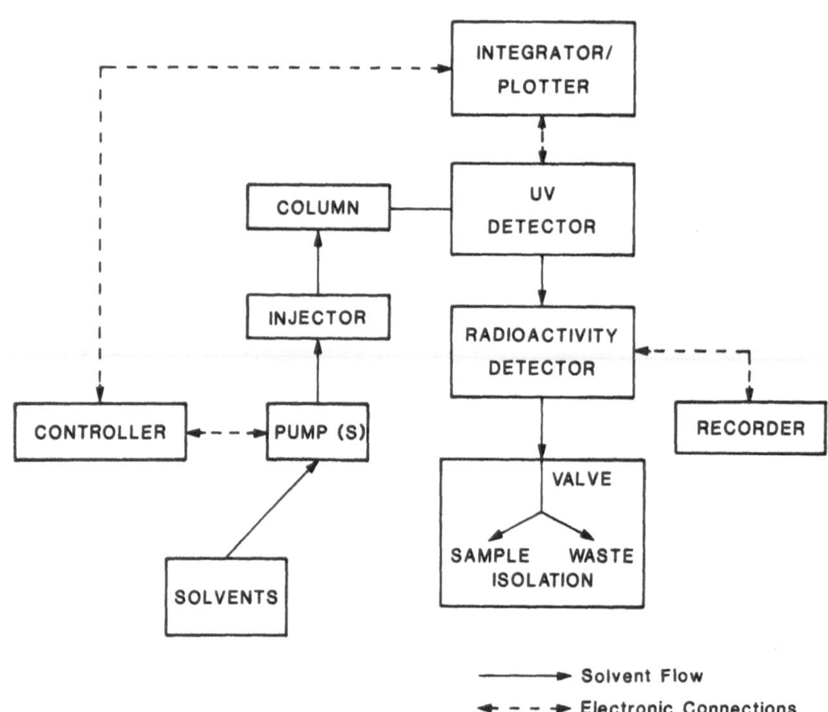

SCHEME I. Typical radio-HPLC setup.

Use of the new small columns that are available, such as the Waters Associates radially compressed columns, helps when columns need to be stored to allow for decay of contaminating radioactivity. Individual use of the radio-HPLC will dictate how much shielding is needed [3] and whether remote control is warranted, such as in the injection of large quantities of radioactivity [4].

One capability that is essential in a radio-HPLC setup to be used for developmental studies is solvent gradient programming. However, because of the difficulty in obtaining reasonable UV detector base lines, my personal preference is to limit the use of gradients. Another particularly important consideration when using gradients is the amount of time necessary to allow the column to equilibrate back to the initial conditions. The solvents used are always HPLC grade (e.g., distilled in glass) and are routinely filtered prior to use (primarily for degassing). As a result of these procedures, it has been possible to run almost daily for five years without replacing the in-line solvent filters. In contrast, samples to be injected are rarely filtered, except when solid particulates are present. This practice causes an increase in the back pressure from the column after a period of time. However, the use of the column can generally be extended by simply reversing the direction of solvent flow through the column. The amount of time and material saved by not filtering every sample when following radiolabeling reactions must more than compensate for premature loss of a column owing to pressure buildup.

Since virtually all of the radionuclides in our studies are gamma emitting, the detector used is simply a 2″ sodium iodide crystal surrounded by adequate lead shielding (a small lead safe works well). Signal detection is accomplished with an appropriate power supply, amplifier, and ratemeter. With this system, detection of nanocurie quantities of radioactivity is possible, but microcuries of activity are routinely used to obtain reasonable base lines on radiochromatographic traces and adequate counts for integration of peaks. Unlike scintillation detectors [5], the flow cell is a straight piece of 1.5-mm teflon tubing (total length of 10 cm) run across the face of the crystal. While other investigators [6] have passed the tubing through a drilled NaI crystal, this has not been necessary in our studies. Interestingly, even though this detector is a distance down the line from the UV detector, there is no appreciable change in resolution of peaks from a 5-μl injection.

Integration of the areas under the peaks in a radiochromatogram is accomplished by connecting the ratemeter to a counter and time and line printer. Attempts to connect a recorder capable of integrating peaks to the ratemeter have not been successful because of the noise in the base line; either many peaks were recorded from the base line, or, after dampening the response, some peaks were not integrated. The method of integration used at present is a manual tripping of the line printer at the beginning and closing of a radioactivity peak. Although this setup requires considerable hands-on effort, the integration can be rapid and reliable. Integration of peaks relative to one another appears to be linear with this setup if background activity is subtracted from the peaks. The linearity of the integrated peaks is lost when large quantities of activity (e.g., millicuries) are in individual peaks.

Because the quantity of radioactivity and the radionuclide used are constantly changing, a quantitative calibration of the detector response has never been carried out. Consequently, it is important to note that without response calibration the radiochromatogram integrated peaks can only show the efficiency of a radiolabeling reaction and cannot give radiochemical yields directly. In fact, all radiochemical yields must be obtained from a measurement of *isolated* activity in a radiolabeled compound in relation to the total activity used in the beginning of the reaction. Isolation of radioactivity shown in peaks on the radiochromatogram can be readily accomplished by using a valve that can switch the effluent from a waste line to sample collection lines. The isolated activity, when compared with a calibrated UV response, can provide a convenient method of determining the specific activity of the radiolabeled compound. Radiochemical purity and stability of the radiolabel can be obtained by re-injecting portions of the isolated samples.

SCALING RADIOLABELING REACTIONS TO nca LEVELS

As a general rule, the radiochemical yield of desired radiolabeled product obtained from a labeling reaction decreases (often dramatically) when the specific activity of a radionuclide is increased in a radiolabeling reaction. Additionally, the decrease in desired product is usually accompanied by an increase in the numbers and quantities of undesired radiolabeled compounds. This change in reaction products is not surprising if the quantities of radionuclide and impurities in the reaction mixtures are considered. As an example, the reactions of nca bromine-77 have been studied in this laboratory. Bromine-77 is produced [7] with a specific activity range of 2500–4500 Ci/mmol after processing [8]. In a reaction with 1 mCi of activity at those specific activities, there is only 18 to 32 nanograms of sodium bromide present. If the reaction is carried out in 50 μl of solvent, the solution is only 10^{-7} *M* in sodium bromide. Considering the concentration of radionuclide, it is apparent that the reaction must be fast to proceed to the desired product. It is also apparent that many impurities are present in much higher concentrations than the radionuclide, even in highly purified reagents and high purity solvents. With this in mind, it is not difficult to imagine how a reaction can become more complex and lower yielding at the nca level (concentration) of radionuclide.

Although reactions can be carried out at the nca level of radionuclide, a careful optimization of radiolabeling reaction conditions is required to achieve the desired results. Prior to radiolabeling studies, the synthesis of the desired compound must be accomplished with stable isotopes of the radioisotope to be used in the radiolabeling. This synthesis is necessary so that a complete purification and characterization of the compound can be achieved. This is absolutely essential when radiolabeling with nca radionuclides, as the retention time of the "cold" compound is the *only* method of identification that is available for the desired radiolabeled product. The next step is to carry out the desired reaction using carrier-added quantities of the radionuclide. While these reactions are essentially the same as those with stable nuclides, this exercise affords the opportunity to evaluate different separation systems and to follow the reactions by radio-HPLC. Furthermore, optimization of reaction conditions can be accomplished more readily, because only those reactions that involve the radioisotope will be monitored in the radiochromatogram. Indeed, the most important aspect of the reactions at this point is the incorporation of the radionuclide.

Once the incorporation of the radionuclide has been optimized at stoichiometric (or near stoichiometric) quantities of nuclide, the amount of added carrier is decreased to increase the specific activity of the radionuclide. If major changes in the radiolabeled products are observed, the causes for the changes are pursued and the radiolabeling reaction yields are optimized again. The reactions are then conducted without adding carrier. Only on a few occasions have the reaction product distributions at this point resembled the carrier-added counterparts. Quite often it has been necessary to completely re-evaluate the reaction conditions. Some reaction variables that have proven to be important considerations when optimizing the nca reaction conditions are: 1) quantities of reagents used, 2) reaction temperatures, 3) purity of reagents, 4) order of addition of reagents, and 5) the reagents themselves.

After conditions are optimized for a desired reaction, the chromatographic conditions are re-examined to obtain the "best" separation. At this point it is important to adjust the UV detector response to the most sensitive setting and attempt to obtain a separation of the UV absorbing species such that a base-line resolution is obtained at the retention time of desired radiolabeled compound. The nca radiolabeling reaction is then repeated to demonstrate reproducibility and to collect samples of the radiolabeled compound. The samples are collected to determine the radiochemical purity, radiochemical yield, and specific activity of the labeled compound.

IDENTIFICATION OF THE nca RADIOLABELED COMPOUND

It should be re-emphasized that the retention time of the radiolabeled compound is the only characteristic that is available to identify it. It is imperative that a method of separation be used that permits high resolution separations so that retention characteristics of the reference compound can be adequately and reproducibly defined. In addition to radio-HPLC, capillary radio-GLC meets these requirements. However, many compounds do not readily pass through a gas chromatograph, and for those that do, the isolation of the purified product from some adsorbent such as charcoal can be difficult with the small quantities of material present in nca radiolabeled compounds. Although it is a personal preference to do all of the developmental work with radio-HPLC, a second radiochromatographic method should be used to help identify the radiolabeled compound. Because most of the HPLC separations are run on reversed-phase columns, one method that can be used is normal-phase radio-TLC [9]. (If another chromatographic system is used, the researcher does not have to be constantly changing the conditions on the HPLC.) The separation afforded by a radio-TLC is most likely adequate as a secondary identification because it is highly unlikely that a different compound or impurity would co-elute under the two separate chromatographic conditions. While the use of two chromatographic systems will not unequivocally identify the radiolabeled species as being the desired radiolabeled compound, it will raise confidence in the identity.

EXAMPLES OF nca RADIOLABELING EXPERIMENTS

The use of organosilanes as intermediates in nca radiohalogenations of compounds has been of interest for some time in this laboratory. A large effort has been spent on developing the reaction conditions necessary to obtain rapid, site-specific introduction of nca radiohalogens into aromatic rings. To exemplify how the radio-HPLC has been used to develop this chemistry,

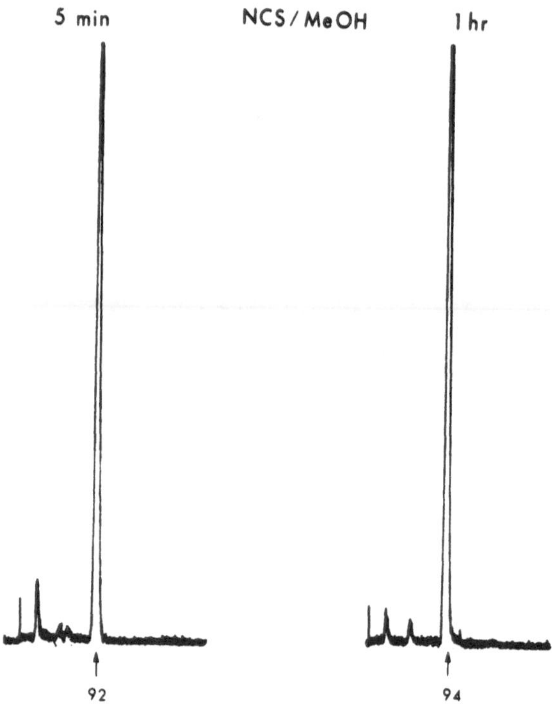

FIG. 11.1. HPLC radiochromatograms for the reaction of ca radiobromine, N-chlorosuccinimide (NCS), and *para*-Trimethylsilyltoluene, *1*.

some radiochromatograms that show the results of varying reaction conditions are given below.

Some of the initial studies of nca radiohalogenations of trimethylsilyltoluenes [10] are particularly exemplary. Figures 11.1 to 11.6 are radiochromatograms obtained in the optimization of these reactions. Figure 11.1 shows the results of a carrier-added radiobromination of *para*-trimethylsilyltoluene, *1*. Figure 11.2 shows the results of the identical reaction using nca bromine-77; no radiolabeled compound was obtained. Figure 11.3 shows the results of changing the oxidant in the reaction to chloramine-T (CAT), where again no product was obtained, even though a good yield had been obtained at carrier-added levels of bromine-77. Further change in the oxidant to *tert*-butylhypochlorite (TBHC) gave a reasonable yield of the nca radiobrominated *1* as shown in Figure 11.4.

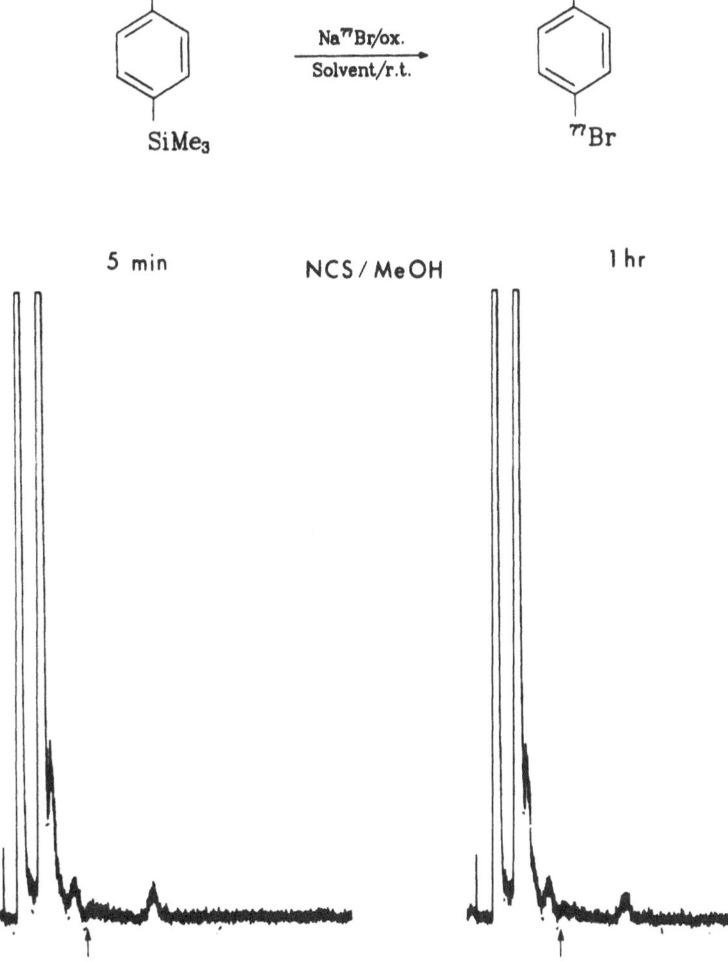

FIG. 11.2. HPLC radiochromatograms for the Reaction of nca radiobromine, NCS, and *1*.

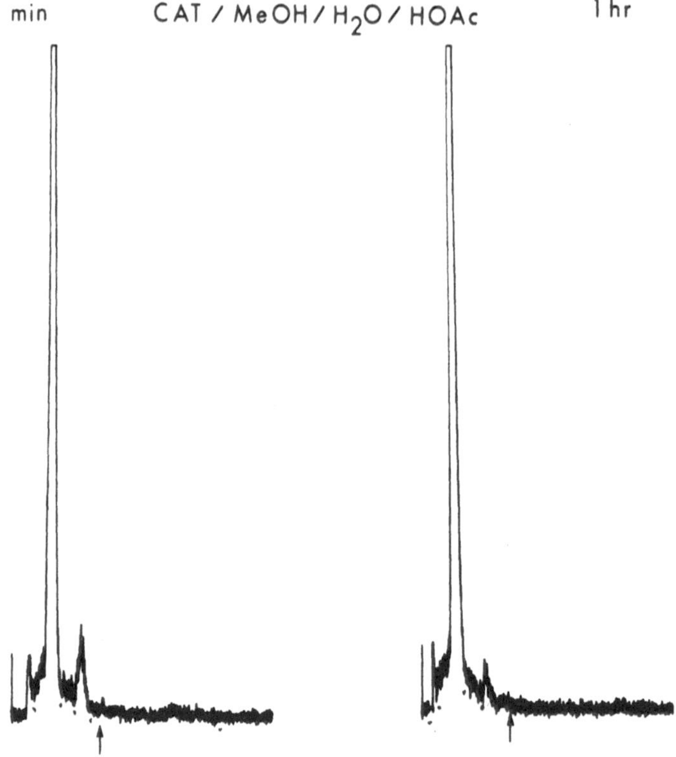

FIG. 11.3. As Fig. 11.2, but the chloramine-T in place of NCS.

Radioiodination of *1* was found to be more difficult than the radiobromination, requiring the use of HOAc as the solvent and elevated reaction temperatures. Figure 11.5 shows radiochromatograms of room temperature reactions of *1* with radioiodine at carrier-added and no-carrier-added concentrations of iodine-131. Figure 11.6 shows the nca radioiodination results when the reaction temperature was raised to 60°C.

CH_3–C_6H_4–$SiMe_3$ $\xrightarrow[\text{HOAc, 1 hr.}]{Na^{131}I/NCS}$ CH_3–C_6H_4–^{131}I

In another series of reactions, a method of regiospecific incorporation of nca radiohalogens into phenols was explored [11]. In Figure 11.7 the results of two nca radiobromination reactions of a phenol derivative, *meta*-trimethylsilylphenyl acetate (*2*) are shown. In the reactions, the order of addition of the oxidizing agent *tert*-butylhypochorite (TBHC) and the radio-

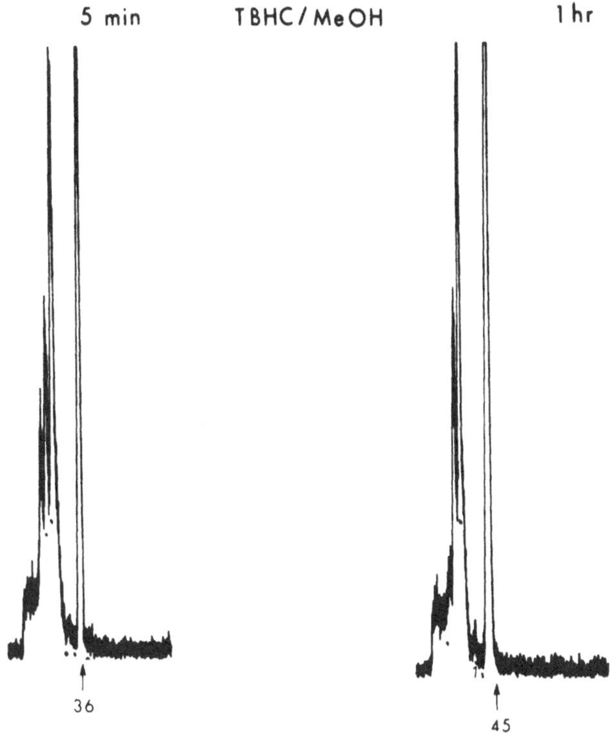

FIG. 11.4. As Fig. 11.2, but with *tert*-butylhypochlorite (TBHC) in place of NCS.

bromine is reversed. In the first reaction, the radiobromine was added prior to the addition of TBHC, in the second reaction the TBHC was added first. The most likely explanation for the observed difference is that in the second reaction TBHC has reacted with impurities prior to producing the electrophilic radiobrominating reagent; thus fewer impurities remain to react with the radiobrominating agent.

2 OAc, $SiMe_3$ $\xrightarrow[\text{HOAc/60°C, 10 min.}]{\text{Na}^{77}\text{Br/TBHC}}$ OAc, ^{77}Br

In another study, the nca radiobromination of *para*-trimethylsilylbenzoic acid (*3*) was explored [12]. Even though very good radiolabeling efficiencies were observed at the carrier-added levels of radiobromine, no yield of the desired product was obtained at nca until the temperature was raised to 50°C as shown in Figure 11.8. The radiochromatogram for the nca reaction indicated that there were many impurities that were reacting with the radiobromine prior to reacting with *3*, even though *3* appeared very pure by

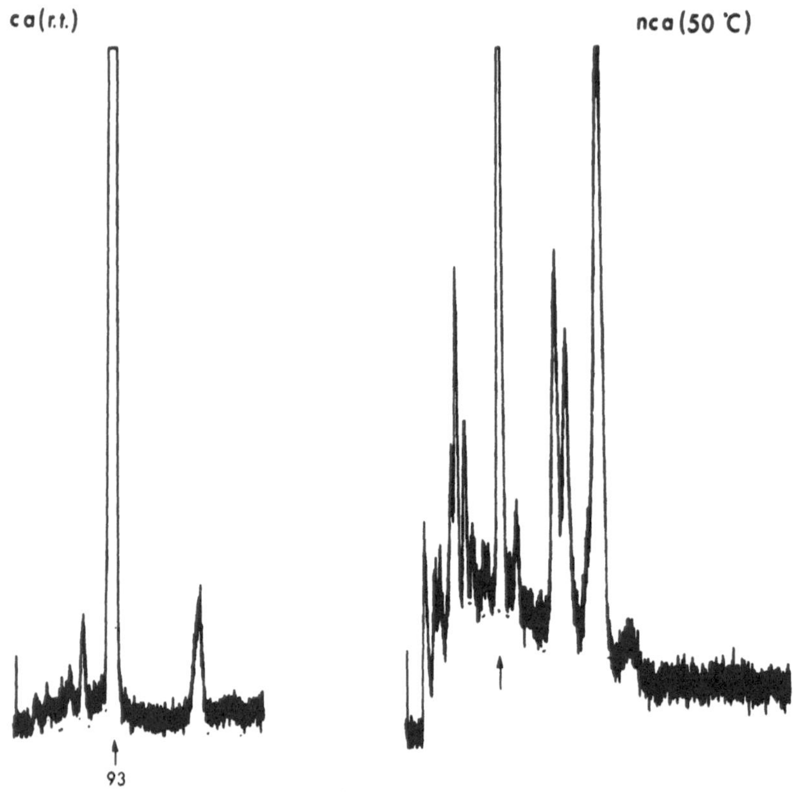

FIG. 11.5. Radiochromatograms for room temperature radioiodinations of *1* using ca and nca iodine-131.

HPLC analysis. Therefore, compound *3* was recrystallized a second time from a different solvent, and the radiobromination was repeated under the same reaction conditions as shown in Figure 11.9.

3

$$p\text{-}(CH_3)_3Si\text{-}C_6H_4\text{-}CO_2H \xrightarrow[\text{HOAc, 10 min.}]{Na^{77}Br/TBHC} p\text{-}{}^{77}Br\text{-}C_6H_4\text{-}CO_2H$$

Other investigations have explored radiobrominations of trimethylsilylbenzylamines and trimethylsilylphenethylamines [13,14]. Simple changes in the reaction conditions, such as the solvent and oxidizing agent used, also make appreciable differences in the radiolabeled product yields. For example, reaction of *para*-trimethylsilyl-N,N-dimethylbenzylamine (*4*) gave much different results when reacted in HOAc and MeOH, as shown in Figure 11.10. As all of the reaction conditions were identical except for the

FIG. 11.6. As nca radioiodination in Fig. 11.5, but with a reaction temperature of 60°C.

solvents, the difference in the numbers of radiolabeled products and radiolabeling yields might be explained as being due to impurities in the MeOH or to a suppression of some of the side reactions of the reagents in HOAc.

4 CH_2NMe_2 (para-$SiMe_3$ benzene) $\xrightarrow[\text{r.t./5 min.}]{\text{Na}^{77}\text{Br/TBHC}}$ CH_2NMe_2 (para-^{77}Br benzene)

Reactions of *para*-trimethylsilylphenethylamine (*5*) with radiobromide oxidized by NCS or TBHC gave quite different results. At carrier-added levels of radiobromine, the radiobrominations gave the desired product in 85% yield using NCS and only 17% yield with TBHC (as shown in Figure 11.11). In the nca radiobrominations of *5*, shown in Figure 11.12, the TBHC reaction gave similar results to the ca reaction, but the reaction of NCS became much more complex. Even with the large number of polar radiobrominated impuri-

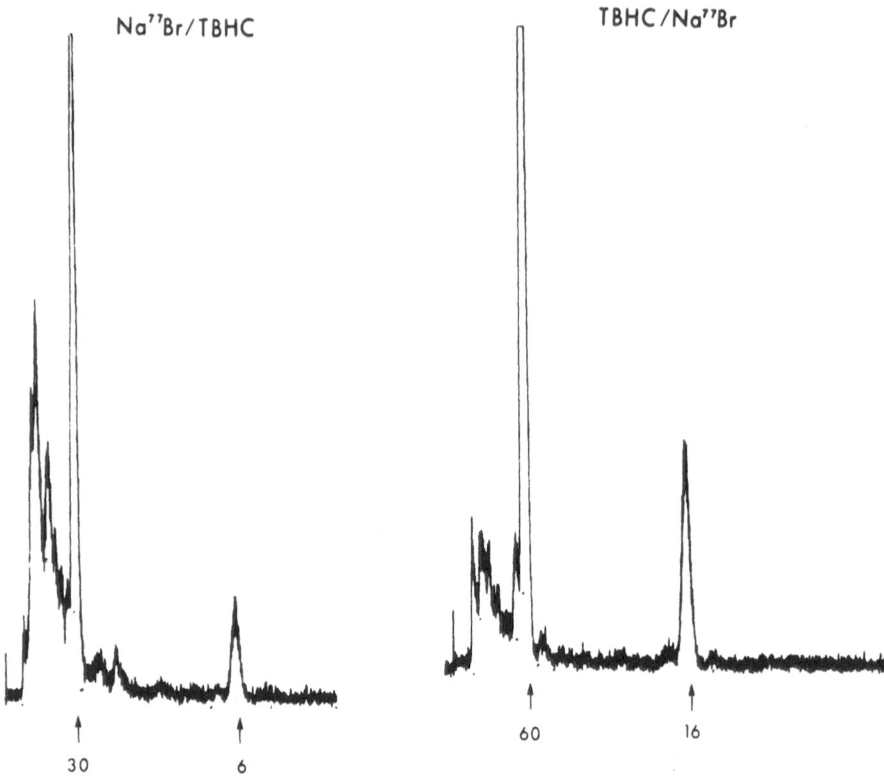

FIG. 11.7. Radiochromatograms for nca radiobromination of *meta*-trimethylsilylphenyl acetate (*2*). In the first reaction, radiobromide was added prior to TBHC; in the second, the order was reversed.

ties present in the nca radiobromination using NCS, the yield was higher for this reaction.

5 $CH_2CH_2NH_2$–C_6H_4–$SiMe_3$ $\xrightarrow[\text{MeOH/r.t. 5 min.}]{^{77}Br^-\text{/ox.}}$ $CH_2CH_2NH_2$–C_6H_4–^{77}Br

The foregoing examples of radiochromatograms were given to illustrate how radio-HPLC is used to modify reaction parameters to optimize nca radiolabeling yields. Usually the isolated radiolabeling yields were only slightly less than those observed on the radiochromatograms. However, in a few examples (e.g., radiobromination of hippuric acid [12]) isolated yields of the products have been considerably less than what the radiochromatogram

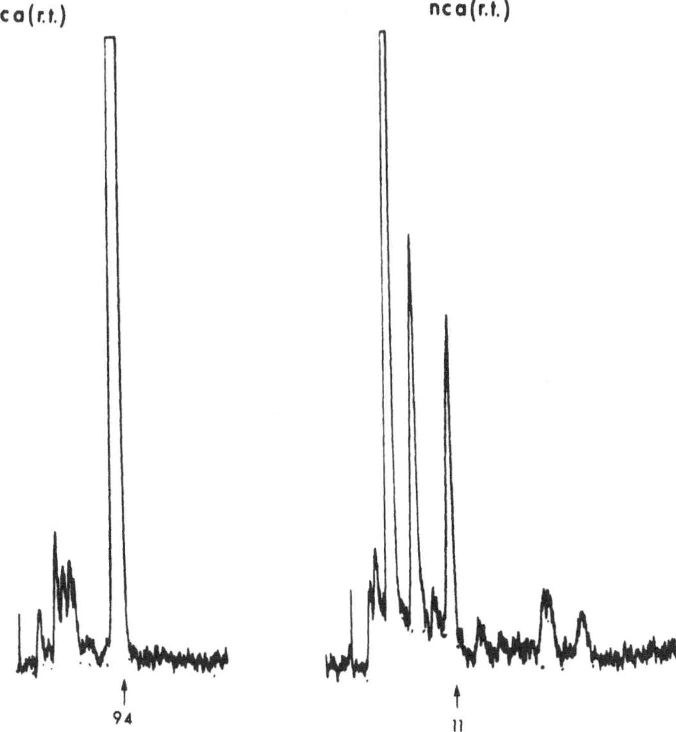

FIG. 11.8. Radiochromatograms for the radiobromination of *para*-trimethylsilylbenzoic acid (*3*) using ca radiobromine at room temperature and nca radiobromine at 50°C.

would indicate. The reasons for this are not readily apparent, but one might speculate that some compounds interact with the column and/or metal surfaces more than others (i.e., through chelation or chemical bonding), and at nca levels an appreciable quantity can be bound up.

CONCLUSIONS

Finding the appropriate chromatographic conditions for radio-HPLC can be the most time-consuming part of a radiolabeling study. However, the time is well spent as the results of a radiolabeling study are probably only as good as the chromatographic separations that have been achieved. This is particularly true in nca radiolabeling, as a multitude of unidentified peaks can be present near the retention time of the desired radiolabeled compound. In fact, the time spent on finding the appropriate chromatographic conditions can be easily regained if the radiolabeling conditions are difficult to develop

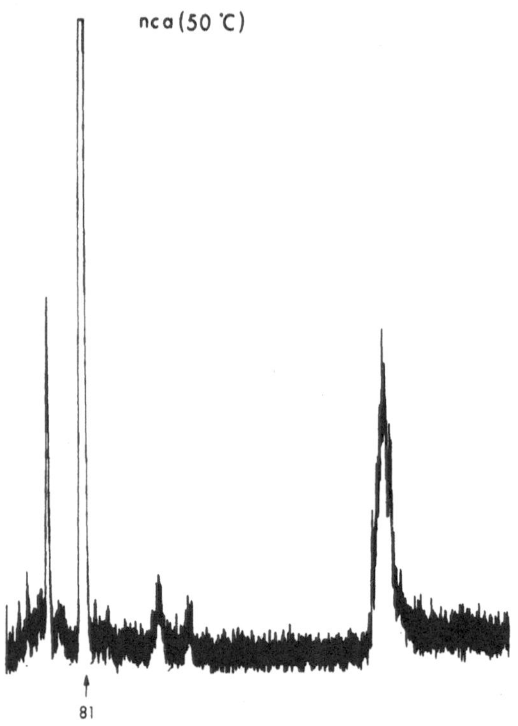

FIG. 11.9. Radiochromatogram for the radiobromination of purified *3* using nca radiobromine at 50 degrees C.

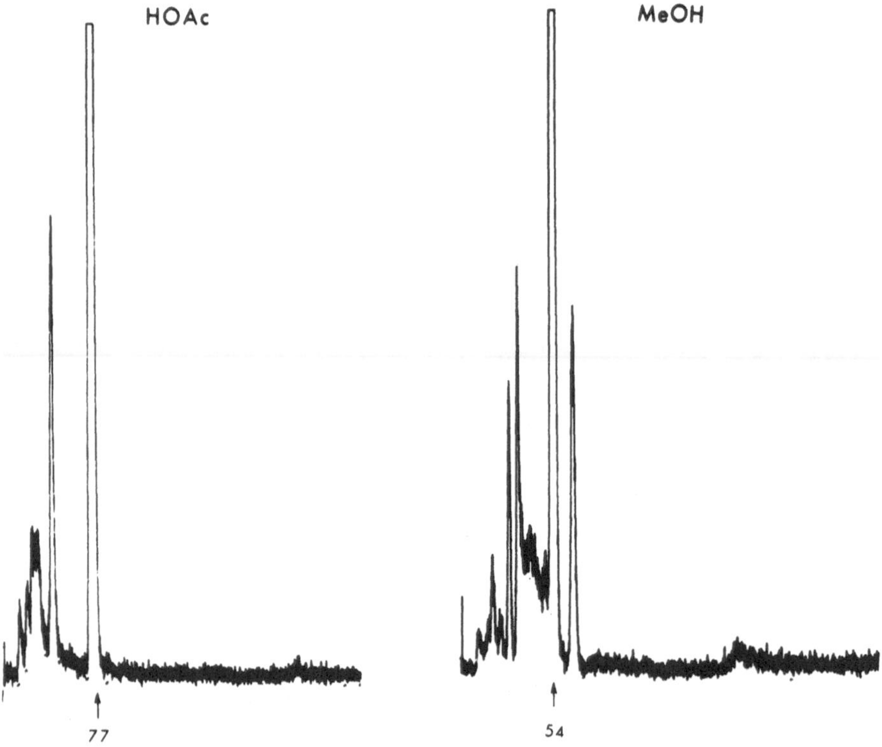

FIG. 11.10. Radiochromatograms for the radiobromination of *para*-trimethylsilyl-N,N-dimethylbenzylamine (*4*) in different solvents.

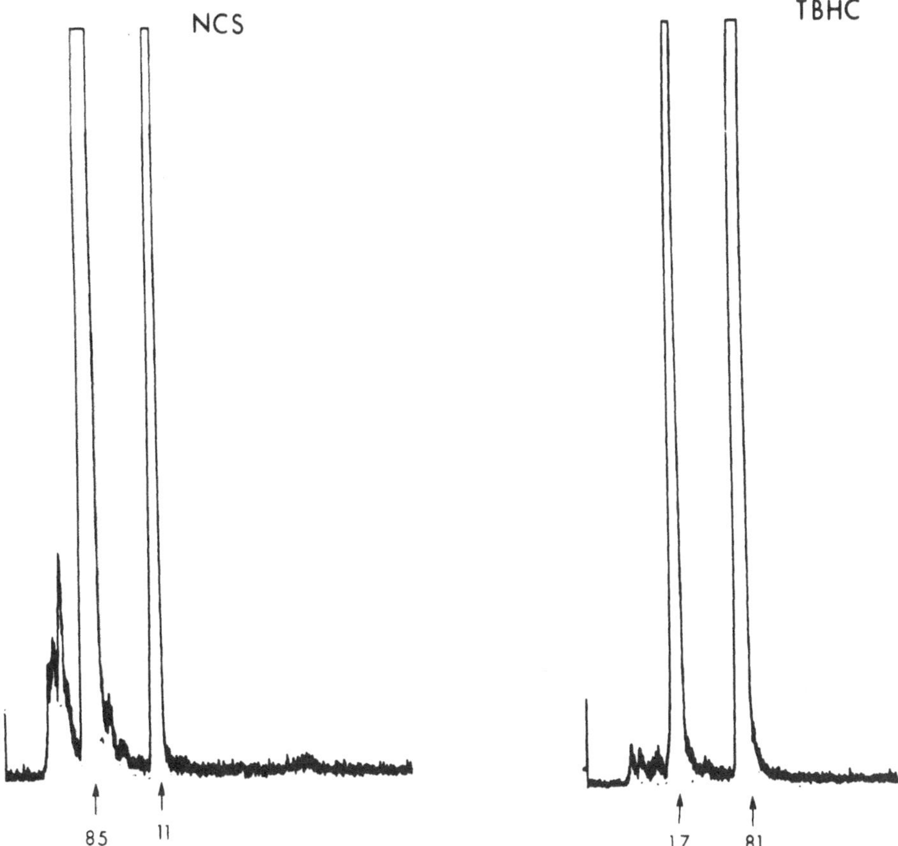

FIG. 11.11. Radiochromatograms for the radiobromination of *para*-trimethylsilylphenethylamine (*5*) using NCS and TBHC.

as the HPLC can quickly determine the effect of changing reaction conditions on the product obtained.

While it may take considerable time to develop a particular separation for the radiolabeling reaction that is being studied, there are many journals and reference books that can help an investigator to get started on a separation problem. Furthermore, many of the commercial companies that supply HPLC equipment have applications departments to demonstrate their products and give advice on particular separation problems. Also, there are always researchers who can be contacted for additional information on particular separations.

A further incentive for the use of radio-HPLC is that the cost of equipment is decreasing. Today it is possible to buy a system at one-half to two-thirds the cost of an equivalent purchased five years ago, and that is not considering the effect of inflation. A basic radio-HPLC can cost from $10,000 to $50,000 depending on what the needs for the instrument are.

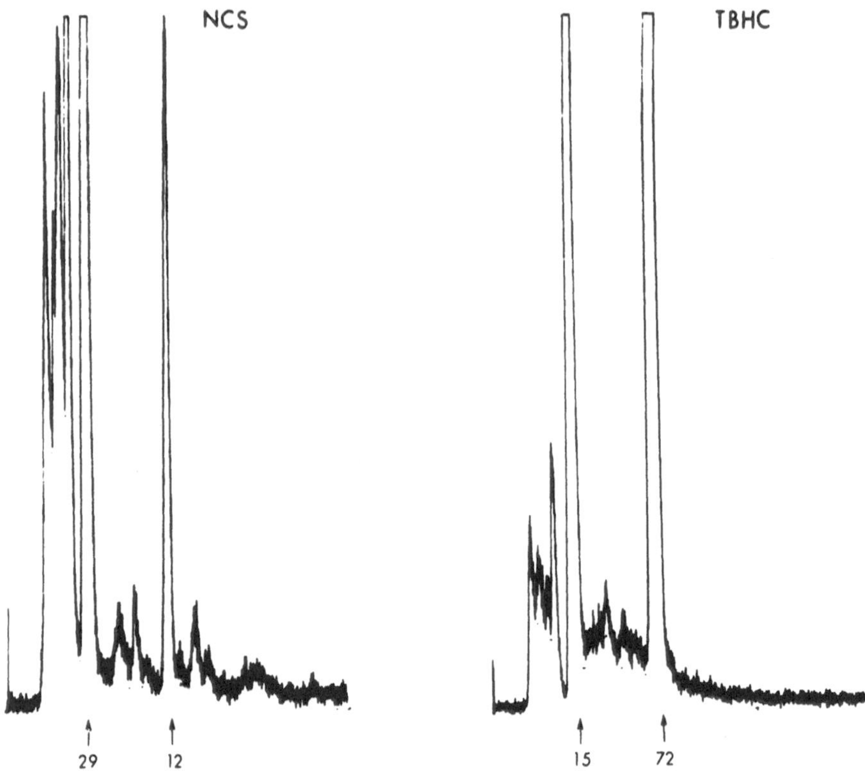

FIG. 11.12. As Figure 11, except both reactions were conducted with nca radiobromine.

If you plan to perform nca radiosyntheses, you should use radio-HPLC. It is an essential tool in the developmental stage of nca radiolabeled compounds. The researcher should also be aware of new technology such as micro-HPLC [15] and supercritical fluid chromatographic [16], as these will have obvious applications in the future nca radiolabeling of compounds.

ACKNOWLEDGMENT

I would like to acknowledge the financial support of the U.S. Department of Energy for studies of nca radiolabeling. I would also like to acknowledge the efforts of Wayne Taylor, Marty Ott, and Neno Segura in making the nca bromine-77 available for the described studies.

REFERENCES

1. Eckelman, W.C., ed.: *Receptor-Binding Radiotracers*. Vols. 1, 2, CRC Press, Boca Raton, 1982.

2. Wong, S. H.-Y.: Roles of high-performance liquid chromatography in nuclear medicine. In *Advances in Chromatography*. Vol. 19, Marcel Dekker, New York, 1981.
3. Berger, G., Maziere, M., Godot, J.J., Prenant, C. and Comar, D. Purification of ^{11}C labeled radiopharmaceuticals by LC. In *Biological/Biomedical Applications of Liquid Chromatography IV*. Chromatographic Science Series, Vol. 20, Hawk, G.L. ed., Marcel Dekker, New York, 1982, pp. 223–230.
4. Berger, G. and Knipper, R.: Remotely controlled injection of a highly radioactive solution into a liquid chromatograph. *J. Radioanalytic Chem.* **56**:251–252, 1980.
5. Roberts, J.R.: Development of continuous flow monitoring in column chromatography. In *Radiochromatography*. Journal of Chromatography Library, Vol. 14, Elsevier, New York, 1978.
6. von Stetten, O. and Schlett, R.: Purification of ^{125}I-labeled compounds by high-performance liquid chromatography with on-line detection. *J. Chromatogr.* **254**:229–235, 1983.
7. Grant, P.M., Whipple, R.E., Barnes, J.W., Bentley, G.E., Wanek, P.M. and O'Brien, H.A.: The production and recovery of Br-77 at Los Alamos for nuclear medicine studies. *J. Inorg. Nucl. Chem.* **43**:2217–2222, 1981.
8. Wilbur, D.S., Garcia, S.R., Adam, M.J. and Ruth, T.J.: An evaluation of the introduction of bromine into high specific activity radiobrominations. *J. Labelled Compd. Radiopharm.* **21**:767–779, 1984.
9. Roberts, J.R.: Radio-thin-layer chromatography. In *Radiochromatography*. Journal of Chromatography Library, Vol. 14, Elsevier, New York, 1978.
10. Wilbur, D.S., Anderson, K.W., Stone, W.E. and O'Brien, H.A.: Radiohalogenation of non activated aromatic compounds via aryltrimethylsilyl intermediates. *J. Labelled Compd. Radiopharm.* **19**:1171–1188, 1982.
11. Wilbur, D.S., Stone, W.E. and Anderson, K.W.: Regiospecific incorporation of bromine and iodine into phenols using (trimethylsilyl)phenol derivatives. *J. Org. Chem.* **48**:1542–1544, 1983.
12. Wilbur, D.S. and Svitra, Z.V.: Electrophilic radiobrominations of hippuric acid: an example of the utility of aryltrimethylsilane intermediates. *J. Labelled Compd. Radiopharm.* **21**:415–428, 1984.
13. Wilbur, D.S., Svitra, Z.V. and O'Brien, H.A.: Applications of aryltrimethylsilanes in radiobrominations. *J. Nucl. Med.* **24**:P43, Abstract, 1983.
14. Wilbur, D.S.: Synthesis and radiobrominations of some trimethylsilylphenethylamines. *J. Nucl. Med.* **25**:P124, Abstract, 1984.
15. Takeuchi, T. and Ishii, D.: Ultra-micro high-performance liquid chromatography. *J. Chromatogr.* **190**:150–155, 1980.
16. Fjeldstred, J.C. and Lee, M.L.: Capillary supercritical fluid chromatography. *Anal. Chem.* **56**:619A–628A, 1984.

CHAPTER 12

From Cyclotron to Patient via HPLC

MICHAEL R. KILBOURN, MICHAEL J. WELCH, CARMEN S. DENCE, and KEITH R. LECHNER

INTRODUCTION

The use of HPLC (high pressure liquid chromatography, also called high performance liquid chromatography) has become an increasingly useful tool in radiopharmaceutical chemistry. Initially a quick method for analysis of radiochemical products, it has evolved into a method now used in many institutions for the preparation of radiopharmaceuticals for human administration.

At the Washington University School of Medicine, numerous radiopharmaceuticals labeled with cyclotron-produced isotopes (carbon-11, oxygen-15, and fluorine-18) are prepared daily for animal and human studies using positron emission tomography (PET). HPLC is used both as part of the preparation of many of these radiolabeled compounds, as well as in the quality control analyses. There are many important aspects of using HPLC for radiopharmaceutical preparations, and many problems to be encountered and overcome. In this chapter we address the practical side of routine use of HPLC in the radiopharmaceutical laboratory (the theoretical aspects are discussed in other chapters), with examples of problems drawn from our laboratories.

HPLC SYSTEMS

A bewildering variety of HPLC systems is available, ranging from the very simple (single pump, manual controls) to the very complex (multiple pumps, microprocessor controlled). Which is appropriate? We have found that no

---

Abbreviations: Positron emission tomography, PET; High pressure liquid chromatography, HPLC; Ultraviolet, UV; Refractive index, RI; Gas chromatography, GC.

single chromatographic system is suitable for all applications, and where a variety of radiopharmaceuticals are to be prepared on a daily basis a multiple-HPLC laboratory is necessary. We have become committed to the extensive use of HPLC in our laboratories; the equipment currently in use is shown in Table 12.1.

Simple, single-pump systems are all that is required when an isocratic separation can be utilized (the design of synthesis to simplify HPLC work is discussed later). For example, our preparation of ^{11}C-glucose [1] via photosynthesis utilizes an isocratic separation of sucrose, glucose, and fructose on a resin column. This allows the use of a small, dedicated pump that is actually placed in the hot cell with the remote synthesis system.

There will be separations where a multiple-pump, gradient system is absolutely necessary. As an example, the purification of the fluorine-18-labeled butyrophenone neuroleptics, ^{18}F-spiroperidol and ^{18}F-haloperidol, requires use of complex solvent mixtures and gradient elution [2] and has required the use of more complex HPLC equipment.

At this point, it is important to discuss the options of self-contained vs. modular HPLC systems. Although a self-contained unit (injector, pump, and detector(s) all in one unit, such as the Spectra-Physics 8000A) is useful in the organic chemistry laboratory, it does not offer the versatility of a modular system, where components (injector, columns) can be easily shielded (porta-

TABLE 12.1. HPLC Equipment in Use at Mallinckrodt Institute of Radiology, Washington University School of Medicine

Chromatographs	
4 Spectra-Physics SP8700	Modular, microprocessor, ternary solvent capability
1 Spectra-Physics SP8000A	Self-contained, microprocessor, ternary solvent capability, variable UV detector
2 Waters Assoc. M6000 pumps	Manual control, single-solvent pumps
1 Waters Assoc. M-660 Solvent Programmer	
1 Waters Assoc. M-45 pump	Manual control, single solvent
1 Altex Model 110A pump	Manual control, single solvent
1 Waters Assoc. System 500 Prep HPLC	Preparatory HPLC
Detectors	
2 Waters Assoc. Model 480 variable UV	
2 Schoeffel variable UV (GM770 monochromator and SF770 monitor)	
3 Waters Assoc. Model R401 differential refractomer	
1 Wescan conductivity detector	
4 NaI(Tl) radioactivity detectors—flow-through loops	
1 CsF radioactivity detector—flow-through loop	

ble lead shielding, or placed in a shielded hood). A modular unit offers additional versatility where multiple-pump systems must share injectors or detectors. Finally, there are numerous modular, microprocessor-controlled units where the pumps, columns, detectors, and injectors are all separate from the controller, an ideal situation for a radiochemistry laboratory. For example, we use the Spectra-Physics 8700 chromatograph, a fully modular unit.

Unfortunately, the various manufacturers of HPLC equipment have not yet agreed to standardize their equipment. For this reason, there are a number of different types of fittings required when a program uses different brands of HPLC equipment (for example, mixing Waters Assoc. and Spectra-Physics equipment). This requires keeping an inventory of numerous small parts of each brand, as well as the adapters necessary for going from one system to another. New workers might be well advised to carefully select and then consistently use one particular brand of HPLC equipment. We would also like to draw attention to the need for maintaining a full complement of spare parts (fittings, pump seals, injector parts, etc.) for the HPLC equipment. It is desirable that an HPLC system not become the weak link in the chain of events extending from the cyclotron to the PET scanner. In other words, the chemists involved in the radiopharmaceutical preparations should be able to repair minor problems with their chromatograph. They should have needed parts on hand at all times and should not have to wait for spare parts to be ordered.

COLUMNS

The variety of solid phases now available in standard columns from various manufacturers has grown tremendously in the past few years. For most applications, the chemist can utilize that HPLC column which is most suitable for the separation at hand; that is, the chemist is more concerned with the analytical ability of the column. For radiopharmaceutical preparations, where one is utilizing the HPLC either as a means of purification or as a means for rapid analysis of material labeled with a short half-lived radionuclide, there are additional considerations in the choice of the HPLC column. For example, it is quite advantageous to use a reversed-phase column and an ethanol–water solvent system in the preparation to radiopharmaceuticals, as the product is obtained in a solvent that is usually (sometimes after reducing the alcohol concentration) compatible with human injection. For this reason, reversed-phased columns have found great favor in radiopharmaceutical preparation.

There will be, however, separations requiring other types of columns, possibly with organic solvent systems (the problems associated with choice of solvents are addressed in the next section). In our laboratories, we make use of a variety of columns in the synthesis of radiopharmaceuticals labeled

TABLE 12.2. HPLC Columns Used in the Routine Preparation and Quality Control of Radiopharmaceuticals Labeled with Carbon-11 and Fluorine-18

Reversed-phase columns (C_8 and C_{18})
^{11}C-diphenylmethanol—preparation and analysis
^{11}C-butanol—analysis
^{18}F-spiroperidol—analysis
^{18}F-1-(2-nitro-1-imidazoyl)-3-fluoro-2-propanol—preparation
Silica gel
^{18}F-spiroperidol—preparation
^{18}F-estrogens—preparation and analysis
Fatty acids analysis column
^{11}C-palmitic acid—analysis
^{11}C-3-methylheptadecanoic acid—analysis
^{11}C-3-methyloctanoic acid—analysis
Carbohydrate analysis column
^{11}C-glucose—analysis
Cation exchange resin
^{11}C-glucose—preparation
Organic acids analysis column
^{11}C-pyruvic acid—analysis

with fluorine-18 and carbon-11 (Table 12.2). In addition to the usual assortment of silica and reversed-phase columns, use is made of commercially available columns where the exact solid phase is not known (Waters Assoc. fatty acid analysis and carbohydrate analysis columns). We also use columns constructed from homemade packing material (proline or hydroxyproline bound to silica gel) and packed by a commercial dealer (Alltech Assoc.) for resolution of amino acid racemate mixtures. Finally, a variety of other columns (cation and anion exchange columns, cyano-substituted silica gel) are utilized in the development and analysis work carried out in our laboratories.

There are few problems in the use of most columns. However, we have found that resin columns (we pack our own using BioRad AG50W-X8) can suffer from microbial contamination. To prevent microbial growth, it is necessary to store the columns with a sodium azide solution. As this is a toxic material, it must be very thoroughly washed out of the columns before use, and the absence of azide ion established by analytical means.

For the preparation of radiopharmaceuticals for human use, it is necessary that HPLC columns remain in the best condition possible. For this reason we have found it necessary to dedicate a column to a particular patient preparation. That column is therefore not available for other radiochemical syntheses, even if the production of the primary radiopharmaceutical is not currently being performed routinely. This eliminates problems of accidental contamination of the column with foreign chemical substances. We have also found it prudent to maintain an inventory of spare columns, and thus a back-up column is always available for use in a radiopharmaceutical preparation. Although seemingly an expensive option, in fact the cost of the spare column is only a small part of the overall program (considering the material and labor costs involved in the production of the radionuclide, its incorporation into a radiotracer, and the subsequent PET study; again, the goal is that the HPLC not be the weak link). Finally, the columns used for patient preparations are routinely checked for retention times and resolution, especially when the radiopharmaceutical is prepared on an irregular or infrequent basis.

Besides differences in packing material, there are also many different sizes of columns. Two particular things in this area should be noted. Most of our analytical work is done using small columns of 0.5 × 25 cm dimensions. Very recently, the use of microbore columns (1 or 2 mm ID, 15–100 mm length) for fast analyses has been described. These innovations should be applied to radiopharmaceutical analyses, as they can result in substantial savings in chromatographic time and solvents (flow rates are less than 1 ml/min and as low as 0.03 ml/min). Second, although most preparative preparations are done with large standard columns (1 × 50 cm), we have also used what are termed *rapid analysis columns*, in particular a Whatman 5 ODS-3 RAC column, which are smaller radially packed columns (9.4 × 100 mm) that allow the shortening of separation times without loss in resolution (3).

SOLVENT SYSTEMS

Although the choice of solvents is usually mandated by the need of the separation, the radiopharmaceutical chemist is restricted by the need for obtaining a final, injectable solution. However, in some cases, there are options, with essentially equivalent separations obtainable in two or more solvent systems. For the use of aqueous, solvents there are problems of high ionic strength, and one is usually limited to the type of buffer that can be injected intravenously (phosphate or acetate, unless the salts can be removed prior to use of the product, as in the removal of calcium ions from preparations of ^{11}C-glucose [1]). In reversed-phase separations, ethanol–water solvent systems are preferable over acetonitrile–water on methanol–water. For organic solvent separations, it is best to use solvents with low boiling points (to facilitate removal by evaporation) and avoid known carcinogenic compounds. For example, one should substitute toluene for benzene, and dichloromethane for chloroform (the latter choice has been made in our preparation of ^{18}F-spiroperidol [2]).

DEDICATED VS. GENERAL INSTRUMENTS

In any large program involved in the preparation of numerous radiopharmaceuticals on a routine basis, along with the development of new methods of radiopharmaceutical chemistry and new radiopharmaceuticals, conflicts eventually (usually sooner rather than later) arise over the use of HPLC instruments. Although ideally one would like to have an HPLC always free for use by any person, for reasons of funding or space this is usually not possible. Therefore, hard choices must sometimes be made. In our laboratories, we have achieved a balance between the various demands of the program. At present, one HPLC pump (Altex pump, single solvent) is totally dedicated to the preparation of ^{11}C-glucose and is located in the cyclotron hood. One of the ternary solvent delivery systems (Spectra-Physics 8700) is dedicated to the preparation of fluorine-18 spiroperidol. A second Spectra-Physics SP8700 system is dedicated to the preparation of fluorine-18 estrogens [4,5]. A third Spectra-Physics SP8700 chromatograph is used for quality control analyses of ^{11}C-butanol [6], ^{11}C-glucose [1], and ^{11}C-palmitic acid [7], all of which use aqueous solvent systems and the refractive index detector (thus facilitating changeover from one column to the next). Finally, four chromatographs (one single-pump isocratic system (Waters M-45 pump), the two-pump, gradient Waters system (6000 pumps and 660 solvent programmer), the Spectra-Physics SP8000A, and one Spectra-Physics 8700) are designated *general* instruments, and their use varies greatly from week to week. However, the SP8000A has been assigned to the preparation of radiopharmaceuticals labeled with longer-lived radionuclides, such as those involving bromine-77 and iodine-131.

One very important question in the assignment of a general instrument is the changeover time. When going from an aqueous to an organic solvent system (or vice versa) significant time is usually needed for system purging and equilibration. The ease of this operation can vary with the type of chromatograph. We have found that this changeover is very simple with the Waters Assoc. chromatographs, but somewhat more difficult and time consuming with the Spectra-Physics systems.

At our institution, where compounds are prepared on a very routine basis, and new compounds and synthetic methods are in continual development, the HPLC units available are usually in continual use. Thus, we have chosen to dedicate instruments to clearly defined roles. The dedication of an instrument to a particular synthesis allows for the chemist to be more involved in the preparative aspects of his work, that is, to be able to do numerous experiments each week, and from week to week, without the constant worry and bother of having to "borrow" the HPLC system for his experiment, with the attendant purging and column equilibration time.

It should be evident from the above discussions that the chromatographs are used with a particular set of radionuclides. The HPLC units used for preparations with the short-lived cyclotron-produced isotopes are never used with longer-lived isotopes, such as iodine-131. This removes all possibilities of contamination. The truly long-lived isotopes—such as tritium and carbon-14—are not even allowed in the laboratory!

DETECTORS

The choice of mass detectors is again dictated by the chemical separations being attempted. Although many compounds have a suitable chromophore for detection by ultraviolet (UV) detectors, compounds such as the fatty acids, carbohydrates, and alcohols are better detected using a refractive index (RI) detector. We have thus found it necessary to provide at least three of our HPLC systems with both UV and RI detectors. For applications involving receptor ligands (such as ^{18}F-labeled spiroperidol and estrogens), where specific activity determination is critical, it is necessary to calibrate the UV detectors, by the injection of known amounts of standard compounds and construction of a standard curve.

We use home-built radioactivity detectors that consist of a scintillator (either thallium-doped sodium iodide or cesium fluoride) around which (or in front of which) is a stainless steel loop through which flows the effluent from the mass detector. These detectors are simple to construct and operate. Each detector is connected to a ratemeter and timer/counter, allowing for the display of the detector results as either a plot on a recorder or a listing of the counts/time. The calculation of the percent activity per peak can be obtained by several methods. First, if one accurately measures the amount of activity injected onto the column, the peaks can be collected as they elute (use of a fraction collector makes this easy), and proportions in each peak can be calculated. If one is certain that all of the activity injected onto the column is eluted, then one can simply sum the counts obtained for all of the peaks (corrected for decay) and then calculate the percentage of the total present in each peak. In our work, when using a reversed-phase column, any peaks with normally long retention times are rapidly eluted by injection of a bolus of dimethylformamide after the peak(s) of interest have eluted. Finally, it is possible to calibrate these simple detectors by insertion of an unpacked column and injecting a known amount of activity at a known flow rate. As an example, if we measure C_m counts from a bolus of M microcuries at a flow rate of 2 ml/min, the number of microcuries in a peak of C counts (integrated) at a flow rate of f ml/min is $MCf/2(C_m)$. Of course, the detector can be calibrated for one positron emitter and used with any other positron emitter.

With many HPLC units in one laboratory, it has proven neither desirable (because of space) or possible (because of cost) to supply each HPLC with a full set of detectors (UV, RI, and radioactivity). Rather, detectors must be shared. In particular, at the present time three HPLC units share the same radioactivity detector, two share UV and RI detectors, and so forth. This does create some conflicts, especially when two chemists need the radioactivity detector, and even worse when different levels of activity (millicuries vs. microcuries) are being handled. This is usually solved by using two different loops in the radioactivity detector, to prevent problems with detector memory (a particular problem with fluorine-18); this is one advantage of our relatively simple flow detectors. Preparations for patients take precedence in all cases, and preparations for PET studies (animals) take precedence over purely chemical work. Finally, the personal utilizing the longer-lived isotope (e.g., fluorine-18 vs. charbon-11) is required to wait (if possible).

PRE-COLUMNS AND FILTERS

It is generally good laboratory practice to have all HPLC columns fitted with appropriate pre-columns, and to use filters to remove particulate matter prior to sample injection. Although not usually a problem, workers must be aware that there are some potential pitfalls. In some cases, we have found that materials can be *caught* on these pre-columns or filters. This can be a particular problem when work is done at the no-carrier-added level with fluorine-18-labeled compounds. Lipophilic labeled products can stick to the materials used in filters (especially lipophilic products in aqueous solution, when filtered through a Millipore filter); we have observed this in the preparation of numerous carbon-11-labeled radiotracers [(3)].

IS HPLC TRULY NECESSARY?

The use of HPLC has grown tremendously in the past few years. In many cases, the use of HPLC is fully justified, but there are examples where simpler, less expensive options would be just as suitable. A few examples of our attempts to limit the use of HPLC might be useful. Our synthesis of ^{11}C-butanol [6] was specifically designed to not use HPLC and employs Waters Assoc. SEP-PAKs® (a form of bonded-phase chromatography) in the isolation and purification step. The product is obtained in suitable purity without any HPLC purification. HPLC is used in the quality control analysis, but as the products and impurities are all volatile, the analysis could just as well be done by gas chromatography (GC). Similarly, SEP-PAKS® are used in the purification of ^{11}C-pyruvic acid [8] with the analysis of the products and impurities achieved using GC, although an HPLC separation was also available. These examples were chosen to illustrate two points. First, there may be simpler methods of purification, such as bonded-phase chromatography [6] or flash column chromatography [9], which can substitute admirably for an HPLC separation. Second, some analyses of products can be done by gas chromatography (using an inexpensive gas chromatography outfitted with a proportional flow counter [10]) which is in general less expensive (columns are much cheaper, and there are no solvent costs).

It should be possible to design synthesis to minimize HPLC use, and if possible this should be pursued. Optional methods of chromatography are usually less expensive and less cumbersome to utilize, and a variety of solid phases is now available commercially. A second option is the use of solid-phase synthesis, recently described for the synthesis of carbon-11-labeled DOPA [11]. Use of an appropriate solid-phase synthesis may obviate problems with the separation of products from starting materials and excess reagents (separation being achieved by simple filtration techniques).

CONCLUSIONS

What pearls of wisdom can we pass on to persons contemplating the large-scale use of HPLC in radiopharmaceutical syntheses? Success requires good planning and coordination of effort. A proper supply of columns and HPLC supplies (solvents, fittings, filters) must always be at hand, chromatographs must be kept in good working order, and personnel should be properly trained in the care and use of the instruments. Preparation of radiopharmaceuticals on short notice (a day or less) is usually not feasible unless a dedicated HPLC system is available. However, with foresight and planning, HPLC can be a very convenient and reliable method for the preparation of radiopharmaceuticals.

ACKNOWLEDGMENTS

The work described herein was supported by grants from the National Institutes of Health (HL13851, HL17646, and NS06833) and the Department of Energy (DE-FG02-84ERG0218.A000).

REFERENCES

1. Dence, C.S., Lechner, K.A., Welch, M.J. and Kilbourn, M.R.: Remote system for the production of carbon-11 labeled glucose via photosynthesis. *J. Labelled Compd. Radiopharm.* **21:**743, 1984.
2. Kilbourn, M.R., Welch, M.J., Dence, C.S., Tewson, T.J., Saji, H. and Maeda, M.: Carrier-added and no-carrier-added synthesis of ^{18}F-spiroperidol and ^{18}F-haloperidol. *Int. J. Appl. Radiat. Isot.* **35:**581, 1984.
3. Dischino, D.D., Welch, M.J., Kilbourn, M.R., Raichle, M.E.: The relationship between lipophilicity and the brain extraction of carbon-11 radiopharmaceuticals. *J. Nucl. Med.* **24:**1030, 1983.
4. Kiesewetter, D.O.: Synthesis of fluorine-18 labeled estrogens for receptor-based imaging of mammary tumors. Ph.D. Thesis, University of Illinois, Urbana, 1984.

5. Landvatter, S.W., Kiesewetter, D.O., Kilbourn, M.R., Katzenellenbogen, J.A. and Welch, M.J.: (2R,3S)-1-(^{18}F)fluoro-2,3-bis(4-hydroxyphenyl)pentane (^{18}F-fluoronorhexestrol), a positron-emitting estrogen that shows highly-selective, receptor-mediated uptake by target tissues in vivo. *Life Sic*. **33:**1933, 1983.
6. Kilbourn, M.R., Dischino, D.D., Dence, C.S. and Welch, M.J.: SEP-PAK preparative chromatography: use in radiopharmaceutical synthesis. *J. Liquid Chrom*. **5:**2005, 1982.
7. Welch, M.J., Dence, C.S., Marshall, D.R., and Kilbourn, M.R.: Remote system for production of carbon-11-labeled palmitic acid. *J. Labeled Compds. Radiopharm*. **20:**1087, 1983.
8. Kilbourn, M.R. and Welch, M.J.: No-carrier-added synthesis of 1-^{11}C-pyruvic acid. *Int. J. Appl. Radiat. Isot*. **33:**391, 1982.
9. Fowler, J.S., Arnett, C.D., Wolf, A.P., MacGregor, R.R., Norton, E.F and Findley, A.M.: [^{11}C]Spiroperidol: synthesis, specific activity determination, and distribution in mice. *J. Nucl. Med*. **23:**437, 1982.
10. Welch, M.J., Withnell, R. and Wolf, A.P.: A new window flow proportional counter for counting radioactive G.L.C. effluents. *Anal. Chem*. **39:**275, 1967.
11. Grierson, J.R., Adam, M.J., Ruth, T.J. and Pate, B.D.: Solid-phase synthesis of ^{11}C-labeled compounds: an expedient synthesis of [3-(C-11)]-L-DOPA. Book of Abstracts, 187th National Meeting of American Chemical Society, St. Louis, 1984.

CHAPTER 13

Potential Artifacts in the Chromatography of Radiopharmaceuticals

THOMAS J. MANGNER

INTRODUCTION

At the foundation of all pharmaceutical research is the assurance that any biological effects observed for a particular pharmaceutical preparation can be attributed to the active compound and not to impurities or components of the drug vehicle. Only when the exact chemical composition of a pharmaceutical preparation or, more specifically, its chemical purity (defined as the proportion, by mass, of a preparation in a specified chemical form) [1] is known can valid correlations between biological activity and chemical structure be made. Assurance of chemical purity and demonstration of sterility and apyrogenicity of a pharmaceutical preparation are particularly important in clinical applications in order for researchers to predict the nature of the biological response and to eliminate the possibility of unwanted pharmacologic or toxic effects.

Radiopharmaceuticals, distinguished primarily by the radioactive atoms they contain, are subject to the same quality controls as any pharmaceutical. The presence of radioactivity, in addition to permitting the use of radiopharmaceuticals as diagnostic imaging agents, can also be used as the basis for very sensitive and selective methods of radiopharmaceutical quality control. Thus, additional standards of purity—radionuclidic purity (defined as the proportion of the total radioactivity as the specified isotope) and radiochemi-

---

Abbreviations: Thin-layer chromatography, TLC; high pressure liquid chromatography, HPLC; *m*-iodobenzylguanidine, MIBG; radioimmunoassay, RIA; gas liquid chromatography, GLC; gel permeation chromatography, GPC; retention time, t_R; capacity factor, k'; *m*-iodohippuric acid, MIHA; high performance thin-layer chromatography, HPTLC; N,N,N′-trimethyl-N′-[2-hydroxyl-3-methyl-5-iodobenzyl]-1,3-propanediamine, HIPDM.

cal purity (defined as the proportion of the total radioactivity as the specified compound) [1]—are applied to radiopharmaceuticals as added quality assurance measures.

The choice of analytical methods for assessment of not only the radiochemical purity of a given radiopharmaceutical preparation, but the chemical purity as well, is influenced by the presence of radioactivity in the sample. This chapter considers two widely used methods of radiopharmaceutical analysis, radio thin-layer chromatography (radio-TLC) and radio high pressure liquid chromatography (radio-HPLC) and focuses on potential erroneous results or artifacts that can occur when these methods are applied especially to radiopharmaceuticals of high specific activity. The intent is to illustrate the potential for artifactual results and not to consider all possible ways and/or instances in which artifacts can occur. The specific illustrative examples presented reflect the author's experience with and bias toward radiopharmaceuticals containing radioiodine; the underlying principles, however, can be applied to the analysis of other classes of radiopharmaceuticals and radiochemicals. The basic principles regarding technique and instrumentation will not be addressed as these are covered in other chapters.

CHOICE OF ANALYTICAL METHODS

In a broad sense, an acceptable assay for a particular compound must be able to selectively determine the amount of the compound contained in a given mixture, even in the presence of potentially interfering or structurally similar compounds. The process usually involves three operations: (a) separation of the desired compound from the mixture, (b) identification or characterization, and (c) quantitation. Depending on the particular application, the method can be rigorously applied to include each and every component in the mixture or merely the principal ingredient. The separation can be either physical (e.g., crystallization, chromatography) or chemical, in cases where the compound, or a derivative thereof, has a unique physical property able to be selectively measured without the need of physical separation. The process of identification involves the comparison of physical properties of the compound of interest to appropriate standard or reference compounds. These properties include, in order of decreasing mass required for analysis, melting or boiling point, solubility behavior, chromatographic behavior, and spectroscopic characteristics, among others. Quantitation methods range from weighing the physically isolated material to measuring physical properties that are proportional to the amount of compound present, as with absorption spectroscopy. Radioactivity, a unique physical property common to all radiopharmaceuticals and radiochemicals, requires a set of identification and quantitation methods separate and distinct from those used to determine chemical identity and mass quantity. Because of its uniqueness, the quantity of radioactivity can be easily determined, generally without need for physical separation of the radioactive species from the mixture. Identification of the radioisotopes responsible for any observed radioactivity is easily accomplished by spectroscopic analysis of the type and energy of the emitted radiation [2]. Because of the level of uncertainty that exists in all separation techniques and measurements of physical properties, no one separation or characterization operation can assure the identity and purity of a given sample. The likelihood that a given determination is accurate is increased when consistent results are obtained using as many different and independent systems as practically possible.

The analysis of most nonradioactive pharmaceuticals is generally straightforward because of the relatively large mass amounts comprising a typical dose and therefore available for full chemical and spectroscopic characterization. In contrast, the mass of a typical radiopharmaceutical dose, to which the required quantity of radioactivity is attached, is kept as low as possible in order to decrease the likelihood of unwanted pharmacologic or toxic effects or to increase the selectivity for targeted receptors. In other words, its specific activity (radioactivity per unit mass), within certain practical constraints, is maximized. Appropriate methods for the chemical analysis of radiopharmaceuticals are limited, therefore, by the relatively small mass amounts available and/or the detrimental effects of the associated radiation.

Although the presence of radioactivity allows for the highly sensitive and selective detection of extremely small quantities of radiopharmaceuticals and radiochemicals, their positive chemical identification is limited by the insensitivity of commonly used methods of mass detection relative to radiation detection, as shown in Table 13.1. Thus, in determinations of purity, stability, and metabolism of radioactive pharmaceuticals, for example, it is often possible to detect radiolabeled compounds in quantities far less than would be even detectable by other means, let alone be enough for full spectroscopic characterization. An illustrative example is provided by the radio-HPLC analysis of ^{131}I-*meta*-iodobenzylguanidine (MIBG) described in Illus-

TABLE 13.1. Approximate Detection Limits for Various Analytical Approaches[a]

Analytical Method	*Detection Limit* (*g*)
Radioactivity detection	10^{-12}–10^{-14}
Mass spectrometry	10^{-10}–10^{-12}
Fluorometry	$\sim 10^{-12}$
Electrochemistry	$\sim 10^{-12}$
Ultraviolet spectroscopy	$\sim 10^{-9}$
Infrared spectroscopy	10^{-6}–10^{-9}
Refractive index	$\sim 10^{-6}$
NMR spectrometry	10^{-4}–10^{-6}

[a] For an ideal compound under ideal conditions. Adapted from [3,4].

trative Examples and depicted in Figure 13.1, Column 2. For each of the analyses represented, the ca. 40 nCi of ^{131}I-MIBG that was injected could easily be detected and quantitated radiometrically. At 50 mCi/mg, this represents 8×10^{-10} g of MIBG sulfate, which is well below the minimum 5×10^{-8} g required for detection by the in-line ultraviolet detector (measured at 254 nm under the HPLC conditions given).

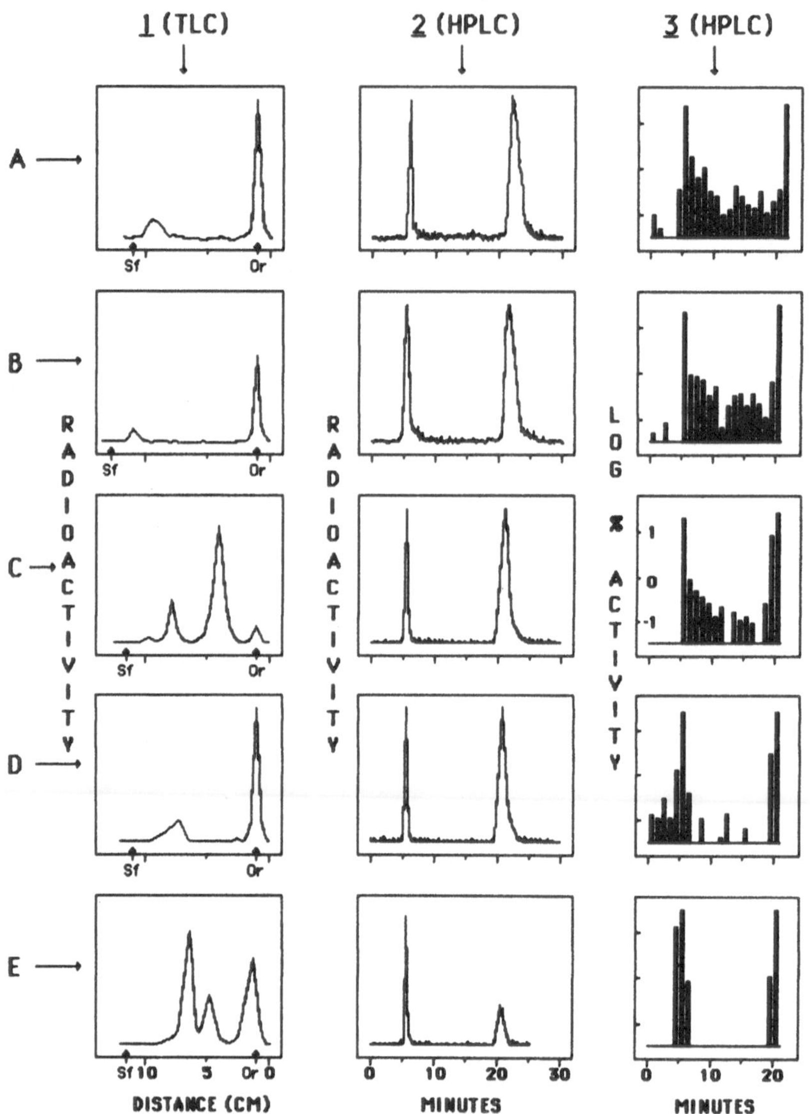

FIG. 13.1. Radio-TLC (Column 1) and radio-HPLC (Columns 2, 3) analysis of a mixture of ^{131}I-iodide and ^{131}I-MIBG contained in (A) water; (B) 13.5 *mM* sodium acetate buffer, pH 4.5; (C) 0.9% bacteriostatic normal saline solution; (D) 10 *mM* sodium iodide solution; and (E) 100 *mM* sodium iodide solution.

Two approaches to counteract the differences in sensitivity between radiation and mass detectors are: (a) to indirectly alter the compound to increase its mass detectability, or (b) to increase the mass available for analysis by using a larger sample. There are several indirect methods available to overcome the small mass problem, but each suffers the disadvantage of requiring prior knowledge of the structure of the compound under examination. One of these is dilution—the addition of *carrier,* a well-characterized unlabeled compound that is structurally identical to the radiochemical being analyzed—with various physical property measurements then being made on the diluted sample. Following carrier addition, the homogeneity of the radiopharmaceutical–carrier mixture can then be assessed through repeated co-crystallizations to a constant specific activity (i.e., reverse isotopic dilution) [5], or by chromatographic means. The chemical structure of an unknown radiopharmaceutical can be determined through proper application of the latter technique, provided appropriate carriers are used and several complementary measurements are made. Dilution methods are not applicable, however, in cases where direct knowledge of the mass of the undiluted radiochemical is required, such as in specific activity determinations of high specific activity radiopharmaceuticals. In these instances, the mass detectability of the radiochemical can be increased by the formation of a derivative that would be more highly detectable by spectroscopic methods. The radioiodination of 2-naphthol, followed by quantitation of the radioiodinated product using HPLC with ultraviolet detection, for example, has been used as an indirect method for the specific activity determination of radioiodide [6]. (See also Chapter 11 for additional examples.) Alternatively, extremely sensitive and selective biochemical methods can be employed, such as radioimmunoassay (RIA) [7], or receptor binding assays [7,8]; each of these, however, requires the labor-intensive development of reagents or systems specific for a single compound or series of closely related compounds. Therefore, these have little utility as general methods of radiopharmaceutical analysis.

One additional consideration in the analysis of especially high specific activity radiochemicals is the presence of radioactivity and its influence, from both a scientific and safety standpoint, on the choice of an appropriate analytical method. For instance, one obvious way to overcome the problem of insufficient mass for analysis is simply to start out with a larger assay sample and isolate the radiopharmaceutical from it for spectroscopic characterization. The high levels of radioactivity that would accompany the required mass, however, would preclude the concentration or total isolation of the radiopharmaceutical, as, at least for radioiodinated compounds, decomposition caused by radiation damage increases with increasing specific activity and concentration [9,10]. The possibility of radiation damage to instruments and equipment not specifically designed for use with radioactive material is another consideration, especially if large amounts of radioactivity are used. Finally, because of the risk of personnel radiation exposure and contamination, methods that require the handling of large amounts of radioactivity are less desirable than those that do not.

Many types of techniques have been used in the analysis of radiopharmaceuticals, some of which are covered by other chapters in this book. The technique that has gained widest acceptance and use is radiochromatography.

RADIOCHROMATOGRAPHY

Introduction

The usefulness of radiochromatography as an analytical technique stems from the fact that it combines the aspects of separation, identification (by chromatographic behavior comparisons), and quantitation (when coupled to various detectors) into a single operation. Almost every type of chromatographic separation technique, including TLC, HPLC, gas-liquid chromatography (GLC), gel permeation chromatography (GPC), etc., has been applied to the chemical analysis of radiopharmaceuticals and radiochemicals [11]. The techniques most widely used today include the complementary ones of TLC and HPLC, for which highly sensitive radiation detectors have been developed (see Chapters 3, 4, 7, 8). Although various chromatographic mass detectors are also available that can be extremely sensitive for specific compounds (e.g., ultraviolet, fluorometric, electrochemical) or that can be applied to a wide variety of compounds (e.g., refractive index), none combines the sensitivity and universality for mass detection that radiometric detectors exhibit for radioactivity [12,13] (see Table 13.1). Possible exceptions are mass spectrometric detectors currently under development which, when coupled to HPLC, may ultimately provide detailed structural data in addition to high detection sensitivity for a variety of different compounds [14].

Use of Reference Compounds and/or Carrier

As a detection method for determining the number of different radioactive components in a given radiopharmaceutical preparation, radio-TLC or radio-HPLC analysis, using appropriate and complementary systems, is generally sufficient. However, to positively assign a chemical structure to the compound represented by a particular peak in a radiochromatogram, additional identifying measurements must be made. Because direct isolation of a high specific activity radiopharmaceutical or radiochemical for subsequent chemical or spectroscopic characterization is at the very least impractical because of the small amount of mass involved, indirect identification by comparison of its chromatographic behavior to a fully characterized reference compound is an often-used technique that can be alternatively employed. Appropriate reference compounds would consist of the radiopharmaceutical and, for any minor radioactive components or impurities that may be present, those compounds most likely to be produced by decomposition or by other conditions to which the radiopharmaceutical is subjected—the chemical environment during its synthesis or the interaction with metabolic enzymes during in vivo metabolism, for example. If its purity could be assured, the ideal reference compound would be the radiolabeled compound with the same specific activity as would be expected in the assay sample. For some applications, detection of free iodide in a preparation of a radioiodinated pharmaceutical, for example, use of the radiolabeled standard (radioiodide) is rather straightforward. If, however, the structure of a minor radioactive component is more complex, a byproduct produced during the synthesis of a radiopharmaceutical, perhaps, or an isolated metabolite, independent synthesis of the appropriate radiolabeled compound for use as a standard would require additional procedures to establish its radiochemical purity. Because this would be prohibitive for all but the most rigorous applications, the retention data (*Rf* for TLC and t_R or k' for HPLC) of the unlabeled reference compound, obtained using quantities great enough to be detected by various mass detectors (e.g., ultraviolet, fluorometric, refractive index), are then compared with that of the radiolabeled compound detected radiometrically. Correlations may be made from separate determinations of the reference compound and assay sample (i.e., as an external standard), or, alternatively, reference compound can be added to a sample of the radiochemical as a carrier, effectively diluting or reducing the specific activity of the radiochemical of interest (provided that the structures of the reference and radiolabeled compounds are identical except for isotopic differences) and allowing concurrent mass and radiometric detection. It is generally assumed that the isotopic differences between the radiochemical and its corresponding unlabeled reference compound are not separated chromatographically, particularly if the isotopic substitution represents only a minor difference in the total mass of the compound. Isotopic separations with HPLC have been reported [15], however, and with the use of HPLC columns of ever-increasing efficiencies, may become more universally observed, particularly with smaller molecules where the difference in mass between isotopic isomers would be more significant.

Methods Validation

Instrumentation

There are many potential sources of variability in quantitative TLC or HPLC analysis that are not exclusive to radiochemical applications. Variations in detector response, sample application, condition of stationary phases and solvents, temperature, etc., can all compromise the accuracy of any qualitative or quantitative results [16]. The handling and treatment of the data obtained from a variety of detectors can also be the source of errors or inconsistencies [17,18]. The best and easiest way to control for these potential variations is to chromatograph a standard mixture of compounds at the time of sample analysis and compare the results to those obtained from previous determinations of the same standard mixture. Mixtures of standard compounds, chromatographed using standard solvent systems, have been used to monitor the efficiency of both silica gel and reversed-phase HPLC columns [19,20]. The best calibration standard for any particular application, however, would be a standard mixture that mimics as closely as possible the components of the assay sample in amounts as close as those expected and within the limitations of the detection system.

Chromatographic Separation

In addition to the separation and identification of any minor radioactive components contained in a radiopharmaceutical preparation, it is also necessary to show that the chromatographic systems being used can separate the radiopharmaceutical from any and all likely contaminants. This allows one to confirm that an apparent single radioactive peak in a chromatogram represents one compound and not several compounds with identical retention behavior in that particular system. Again, the choice of likely contaminants would depend on the nature of the radiopharmaceutical and conditions to which it is subjected. To further confirm the radiochemical purity of a given sample, it is generally necessary to establish its radiochemical homogeneity using several different chromatographic systems. However, radiochemical purity can be assured only if it can be demonstrated that the chromatographic systems can, either individually or in combination, separate all likely impurities.

The example depicted in Figure 13.2 illustrates the co-elution of two structurally very dissimilar compounds, MIBG and its major metabolite, *m*-iodo-

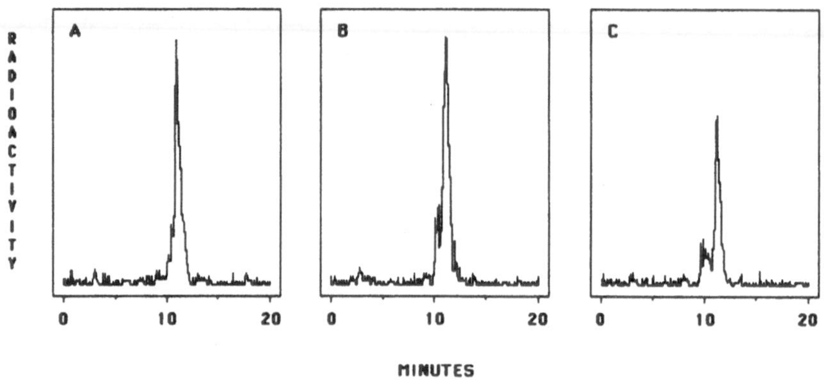

FIG. 13.2. Radio-HPLC chromatograms of ^{131}I-MIBG and metabolites in a sample of dog urine. (A) In the absence of any carrier MIBG; (B) in the presence of 0.16-*mM* carrier MIBG; (C) in the presence of 0.5-*mM* carrier MIGB.

hippuric acid (MIHA) (see Radio-HPLC of ^{131}I-MIBG in Urine—Effect of Added Carrier), even though the HPLC system used was previously shown to be very efficient in separating a variety of compounds closely related to MIBG [21]. An HPLC system with greater selectivity for this particular separation was subsequently used in the study of MIBG metabolism [22]. Other examples from recent radiopharmaceutical literature include the separation and detection of a number of ^{99m}Tc species in experimental ^{99m}Tc-radiopharmaceutical preparations, which were not evident by analysis with the normally used TLC methods [23], and the case of [^{18}F]2-deoxy-2-fluoro-D-glucose in which chromatographic separation from its anomer, [^{18}F]2-deoxy-2-fluoroD-mannose, has thus far not been accomplished despite the various methods that have been tried [24]. (See also Chapter 11). Again, each of these examples emphasizes the necessity of confirming the selectivity of the chromatographic system.

Once all the components of a radiopharmaceutical preparation have been identified by comparison to reference standards, the detection limit and linearity for each can then be established. For minor radioactive components, because the lower limit of detectability is most likely governed by the amount of radioactivity and not mass present, the detection limit can be estimated based on the efficiency of the radioactive detector under similar chromatographic conditions and the approximate specific activity of the compound. In any quantitative assay, including the radiochromatography of radiopharmaceuticals, it is also necessary to establish the range of sample size, in terms of both radioactivity and mass, within which consistent results can be expected. Researchers can do this by analyzing varying amounts or dilutions of a sample and noting the detector response. A calibration curve established in this way is used to test both detector linearity and the linear capacity of the chromatographic system [25]. Deviations from linearity in the sample amount/detector response relationship are more likely to occur at either end of the sample size range. The effects of exceeding the capacity of, or overloading a TLC plate or HPLC column are: (a) loss of resolution, and (b) alteration in retention behavior (i.e., a deviation in established TLC R_f or HPLC k' values) [26]. Because the amount of a particular compound that would exceed the capacity of most chromatographic systems in use today is well within the detectability range of virtually all mass detectors, the proper amount of carrier compound to be added to an assay sample can be easily determined from a calibration curve established using mass detection. On the low end of the scale, between the lower detectability limit of radiometric and mass detectors, however, examination of any deviation from linearity is limited again by the mass of the radiolabeled compound.

Of all the requirements for an accurate analytical method, perhaps none is more important than reproducibility. This is especially true of high specific activity radiopharmaceuticals where minor variations in the chromatographic conditions can lead to inconsistent chromatographic behavior and/or quantitative results. One way to avoid these inconsistencies is to define the chromatographic conditions as specifically as possible and keep tight controls over the materials (i.e., HPLC column or TLC plates, elution solvents, etc.) and procedures used in the assay. Reproducibility also means that identical results, within small, well-defined limits of experimental error, should be obtained when a particular sample is subjected to multiple chromatographic analyses performed at the same time under identical conditions. For example, in all but the most routine applications, at least three TLC plates or three consecutive HPLC injections of a given sample should be run to demonstrate consistency of results. Occasionally, however, consistent results are not obtained during side-by-side chromatography of the same sample and, in many instances, these inconsistent results, or artifacts, are complications resulting from the small amount of mass involved.

POTENTIAL ARTIFACTS/INCONSISTENCIES IN RADIOCHROMATOGRAPHY

Introduction

The chromatographic behavior of a particular compound can be dependent on the sample size, and direct comparison of radiochromatograms of no-carrier-added material to absorbance-derived chromatograms of the unlabeled compound may, in certain cases, not be valid. Similarly, chemical interaction of the radiolabeled compound with added carrier can significantly alter its chromatographic behavior. The remainder of this chapter illustrates the types of artifacts that can occur during radio-TLC or radio-HPLC analysis of radiopharmaceuticals, with particular emphasis on those that are a consequence of small sample size.

Artifacts in TLC Analysis

General

Thin-layer chromatography is a widely used and ideally suited technique for the analysis of radiopharmaceuticals because of its resolution, simplicity, and the fact that there is little potential for loss of material during sample preparation and analysis (except for potential loss by volatilization), allowing the analysis of the entire sample applied to a TLC plate. Additionally, the easy disposal of the radioactive plates following analysis minimizes the radiation hazard potential. The potential for artifactual or inconsistent results during radio-TLC analysis of radiopharmaceuticals exists and, indeed, is not specifically limited to this application. The most likely causes of inconsistent results during TLC analysis can generally be attributed to use of an unsuitable system, improper technique in sample application or plate development, or to inconsistencies in the quality and uniformity of the TLC plates and developing solvent [27,28] (see also Chapters 2 and 3).

Radiopharmaceuticals

The nature of radiopharmaceuticals that makes them especially susceptible to inconsistent results is, as previously mentioned, the small mass amounts involved. As pointed out by Sheppard [18,29], radio-TLC analysis of high specific activity radiochemicals in the absence of carrier often results in "streaking," or excessive peak broadening resulting from secondary interactions between the compound and the TLC plate, such as complexation of the compound with trace metal ions present in the sorbent material. Since the capacity of this type of minor secondary interaction is usually very limited, these effects are accentuated with samples of low mass in which the number of molecules involved can be a significant percentage of the total. The use of TLC plates of higher quality, such as those available for high performance TLC (HPTLC) applications, might be the answer to some of these problems. A similar type of secondary effect can be observed during reversed-phase TLC and HPLC in which mixed retention mechanisms are possible owing to the presence of both the liquid bonded phase and unbonded silanol groups of the silica gel support [30]. Additionally, if a radioactive component of a sample mixture is marginally unstable under the chromatographic conditions being used, as, for example, in the general susceptibility of radioiodide to air oxidation, overall decomposition will be enhanced by the increased exposure afforded a smaller sample mass. Effects of these types are more easily demonstrated with radiopharmaceuticals and radiochemicals of high specific activity where the small amount of mass favorable to these secondary interactions is coupled to the acute detectability of any radioactive species involved. Proper use of carrier can increase consistency by overwhelming any of these easily saturated and often variable secondary effects.

Example 1, described in the following section and depicted in Figure 13.1, illustrates the positive and, in this case, negative effects the addition of carrier can have on radiochromatographic results. The R_f values for both ^{131}I-iodide and ^{131}I-MIBG were significantly altered when unlabeled carrier sodium iodide was added to the TLC sample containing a 1 : 3 mixture of the two radioactive species. The R_f value for radioiodide decreased by as much as 40% in the presence of not only sodium iodide, but sodium chloride as well, and in both cases the radioiodide peak was much sharper (Figure 13.1, Column 1) than with samples in which these anions were absent. The effect of sodium iodide or chloride on the R_f of ^{131}I-MIBG was often manifested as two separate peaks of widely varying proportions, but, when added together, they represented a constant proportion of the total sample radioactivity. This apparent ion-pair effect between MIBG and the added anions was not seen when the samples were analyzed by a selective HPLC system [22] in which identical results, showing only ^{131}I-iodide and ^{131}I-MIBG, were obtained regardless of the presence or absence of carrier iodide or chloride (Figure 13.1, Column 2).

Artifacts in HPLC Analysis

General

Although the technique of HPLC is not without its drawbacks (see Chapter 5), its high sensitivity, resolution, and general applicability to a wide variety of different compounds, from inorganic to labeled organic and even to labeled antibodies and proteins, make it an almost essential tool in radiopharmaceutical analysis. The components of an HPLC system (e.g., pumps, injectors, detectors, etc.), which, by virtue of their increased precision and sophistication have led to the increased usefulness of HPLC as an analytical technique, are often the source of inconsistent or artifactual results [31]. An excellent summary of the many types of artifacts that can generally occur in HPLC analysis is provided by Snyder and Kirkland [32]. Again, regular calibration and performance evaluation of the HPLC system using standard mixtures could detect any inconsistencies brought on by instrumentation problems.

The most important component of any HPLC system is the column, and the choice of the proper column and elution solvent contributes most to generation of consistent results. Because, unlike TLC plates, HPLC columns are used for multiple analyses, any change in the characteristics of the column over time can have an adverse effect on the efficiency of the assay. In addition to the gradual decline in column efficiency brought on by a gradual buildup of strongly retained material and/or dissolution of the stationary phase [33], radiation-induced damage of the stationary phase, as previously mentioned, may also play a role when large amounts of radioactivity are used. The influence of conditions of secondary importance, such as column temperature, can vary among systems and have a larger than expected effect in certain cases. The t_R of MIBG, when analyzed under the chromatographic conditions of Example 1, for instance, is shortened by as much as 10% by a 5°C rise in column (ambient) temperature. As a further complication, the magnitude of the effect of column temperature on t_R values can vary widely among components of the same sample, with some greatly affected and others not at all [34].

Radiopharmaceuticals

As with TLC, there are a number of ways sample size or concentration can influence an HPLC separation, and if uncontrolled, can lead to errors in quantitation. Considerations of sample size weigh particularly heavily in the decision about whether to add carrier to the radiopharmaceutical assay sample prior to HPLC analysis. In many cases, consistent and reproducible results may be obtained with the undiluted radiopharmaceuticals, and addition of carrier would then not be needed. The goal of adding carrier would be to adjust the amount of compound contained in the sample in order to be within the established linearity range of the system. Again, this would be to counteract any low capacity secondary interactions that would have a greater influence with a smaller sample size. The effect of added carrier iodide on the elution of radioiodide using a reversed-phase HPLC system is illustrated in Figure 13.1, Columns 2 and 3. In the presence of NaI (and to a certain extent NaCl), the elution of iodide was much sharper. This is only evident, however, upon examination of the base-line "noise" following the radioiodide peak. If the base-line noise is due to the incomplete elution or slow washout of radioiodide, as it appears, this would introduce considerable error into the quantitation of radioiodide if not counted. Indeed, in the samples illustrated in Figure 13.1, the proportion of radioiodide represented by the "wash-out" activity between 7 and 19 min ranges from ca. 16% with no carrier added (A and B) to ca. 10% with NaCl added (C) to less than 1% in the presence of carrier iodide (D and E). Although, by personal experience, addition of carrier iodide to a variety of radioiodide-containing samples generally increases consistency in its analysis, this practice can sometimes have an adverse effect. One such effect is illustrated by Example 2 (Figure 13.3) in which added carrier iodide underwent exchange with the radioiodinated product resulting in an erroneously high level of radioiodide.

Of equal importance to adding the right amount of carrier to an assay sample is not adding too much so that the linear range of the column and detectors are exceeded. As already mentioned, one result of exceeding the capacity of an HPLC column is a decrease in t_R. This effect can vary among components of the same sample, being more pronounced with compounds of greater k' values [35]. For example, with a reversed-phase HPLC system, the k' value for 3-bromo-α-methyltyrosine decreased from 6.5 with $<100\ \mu g$ total injected to 1.2 with a sample mass of >1 mg, while the decrease in k' for α-methyltyrosine was from 2.2 to 0.8 over the same mass range [36]. An

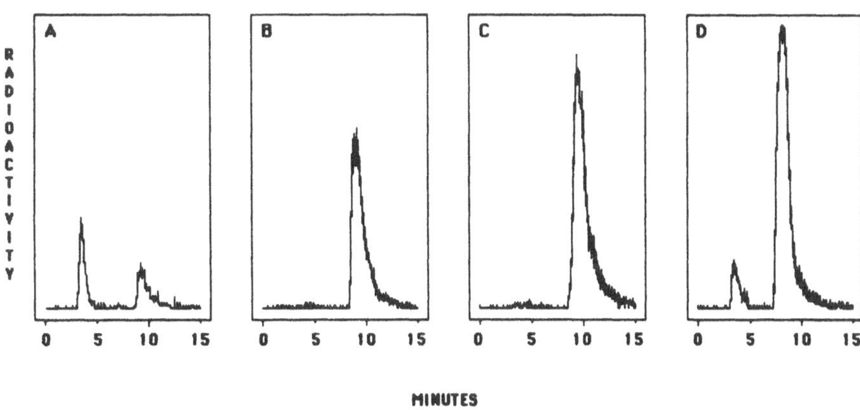

FIG. 13.3. Radio-HPLC chromatograms of ^{123}I-HIPDM. (A) Before radioiodide removal; (B) after radioiodide removal in the presence of 1 *mM* carrier HIPDM and NaI; (C) same sample as in B 3 hr later; and (D) same sample 10 min after addition of more NaI to a concentration of 3.4 *mM*.

additional complication one must consider is the effect on the t_R of a compound brought on by the overloading of other components in the mixture. For instance, in Example 3, the t_R for ^{131}I-MIBG that was contained in a urine sample was decreased from ca. 11 min (coincident with the second peak, *m*-iodohippuric acid) to ca. 10 min by the addition of carrier MIBG (see Figure 13.2). Since the total amount of MIBG (ca. 8 μg) that caused the 1-min decrease in t_R (Figure 13.2C) is well within the upper limit of 50 μg previously established for the system with a standard calibration curve, the observed decrease in t_R is a result of the influence of other components contained within the urine sample.

ILLUSTRATIVE EXAMPLES

Effect of Carrier Iodide on Radiochromatography of ^{131}I-MIBG

Introduction

In our procedure for ^{131}I-MIBG synthesis [21], a simple TLC system is used to confirm the absence of unbound radioiodide in the final product. The system provides an efficient separation of ^{131}I-MIBG (R_f 0–0.05) and free iodide (R_f 0.6–0.8) (see Figure 13.1). An additional radioactive peak at R_f 0.2–0.3 has been occasionally detected when: (a) performing stability studies of ^{131}I-MIBG formulated for human use with normal saline, or (b) performing TLC analysis of ^{131}I-MIBG in the presence of unlabeled carrier NaI, added to reduce radioiodide oxidation and volatilization. Upon subsequent radio-HPLC analysis of the sample exhibiting the extraneous TLC peak, only MIBG and iodide, if present, were detected. The experiment depicted in Figure 13.1 was used to confirm the fact that the extraneous peaks were indeed artifactual.

Experimental Conditions

TLC samples: ^{131}I-MIBG(specific activity = 15 Ci/mmol) containing ca. 25% free ^{131}I-iodide was dissolved to a concentration of 115 to 120 μCi/ml in: (A) water, (B) 13.5 *mM* sodium acetate buffer, pH 4.5, (C) 0.9% bacteriostatic sodium chloride solution, (D) 10 *mM* sodium iodide solution, and (E) 100 *mM* sodium iodide solution (*Note*: A different ^{131}I-MIBG sample was used here than for A–D, containing ca. 40% free ^{131}I-iodide). *TLC conditions*: Duplicate TLC plates (Whatman K6F silica gel-coated glass plates) were spotted with approximately 3–7 μl of each of the ^{131}I-MIBG solutions and developed with a solvent of EtOH/EtOAc/conc. NH_4OH, 20/20/1. The plates were developed to ca. 10 cm, allowed to air dry at room temperature, and immediately scanned using a Packard model 7201 radiochromatogram scanner. The results are depicted in Figure 13.1, Column 1.

HPLC samples: 10 μl of each of the TLC samples A–D were diluted to 500 μl with the appropriate solvent as listed above. HPLC analysis (duplicate injections) was performed on 20 μl aliquots of the diluted samples. *HPLC conditions*: A Waters HPLC system, equipped with a Model 440 ultraviolet detector (254 nm) and a Radiomatic Model Flo-One DR radiometric detector (200 μl solid scintillant cell), was used. Two Waters μBondapak C18 columns (4.6 × 250 mm) connected in series and protected by a 3-cm C18 pre-column cartridge (Brownlee) were eluted with 0.2 *M* ammonium phosphate (pH 7.0)/THF, 80/20, at a flow rate of 1.5 ml/min. The HPLC results are depicted in Figure 13.1, Column 2. Each chromatogram depicted in Column 2 was replotted in Column 3 as the log of the percent of total radioactivity per 1-min elution interval up to the elution of the ^{131}I-MIBG peak to emphasize differences in the magnitude of base-line noise between the iodide (t_R = 5.5 min) and MIBG (t_R = 21 min) peaks.

Observations

Several relevant observations can be made regarding the results illustrated in Figure 13.1:

1. There are obviously three peaks in the TLC scans of the NaCl and 100 *mM* NaI samples (C and E), a hint of a third peak from the 10 *mM* NaI sample (D), and only two peaks from the water and acetate buffer samples (A and B) (Figure 13.1, Column 1).
2. The R_f value for radioiodide is shifted from 0.84, found in samples A and B, to 0.62 for D, 0.55 for E, and 0.65, with a shoulder at 0.82, for C.
3. Reversed phase HPLC analysis of all the samples, including those displaying multiple TLC peaks, was essentially identical, showing only the presence of radioiodide and MIBG (Figure 13.1, Column 2).
4. The elution of radioiodide was much cleaner for the NaI-containing samples D and E than for the carrier-absent ones. This is reflected in the base-line noise between the radioiodide and MIBG peaks in a logarithmic plot of the HPLC chromatograms (Figure 13.1, Column 3).

Effect of Added Carrier Iodide on the Stability of ^{123}I-HIPDM

A sample of ^{123}I-HIPDM was labeled by iodide exchange [37] to a specific activity of 13 mCi/mg. Following removal of free radioiodide by anion exchange (Cellex-D), HPLC analysis (sample solution: 5–10 μCi of ^{123}I-HIPDM in a solution of 1 *mM* unlabeled HIPDM and 1 *mM* NaI; column: μ Bondapak phenyl 250 × 4.6 mm; elution solvent: 1 % ammonium acetate (pH 5.5)/methanol, 30/70; flow rate: 1 ml/min) revealed that the radioiodide present at t_R = 4 min before purification (Figure 13.3A) was completely removed (Figure 13.3B). Furthermore, on analysis 3 hr later, the same HPLC sample showed little change (Figure 13.3C). However, when additional carrier NaI was added to the sample to give a final concentration of 3.4 *mM*, HPLC analysis 10 min later showed the presence of ^{123}I-iodide, comprising 8% of the activity in the sample (Figure 13.3D).

Radio-HPLC of ^{131}I-MIBG in Urine—Effect of Added Carrier

When beginning a study on the metabolism of ^{131}I-MIBG, urine obtained from a dog given a 0.5 mCi dose was subjected to HPLC analysis (conditions as per Figure 13.1 except that only one C18 column was used and the pH of the elution solvent was 4.6). When a 35-μl sample of urine was directly analyzed, a single radioactive peak with a retention time roughly corresponding to that observed for a measurable amount of unlabeled MIBG was the result (see Figure 13.2A). However, when unlabeled MIBG was added to the urine to a concentration of 0.16 *mM*, a small peak with a shorter retention time appeared (Figure 13.2B). By further increasing the added carrier to 0.5 *mM* (with a 50-μl injected volume), the separation of the two peaks became more pronounced (Figure 13.2C), with the more quickly eluted peak corresponding to MIBG. The other peak was subsequently identified as the MIBG metabolite, *m*-iodohippuric acid [22].

SUMMARY

The chemical characterization of radiopharmaceuticals is complicated by the small mass amounts available for analysis. Radio-TLC and radio-HPLC are useful techniques for radiopharmaceutical analysis that can overcome this difficulty, especially when appropriate reference compounds are used. The small sample mass associated with radiopharmaceuticals also increases the potential for inconsistent or artifactual results during radiochromatographic analysis. Because of this, multiple analyses using several complementary systems should be used to determine the radiochemical purity of a radiopharmaceutical sample.

REFERENCES

1. Cohen, Y.: Purity criteria and general specifications of radiopharmaceuticals. In *Analytical Control of Radiopharmaceuticals*, International Atomic Energy Agency, Vienna, 1970, pp. 1–30.
2. Knoll, G.F.: *Radiation Detection and Measurement*. John Wiley and Sons, New York, 1979.
3. Breimer, D.D.: Rational selection of methods for therapeutic drug monitoring. In *Therapeutic Relevance of Drug Analysis*, DeWolff, F.A., Mattie, H. and Breimer, D.D., eds., Martinus Nijhoff, The Hague, 1979, pp. 9–21.
4. Dorn, H.C.: ^{1}H-NMR—A new detector for liquid chromatography. *Anal. Chem.* **56**:747A–758A, 1984.
5. Gorsuch, T.T.: *Radioactive Isotope Dilution Analysis, Review 2*. The Radiochemical Centre, Amersham, 1968.
6. Kloster, G. and Laufer, P.: Determination of specific activity of radiohalide preparations (^{75}Br, ^{77}Br, ^{123}I, ^{131}I) by HPLC-UV detection following chemical

derivatization to 1-halonaphthol-2. *J. Label. Compd. Radiopharm.* **20:**1305–1315, 1983.

7. Odell, W.D.: Principles of competitive binding assays: In *Radiopharmacy,* Tubis, M. and Wolf, W., eds., John Wiley and Sons, New York, 1976, pp. 833–859.
8. Gibson, R.E., Eckleman, W.C., Francis, B., O'Brien, H.A., Mazaitis, J.K., Wilbur, S. and Reba, R.C.: [^{77}Br]-17α-Bromoethynylestradiol: in vivo and in vitro characterization of an estrogen receptor radiotracer. *Int. J. Nucl. Med. Biol.* **9:**245–250, 1982.
9. Bayly, R.J. and Evans, E.A.: *Storage and Stability of Compounds Labelled with Radioisotopes, Review* 7. The Radiochemical Centre, Amersham, 1968.
10. Galatzeanu, I. and Cook, G.B.: Self decomposition of some ^{131}I-labelled radiopharmaceuticals. In *Analytical Control of Radiopharmaceuticals,* International Atomic Energy Agency, Vienna, 1970, pp. 131–152.
11. Roberts, T.R.: *Radiochromatography.* Elsevier, Amsterdam, 1978.
12. Scott, R.P.W.: *Liquid Chromatography Detectors.* Elsevier, Amsterdam, 1977.
13. Vickrey, T.M., ed.: *Liquid Chromatography Detectors.* Marcel Dekker, New York, 1983.
14. Crowther, J.B., Covey, T.R., Silvestre, D. and Henion, J.D.: Direct liquid introduction into LC/MS: four different approaches. *Liquid Chromatogr.* **3:**240–254, 1985.
15. Jefferey, A.M. and Fu, P.P.: *Anal. Biochem.* **77:**298, 1977.
16. Poole, C.F. and Schuette, S.A.: *Contemporary Practice of Chromatography.* Elsevier, Amsterdam, 1984.
17. Mellish, C.E.: Limits of accuracy in the determination of purity by thin layer chromatography and paper chromatography. In *Analytical Control of Radiopharmaceuticals,* International Atomic Energy Agency, Vienna, 1970, pp. 115–125.
18. Sheppard, G.: *The Radiochromatography of Labelled Compounds, Review 14.* The Radiochemical Centre, Amersham, 1972.
19. Atwood, J.G. and Goldstein, J.: *J. Chromatogr. Sci.* **18:**650–654, 1980.
20. Smith, R.M.: *J. Chromatogr.* **236:**313–320, 1982.
21. Mangner, T.J., Anderson-Davis, H., Wieland, D. and Swanson, D.P.: Synthesis of I-131- and I-123-metaiodobenzylguanidine for diagnosis and treatment of pheochromocytoma. *J. Nucl. Med.* **24:**P118, 1983.
22. Mangner, T.J., Tobes, M.C., Wieland, D.M. and Sisson, J.C.: Metabolism of I-131-*m*-iodobenzylguanidine (MIBG) in humans. *J. Nucl. Med.* **25:**P123, 1984.
23. Pinkerton, T.C., Heineman, W.R. and Deutsch, E.: Separation of technetium hydroxyethylidene diphosphonate complexes by anion exchange high performance liquid chromatography. *Anal. Chem.* **52:**1106–1110, 1980.
24. Bida, G.T., Satyamurthy, N. and Barrio, J.R.: The synthesis of 2-[F-18]fluoro-2-deoxy-D-glucose using glycals: a reexamination. *J. Nucl. Med.* **25:**1327–1334, 1984.
25. Delaney, M.F.: Chromatographic calibration. *Liquid Chromatogr.* **3:**264–268, 1985.
26. Snyder, L.R.: *Anal. Chem.* **39:**698, 1967.
27. Touchstone, J.C. and Dobbins, M.F.: *Practice of Thin Layer Chromatography.* 2nd ed., John Wiley and Sons, New York, 1983.

28. Janchen, D.: Considerations on the reproducibility of thin-layer chromatographic separations. In *HPTLC-High Performance Thin Layer Chromatography,* Zlatkis, A. and Kaiser, R.E., eds., Elsevier, Amsterdam, 1977, pp. 129–145.
29. Sheppard, G. and Thomson, R.: Quality control and analysis of radiotracer compounds. In *Radiotracer Techniques and Applications,* Evans, E.A. and Muramatsu, M., eds., Marcel Dekker, New York, 1977, pp. 171–235.
30. Brinkman, V.A.T. and DeVries, G.: *J. Chromatogr.* **258:**43–55, 1983.
31. Walker, J.Q., Jackson, M.T. Jr. and Maynar, J.B.: *Chromatographic Systems: Maintenance and Troubleshooting.* 2nd ed., Academic Press, New York, 1977.
32. Snyder, L.R. and Kirkland, J.J.: *Introduction to Modern Liquid Chromatography.* 2nd ed., John Wiley and Sons, New York, 1979, pp. 781–823.
33. Snyder, L.R. and Kirkland, J.J.: *Introduction to Modern Liquid Chromatography.* 2nd ed., John Wiley and Sons, New York, 1979, pp. 226–245.
34. Wong, S.H.-Y.: Roles of high performance liquid chromatography in nuclear medicine. *Adv. Chromatogr.* **19:**1–36, 1981.
35. Snyder, L.R. and Kirkland, J.J.: *Introduction to Modern Liquid Chromatography.* 2nd ed., John Wiley and Sons, New York, 1979, pp. 541–574.
36. Kloster, G. and Laufer, P.: Identification of radiopharmaceuticals by their retention on HPLC: a caveat. *Int. J. Appl. Radiat. Isot.* **35:**545, 1984.
37. Mangner, T.J., Rogers, W.L., Wieland, D., Clinthorne, N., Juni, J., Koral, K. and Mayans, R.: Solid phase exchange labeling of I-123-HIPDM. *J. Nucl. Med.* **25:**P123, 1984.

CHAPTER 14

HPLC of Radiolabeled Antibodies

DONALD J. HNATOWICH

INTRODUCTION

Widespread clinical use of radiolabeled proteins in nuclear medicine is currently restricted to human serum albumin, either in its native state for blood pool imaging or denatured for perfusion imaging. It now appears that radiolabeled antibodies may be the second protein to enjoy extensive clinical use, particularly for, but not restricted to, tumor imaging and therapy. Fortunately, the development of monoclonal antibodies with superior properties for these applications has coincided with the development of new analytical techniques for protein analysis.

Although improvements have been made in traditional methods of protein analysis such as polyacrylamide gel electrophoresis and thin-layer chromatography, the technique that shows the most promise is high performance liquid chromatography (HPLC). Whereas, gel electrophoresis and thin-layer chromatography may often achieve the resolution of HPLC while precipitation techniques may achieve the speed of HPLC, only HPLC itself combines both of these desirable features. Being able to separate radiolabeled proteins from radiocontaminants and other species and do so quickly is key to the needs of nuclear medicine.

The field of protein separation by HPLC is relatively new (such that uniform nomenclature has not yet evolved) and is changing rapidly. Several review articles [1–3], book chapters [4,5], and other references [6] are concerned with the separation by HPLC of proteins and peptides, although not

---

Abbreviations: High performance liquid chromatography, HPLC; High performance ion exchange chromatography, HPIEC; High performance reversed-phase chromatography, HPRPC; High performance size exclusion chromatography, HPSEC; Ultraviolet, UV; Diethylenetriaminepentacetic acid, DTPA; Isoelectric pH, pI.

with the analysis of radiolabeled proteins for nuclear medicine applications [7]. This chapter attempts to fill a void by briefly discussing HPLC as it has been applied in one laboratory for the analysis of radiolabeled antibodies. It is beyond the scope of this chapter to enter in depth into the many aspects of HPLC that determine resolution, recovery, etc.; these are, however, discussed in the above-mentioned references.

THEORY

The object of HPLC, to provide adequate separation and to do so rapidly, is achieved through the use of column packings consisting of microparticles with diameters of 10 μ or less. Because diffusion distances are decreased in column packings comprised of these particles, both the equilibrium time and the separation time are also decreased. However, in order to maintain the high flow rates through these columns required for rapid separations, high pressures (1000 to 3000 psi) are usually necessary. Organic-based polymers, such as polyacrylamides and polydextrans, useful in low pressure open-column chromatography collapse under these pressures and are therefore unsuitable for HPLC. Because of this need for rigidity under pressure, with only few exceptions [8,9], packing materials used in HPLC are rigid inorganic gels usually based on silica. It has been possible to prepare silica-based column packings with the proper particle and pore sizes and derivatized with the many functional groups required for various HPLC applications.

The use of silica is, however, not without its drawbacks. In particular, column life may be reduced because of the appreciable aqueous solubility of this material. In addition to particle size, temperature, and electrolyte concentration, the solubility of silica is dependent on eluant pH. Consequently, the pH range of eluants used with these columns is limited to about 2 to 8 [10]. In addition, silica is negatively charged at pH values greater than approximately 2; therefore, unwanted ionic interactions between sample and packing may occur even in the case of well-derivatized silica.

With the exception of specialized columns such as those performing affinity chromatography [3], commercially available HPLC columns designed for use with macromolecules fall into three catagories: those that principally perform size exclusion, reversed-phase, or ion exchange chromatography.

Size Exclusion Chromatography

HPLC employing size exclusion (often referred to as HPSEC) separates according to molecular size and shape. Historically, the technique was developed as gel filtration (aqueous eluants) and gel permeation (nonaqueous eluants) chromatography. The column packing consists of particles with pores of a particular size. When a sample containing different-sized molecules is passed into a column of these particles, the smaller molecules will experience the greatest ease in entering these pores, while the largest molecules may be excluded entirely. Thus, the effective path length of the small molecules will be much greater than that of the larger molecules in their passage through the column, and, as a result, elution from the column will be in order of decreasing molecular size. Those molecules which, because of their large size, are totally excluded from the pores, will co-elute in a volume of eluant known as the void or excluded volume (V_o). Likewise, all molecules which, because of their small dimensions, are totally permeable within the pores will also co-elute, but in a volume equal to the total volume of the column, referred to as the total included volume (V_t). Separation, therefore, only occurs at elution volumes within these limits (i.e., $V_t - V_o$). Since the dimensions of the pores determine the size of the molecules eluting within these limits, it is possible to separate molecules over widely different molecular size ranges by using columns packed with particles of different pore dimensions. As an example, the separation ranges of three HPSEC columns from one manufacturer are listed in Table 14.1 [11]. The separation ranges are clearly dependent on molecular shape such that macromolecules with a random coil configurations appear larger than those with globular configurations for the same molecular weight.

Although protein separation by HPSEC is predominantly determined by molecular size and shape, ionic and hydrophobic interactions between the protein and column packing may influence the separation. The pH of eluants used in protein separations is typically near that of physiologic pH (i.e., 6.6 to 8.0), and, at this pH, silanol groups that have escaped derivatization are negatively charged. Consequently, basic proteins may experience retention owing to charge attraction and elute later than expected, while acidic proteins may experience ionic exclusion and elute early. Likewise, proteins with hydrophobic regions may be attracted to alkyl or other hydrophobic groups on the derivatized silica packing material.

These interactions must be minimized if information concerning molecular size and shape of proteins is of interest. Ionic interactions may be minimized by adjusting the eluant pH to be equal to the pI of the protein of interest or, more usefully, by increasing the ionic strength of the eluant. Hydrophobic interactions may be reduced by decreasing the hydrophilicity of the eluant through the addition of a miscible organic solvent [12].

A characteristic of HPSEC is that resolution cannot usually be improved by changing the eluant because the composition of the eluant often has little

TABLE 14.1. Molecular Weight Separation Ranges for Several Commercially Available (Waters) HPSEC Columns[a]

Column	*Native Globular (daltons)*	*Random Coil (daltons)*
I-60	1,000–30,000	600–8,000
I-125	2,000–80,000	1,000–30,000
I-250	10,000–500,000	2,000–150,000

[a] See [11].

effect on a separation. Improvements in separation may be achieved by adding additional columns in series, reducing the flow rate, or by recycling. Nevertheless, a twofold difference in molecular weight is usually necessary for adequate resolution [2].

Reversed-Phase Chromatography

Reversed phase refers to adsorption chromatography in which the eluant is more polar than the column packing. Separations occur because polar molecules experience less attraction for the nonpolar packing and therefore elute ahead of nonpolar molecules. The packing material used in reversed-phase HPLC (HPRPC) usually consists of silica derivatized with long chain alkyl or other hydrophobic groups to reduce charge interactions, but primarily to maximize hydrophobic interactions.

HPRPC has seen its greatest use in the separation of small molecules, however it is being increasingly employed for macromolecular separations [13]. Because most proteins are strongly retained by the hydrophobic column packings used in HPRPC with aqueous eluants, protein separations are rarely performed without gradient elutions [14]. In contrast to isocratic elution where the eluant composition is unchanged during analysis, during gradient elution the composition of the eluant is continually changed. In the case of HPRPC, the eluant is usually made increasingly less polar by the addition of the miscible nonpolar organic solute such as acetonitrile, thereby encouraging protein elution.

One strength of HPRPC is its versatility. Protein retention may be altered by changes in the pH, ionic strength, and solvent strength of the eluant, and these changes may be programmed by gradient elution to occur at any point during the analysis. Therefore, with the appropriate choice of conditions, it is usually possible to achieve a desired separation. However, depending on conditions, protein recovery may be reduced and the recovered proteins are likely to have been denatured by organic solvents in the eluant [15].

Ion Exchange Chromatography

Whereas charged interactions between sample and column packing material are usually undesirable in the case of HPSEC and HPRPC, in the case of high performance ion exchange chromatography (HPIEC), these interactions are encouraged. Accordingly, the packing employed in HPIEC is derivatized with charged groups such as tertiary amines in the case of anion exchange chromatography, and sulfonates in the case of cation exchange chromatography. Since proteins are positively charged below and negatively charged above their pI, anion exchange columns will retain proteins above their pI, while cation exchange columns will retain proteins below their pI [16].

HPIEC is the most recent of the HPLC techniques to be applied to protein separations and is not thoroughly evaluated. Separations are achieved by gradient elution as in the case of HPRPC, but by altering the pH and/or the ionic strength of the eluant rather than the concentration of an organic solvent. Consequently, HPIEC has the advantage over HPRPC of denaturing proteins to a lesser extent [17]. Furthermore, several commercially available HPIEC columns use organic-based packings, and therefore may be used over an extended pH range.

AN HPLC SYSTEM FOR THE ANALYSIS OF RADIOLABELED ANTIBODIES

Perhaps the most important decision in the design of an HPLC system is the choice between HPSEC, HPRPC, and HPIEC. The decision about which of these chromatographic techniques to use for the analysis of radiolabeled antibodies was made in this laboratory in favor of HPSEC, primarily because of its simplicity. A major advantage of HPSEC is that it is not necessary to experiment with different eluant compositions to improve resolution because this will not normally influence the separation and because virtually all water soluble species will eventually elute, regardless. For this reason, HPSEC is less versatile than HPRPC or HPIEC, however, provided proteins differ sufficiently in molecular weight, adequate separations are possible with minimal understanding of or attention to separation conditions. As a further advantage, the HPLC system can be less expensive because elutions are performed isocratically and therefore without the additional pump and controller unit required for gradient elutions. Furthermore, because of the isocractic conditions, repeat analysis may be performed more quickly than in the case of gradient elutions, where initial conditions need to be re-established prior to each analysis.

An essential detector for use with radioactive samples is, of course, the flow-through radioactivity detector. Several flow-through radioactivity detectors designed for HPLC applications are commercially available. However, virtually all use liquid scintillation, and therefore detect beta particles with good sensitivity, but detect gamma rays only poorly. Because all radionuclides in nuclear medicine decay with the emission of gamma rays (although beta rays may sometimes be involved in their decay as well), more sensitive detectors for these radionuclides employ inorganic scintillators such as NaI(Tl). At least one such detector is commercially available (Flow-one Model GA, Radiomatic Instruments and Chemical Co., Tampa, FL). Fortunately, a flow-through gamma-ray detector may be constructed fairly cheaply. Figure 14.1 illustrates one such detector used in this research, along with its associated electronics. The detector itself consists of a $1 \times 1\frac{1}{2}''$ NaI(Tl) crystal-photomultiplier preamplifier combination wrapped once with a small diameter tube (A) from the outlet of the UV detector. High voltage

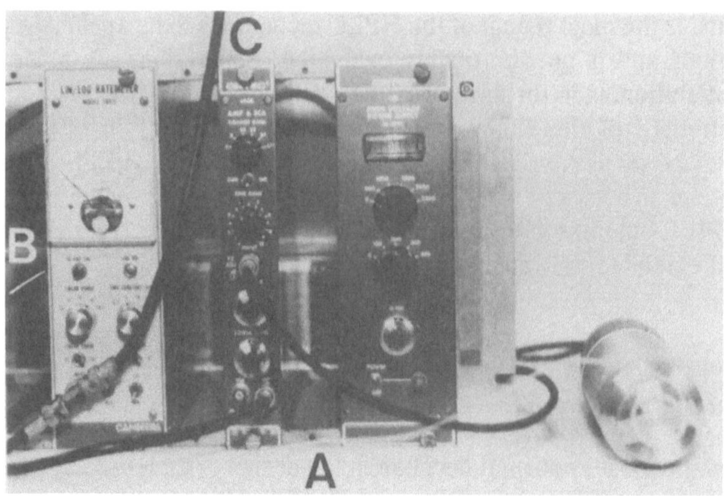

FIG. 14.1. A photograph of a flow-through radioactivity detector and associated electronics. Effluent carried in a small diameter plastic tube (A) flows in a loop in front of an NaI(Tl) detector. The amplified signal from the detector is routed to a ratemeter connected via a cable (B) to a chart recorder. Output from the amplifier is also routed via a coaxial cable (C) to a multichannel analyzer.

for the photomultiplier is provided at the rear of a high voltage power supply such as the Ortec Model 456 (Ortec Inc., Oak Ridge, TN). The signal is amplified by an Ortec Model 490B amplifier or the equivalent, and the amplified signal sent to a linear ratemeter such as the Canberra Model 1418L Lin/Log ratemeter (Canberra Industries, Meriden, CT). Output from the ratemeter is conducted via a cable (B) from the rear of the unit to a dual-pen chart recorder. All three units are housed and powdered by a Canberra Model 2000 Nim Bin or the equivalent.

Another important decision in the design of an HPLC system concerns the choice of a second detector. Since radiolabeled antibodies are to be analyzed, a flow-through radioactivity detector may be all that is needed. Nevertheless, a second detector sensitive to some property of the protein itself can be very useful. For example, only by demonstrating that the radioactivity and protein co-elute can one say that the two may be associated. A useful detector for HPLC applications involving the analysis of proteins is an ultraviolet (UV) detector. Most proteins absorb strongly in the UV at about 280 nm because of the aromatic amino acids, tryptophane, tryosine, and phenylalanine [18], and, as a consequence, UV detection tends to be sensitive.

The dual-pen recorder charts both the radioactivity and UV traces. In addition, the output from the UV detector (after digitizing using a Canberra Model 6271 digitizer) and the output from the ratemeter (C) are routed via a coaxial cable to a Canberra Model 8100 multichannel analyzer capable of operating in multiscale mode. In this mode, all counts received for a fixed time interval are accumulated in the first channel, all counts received during a second time interval of the same duration are accumulated in the second channel, and so on. The result is a chromatogram with features identical to that appearing on the chart recorder, but stored in the analyzer memory and displayed on a display scope. In this manner, the radioactivity and UV absorbance traces can be transferred via the display scope to polaroid film, retrieved by teletype, or transferred to magnetic tape for long-term storage and for analysis by off-line computer. Most data analysis, such as the calculation of peak areas, can also be performed, if desired, using the microprocessor-controlled features of the analyzer.

The HPLC used in the studies to be described below consisted of a Waters Model 45 or 6000A solvent delivery system (flow rates 0.1 to 9.9 ml/min), a Model U6K universal injector (injector loop capacity 1 μl to 2.0 ml), and a dual-pen chart recorder. Absorbances at 280 nm were monitored with a flow-through Model 440 constant-frequency or Model 481 variable-frequency detector. The flow-through radioactivity detector and associated electronics have been described above. Separations were achieved on a single 7.8 mm $\times$ 30 cm I-125 or TSK 3000 SW (Bio-Rad, Richmond, CA) HPSEC column using 0.1 *M* phosphate buffer, pH 7.0, as eluant. The buffer was degassed under water aspirator vacuum to remove dissolved gases that can affect the base line of the absorbance detector. Both the eluant and sample were filtered through a 0.45-μ Millipore cellulose acetate filter to remove particulates that can plug the column inlet and outlet filters. A guard column (Waters Model 84550 with bulk packing 85290) was placed between the injector and column to protect the column from contaminants, that can shorten its life. Between uses, the entire system was stored in distilled water containing 0.05% NaN_3 as a bacteriostat.

APPLICATIONS

Evaluation of Radiochemical Purity

One of our least demanding applications of HPLC requires only that labeled protein be resolved from radiolabeled low molecular weight contaminants. In our laboratory, the first step in the labeling of proteins involves covalently attaching to the protein the strong chelator diethylenetriaminepentaacetic acid (DTPA) using a cyclic anhydride [19]. Because hydrolysis of the anhydride competes with the coupling reaction, the coupled protein is always contaminated with some free DTPA that is removed by preparative open-column gel filtration chromatography. The coupled and purified proteins are then stored and eventually labeled, usually with ^{111}In. Throughout, the percentage of radioactivity present as labeled protein and as labeled free DTPA must be determined. In the case of coupled but unpurified protein, the analy-

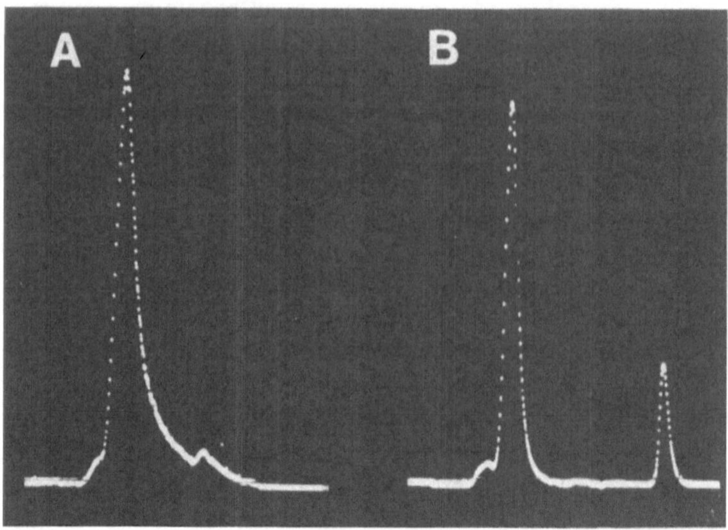

FIG. 14.2. A composite image obtained from the display scope of a multichannel analyzer. Both the radioactivity trace (A) and UV absorbance trace (B) are shown for a patient preparation of an ^{111}In-labeled F(ab′)$_2$ antibody fragment. Abscissa: channel number (retention time); ordinate: counts (A) and UV absorbance (B).

sis of a small aliquot of the preparation after labeling with ^{111}In permits the average number of DTPA groups per protein molecule to be determined [19]. In the case of purified protein preparations, analysis after labeling determines the radiochemical purity. Figure 14.2 is a composite prepared from two images of the display scope of the multichannel analyzer and shows the radioactivity trace (A) and UV absorbance trace (B) of one patient preparation of the F(ab′)$_2$ antibody fragment labeled with ^{111}In. The UV absorbance trace shows only the presence of human serum albumin and a preservative added as a stabilizer; however, since albumin and the antibody co-elute in this system, the absorbance trace is useful to determine the retention time of the antibody. It is then apparent from the radioactivity trace that greater than 95% of the radioactivity elutes with the antibody. A small percentage of radioactivity elutes at a longer retention time, probably as labeled free DTPA, and a small percentage of activity elutes at an earlier retention time, probably as labeled antibody dimer. Since this analysis took only about 15 min, the clinical study was not substantially delayed by the determination of radiochemical purity.

Pharmacokinetic Studies

In our laboratory, we are interested in determining the chemical forms of ^{111}In circulating in serum and present in urine following administration of

^{111}In-labeled antibodies to patients. In the following study, the TSK 3000 SW column was used in place of the I-125 column for improved resolution at high molecular weights in a search for antigen-antibody complexes.

Serum and urine samples obtained from a patient administered an ^{111}In-labeled $F(ab')_2$ fragment were injected directly into the HPLC without prior sample preparation other than filtration. No more than 250 μl (about 0.25 μCi) was injected to minimize peak broadening resulting from sample loading. The flow-through radiation detector was too insensitive to monitor radioactivity, so that, in this case, fractions were collected and were counted in a Na(Tl) well counter. Figure 14.3 presents a radiochromatogram obtained by the analysis of a serum sample obtained at 30 min post-injection (trace A). For comparison, a radiochromatogram of the injectate is also presented (trace B). Both radiochromatograms show the majority of activity to be

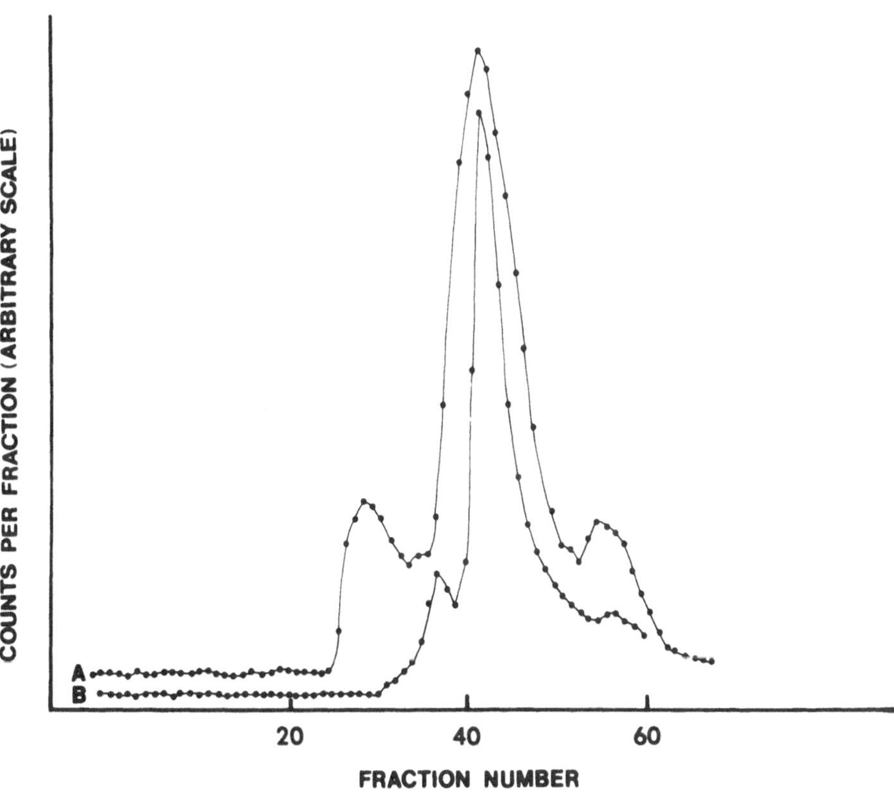

FIG. 14.3. Radiochromatograms obtained for a labeled antibody (B) and for a serum sample from a patient administered the labeled antibody 30 min previously (A). Conditions: one TSK 3000 SW column, 0.1 *M* phosphate pH 7.0 buffer, 0.8 ml/min flow rate. Radioactivity was measured by collecting fractions for counting in a well counter. The effect of sample size on peak broadening is apparent; trace A was obtained following injection of 250 μl, while only 25 μl were injected in the case of trace B.

present as labeled protein eluting in fractions 40 to 50. The injectate is seen to contain a small percentage of a low molecular weight radiocontaminant, probably labeled free DTPA, eluting in fractions 55 to 60. A similar, but larger, peak appears in the radiochromatogram of the serum sample, suggesting that it may be the result of antibody catabolism. The injectate also contains a small percentage of antibody dimer appearing in fractions 35 to 39. The radiochromatogram of the serum sample shows a pronounced peak at an even higher molecular weight, eluting in fractions 26 to 34, which is not present in the injectate, suggesting the presence of antigen–antibody complex.

Evaluating the Properties of New Radiolabels for Proteins

One of the advantages of attaching DTPA groups to proteins is that these proteins may be radiolabeled with a variety of metallic radionuclides of which ^{111}In is only one. Whenever a new radionuclide is employed for this purpose, however, it is imperative to establish: (a) that the label is attached to the DTPA groups on the protein and is not nonspecifically bound to other regions of the protein, and (b) that the label does not dissociate from the protein in aqueous solution and, in particular, in 37°C serum.

In our laboratory, the absence of nonspecifically bound radioactivity is established by a hydrolyzed-control study in which the cyclic anhydride is deliberately hydrolyzed to free DTPA before adding to the protein [19]. Thus, the only differences between the coupled protein sample and the hydrolyzed control sample is that, in the case of the latter, the protein is not coupled with DTPA. Figure 14.4 shows two radioactivity traces obtained on the I-125 column by the analysis of both a coupled and control sample following radiolabeling with 90Y and without purification from free DTPA. The radioactivity trace for the coupled sample (bottom trace) shows that the majority of activity elutes at early retention times as labeled protein and somewhat less radioactivity elutes later as labeled free DTPA. The radioactivity trace for the hydrolyzed control sample also shows radioactivity eluting as labeled free DTPA but, in this case, no radioactivity co-elutes with the protein.

The determination of serum stability of the label is illustrated below for the case of an IgG antibody radiolabeled with ^{99m}Tc. Using the TSK 3000 SW column and the flow-through radioactivity detector, the radioactivity trace (A) presented in Figure 14.5 was obtained and shows that approximately 95% of the radioactivity is protein bound in a form that is stable to analysis in this system. A small percentage of radioactivity elutes in the V_o and is probably ^{99m}Tc-labeled radiocolloids; the presence of radiocolloids in these samples was also indicated by the poor recovery of radioactivity in these analyses (probably resulting from trapping of radiocolloids in the column packing). After incubation for 1 hr in 37°C serum, the radioactivity trace (B)

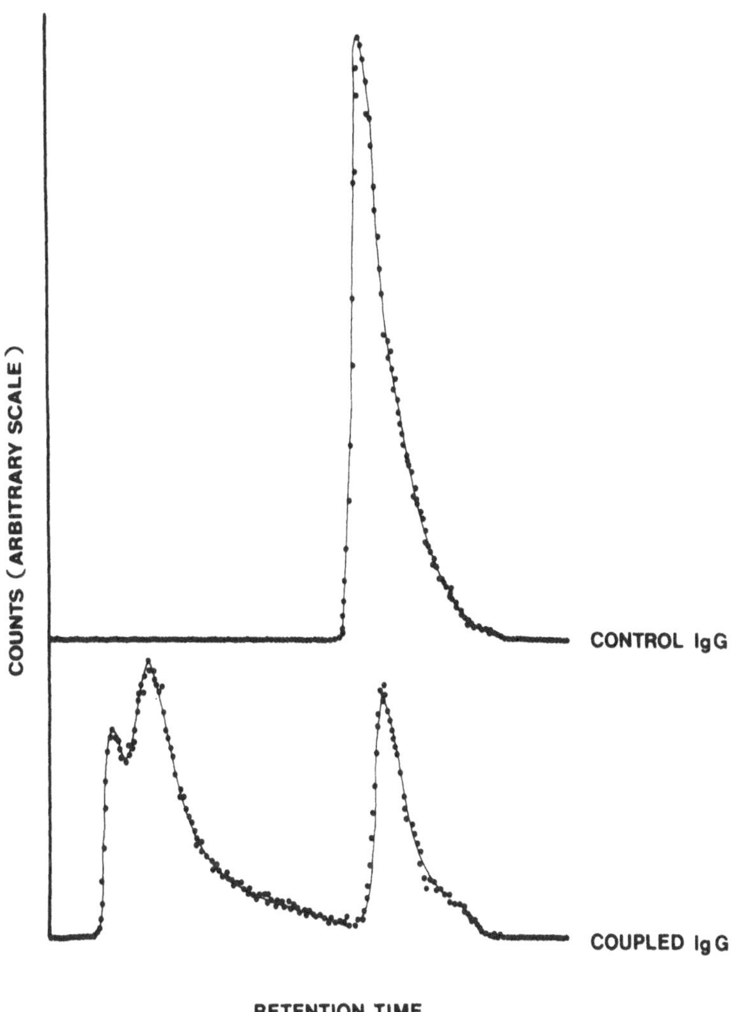

FIG. 14.4. Two radioactivity traces of an IgG antibody coupled with DTPA (bottom) and its corresponding hydrolyzed control (top) after labeling with ^{90}Y and without purification from free DTPA. Conditions: one I-125 column 0.1 *M* phosphate pH 7 buffer, 1.0 ml/min flow rate.

shows increased radioactivity present as the high molecular weight species. After incubation for 3 hr (trace C), increased radioactivity is again present in the high molecular weight species, but two other changes are now apparent: ^{99m}Tc-pertechnetate, resulting from oxidation of reduced ^{99m}Tc, appears at an elution volume of 12.5 ml, and the effects of transcomplexation of the radioactivity from antibody to another, lower molecular weight protein, are apparent in the appearance of an additional peak at an elution volume of 8.5 ml.

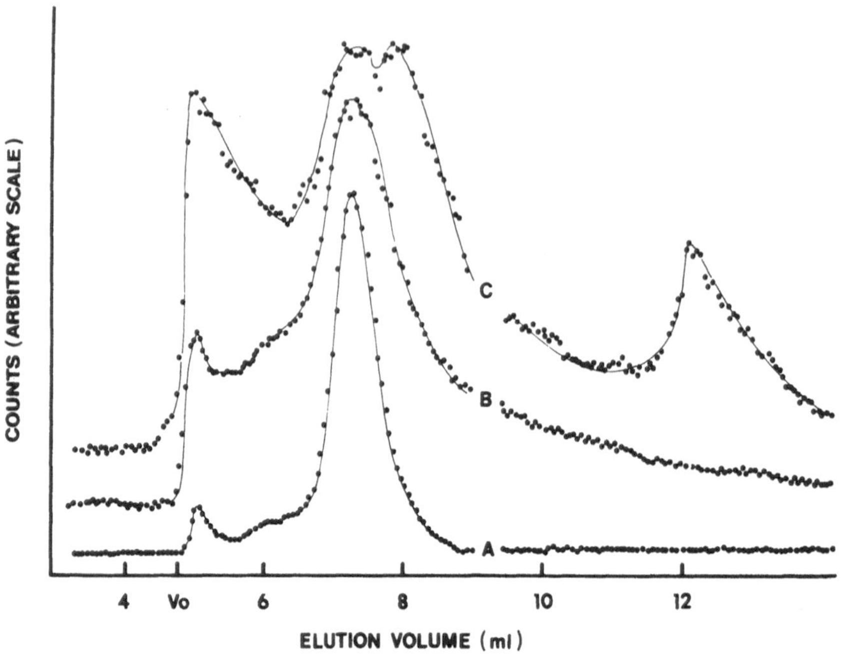

FIG. 14.5. Three radiochromatograms obtained by the analysis of an IgG antibody preparation radiolabeled with ^{99m}Tc (A), analysis of the labeled antibody preparation after a 1 hr serum incubation (B), and analysis of the labeled antibody preparation after a 3 hr serum incubation (C). Conditions: one TSK 3000 SW column, 0.1 *M* phosphate pH 7.0 buffer, 0.8 ml/min flow rate.

DISCUSSION

Each of the separations described could have been performed by methods more conventional than HPLC. Both open-column gel filtration chromatography and gel electrophoresis are capable of resolving macromolecules according to their molecular weight; free activity may be separated from protein-bound activity by paper or thin-layer chromatography or by dialysis. Furthermore, these techniques are far less costly than HPLC. The principal advantage of HPLC is its ability to perform these analyses rapidly.

There are, however, drawbacks to HPLC, not the least of which is cost. The HPLC system used in these studies currently costs approximately $13,000 to 15,000, not including the multichannel analyzer. If gradient elution capability is important, then a second pump and a controller unit is required at additional cost. The columns employed in this work cost about $600 each, and many separation applications require multiple columns in series. In addition, the columns may be expected to last only about one year under normal use. Fortunately, protein separations are usually performed with aqueous solvents, so that solvent costs are negligible.

An additional disadvantage is that the resolution obtained by HPSEC may be disappointing in comparison to electrophoresis or open-column gel filtration. For example, it was not possible in the study presented in Figure 14.3 to resolve the antibody fragment (MW about 100,000 daltons) from that of transferrin (MW 80,000 daltons). The resolution can likely be improved by the use of multiple columns in series, slower flow rates, and possibly by recycling through the column(s).

It may also be difficult to determine accurately the molecular weight of macromolecules using HPSEC. As mentioned above, interactions other than size exclusion may become important under certain conditions, leading to shifts in elution volume. For example, it was not possible to calibrate accurately the chromatographic system used to obtain Figure 14.2 since two of the molecular weight standards, trypsinogen (MW 18,400 daltons) and beta-lactoglobulin (MW 24,000 daltons) eluted in the reverse order to that expected from their molecular weights.

Finally, HPLC separations may be greatly influenced by contamination of the column from previous injections; as a consequence, it is wise to analyze regularly a series of protein standards as a check on the performance of the column. It is also wise to monitor the fraction of injected protein activity that is recovered from the column because increased retention may indicate increased adsorption resulting from changes in the column. In cases where such changes are observed, it may be necessary to wash the column with organic solvents (after first rinsing with water to remove salts) to remove adsorbed lipids. In difficult cases, such as that following multiple injections of serum, it may be necessary to resort to 6 *M* guanidine-HCl or urea to regenerate the column, even though this treatment may shorten column life [20].

CONCLUSIONS

The applications of HPLC to the analysis of radiolabeled proteins in our laboratory fall into three main categories: 1) the routine quality assurance of patient preparations prior to administration, 2) pharmacokinetic studies, and 3) in vitro studies including serum stability determinations for proteins radiolabeled with new radionuclides. The principal advantage of HPLC in connection with quality assurance is that the determination requires only a few minutes and does not, therefore, delay an investigation. In the case of pharmacokinetic studies, although speed of analysis is not directly a concern, the numbers of samples requiring analysis can be so large that the use of any time-consuming analytical technique would be impractical. We have also observed that urine and even serum samples may be analyzed directly by HPLC without any sample preparation other than filtration, and as a consequence, the results obtained in our pharmacokinetic studies and serum stability studies have been unaffected by sample manipulation. Finally, it is obvious that proteins radiolabeled with new short-lived radionuclides must be and can be analyzed quickly by HPLC. In conclusion, we have found HPLC to be very useful and generally superior to more convention analytical techniques.

ACKNOWLEDGMENTS

The author expresses his appreciation to David Lanteigne, Richard Childs, and Cynthia Kosciuczyk for performing the separations discussed herein, to Centocor for providing the antibody, and to Medi-Physics for providing the ^{111}In used in these studies. This work was supported in part by DOE contract, no. DE-AC02-83ER60175 and NIH grant no. CA 33029.

REFERENCES

1. Hearn, T.W., Regnier, F.E. and Wehr, C.T.: HPLC of peptides and proteins. *Amer. Lab.* **14**(10):18–39, 1982.
2. Regnier, F.E.: High-performance liquid chromatography of proteins. In *Methods in Enzymology 91,* Hirs, C.H.W. and Timasheff, S.N., eds., Academic Press, New York, 1983, pp. 137–190.
3. Stein, S.: Protein separations. *Trends in Analyt. Chem.* **3**:99–101, 1984.
4. Saunders, D.L.: Techniques of liquid column chromatography. In *Chromatography: A Laboratory Handbook of Chromatographic and Electrophoretic Methods,* Heftmann, E., ed., 3rd ed., Van Nostrand Reinhold, New York, 1975, pp. 77–109.
5. Tomono, T., Suzuki, T. and Tokunage, E.: Analysis of peptic fragmentation of human immunoglobulin G using high-performance liquid chromatography. In *High Performance Liquid Chromatography of Proteins and Peptides, Proceedings of the First International Symposium,* Hearn, T.W., Regnier, F.E. and Wehr, C.T., eds., Academic Press, New York, 1983, pp. 9–16.
6. *CRC Handbook of HPLC for the Separation of Amino Acids, Peptides, and Proteins.* Vol. I and II, Hancock, W.S., ed., CRC Press, Boca Raton, 1984.
7. Liquid column chromatography, bibliography. *J. Chromatogr.* **304**:B19–21 and B139–141, 1984.
8. Benson, J.R. and Woo, D.J.: Polymeric columns for liquid chromatography. *J. Chromatogr. Sci.* **22**:386–399, 1984.
9. Richey, J.: FPLC: a comprehensive separation technique for biopolymers. *Amer. Lab.* **14**(10):104–129, 1982.
10. Unger, K.: The structure of microparticulate silica. In *CRC Handbook of HPLC for the Separation of Amino Acids, Peptides, and Proteins,* Vol. I, Hancock, W.S., ed., CRC Press, Boca Raton, 1984, p. 21.
11. Pollak, J.K., and Campbell, M.T.: Waters protein columns. In *CRC Handbook of HPLC for the Separation of Amino Acids, Peptides, and Proteins,* Vol. II, Hancock, W.S., ed., CRC Press, Boca Raton, 1984, p. 372.
12. Wehr, T.C.: Commercially available columns. In *CRC Handbook of HPLC for the Separation of Amino Acids, Peptides, and Proteins,* Vol. I, Hancock, W.S., ed., CRC Press, Boca Raton, 1984, p. 54.

13. Hancock, W.S. and Harding, D.R.K.: Review of separation conditions. In *CRC Handbook of HPLC for the Separation of Amino Acids, Peptides, and Proteins,* Vol. II, Hancock, W.S., ed., CRC Press, Boca Raton, 1984, p. 303.
14. Hancock, W.S. and Harding, D.R.K.: Review of separation conditions. In *CRC Handbook of HPLC for the Separation of Amino Acids, Peptides, and Proteins,* Vol. II, Hancock, W.S., ed., CRC Press, Boca Raton, 1984, p. 308.
15. Regnier, F.E.: High performance liquid chromatography of proteins. In *Methods in Enzymology 91,* Hirs, C.H.W. and Timasheff, S.N., eds. Academic Press, New York, 1983, p. 183.
16. Regnier, F.E.: High performance liquid chromatography of proteins. In *Methods in Enzymology 91,* Hirs, C.H.W. and Timasheff, S.N., eds. Academic Press, New York, 1983, p. 171.
17. Welinder, B.S., and Linde, S.: High performance ion-exchange chromatography of insulin and insulin derivatives. In *CRC Handbook of HPLC for the Separation of Amino Acids, Peptides, and Proteins,* Vol. II, Hancock, W.S., ed., CRC Press, Boca Raton, 1984, p. 362.
18. Metzler, D.E.: In *Biochemistry: The chemical reactions of living cells.* Academic Press, New York, 1977, p. 757.
19. Hnatowich, D.J., Childs, R.L., Lanteigne, D. and Najafi, A.: The preparation of DTPA-coupled antibodies radiolabeled with metallic radionuclides: an improved method. *J. Imm. Method* **65:**147–157, 1983.
20. Waters Associates: *The protein column series, care and use manual.* 1983.

Index